Design and Analysis of Thermal Systems

Thermal systems are essential features of all domestic and industrial applications involving heat and fluid flow. Focusing on the design of thermal systems, this book bridges the gap between the theories of thermal science and design of practical thermal systems. Further, it discusses thermodynamic design principles, mathematical and CFD tools that will enable students as well as professional engineers to quickly analyze and design practical thermal systems. The major emphasis is on practical problems related to contemporary energy- and environment-related thermal systems including discussions on computational fluid dynamics used in thermal system design.

Features

- Exclusive book integrating thermal sciences and computational approaches

- Covers both philosophical concepts related to systems and design, to numerical methods, to design of specific systems, to computational fluid dynamics strategies

- Focuses on solving complex real-world thermal system design problems instead of just designing a single component or simple systems

- Introduces usage of statistics and machine learning methods to optimize the system

- Includes sample PYTHON codes, exercise problems, special projects

This book is aimed at senior undergraduate/graduate students and industry professionals in mechanical engineering, thermo-fluids, HVAC, energy engineering, power engineering, chemical engineering, nuclear engineering.

Design and Analysis of Thermal Systems

Malay Kumar Das
Pradipta K. Panigrahi

CRC Press
Taylor & Francis Group
Boca Raton London New York

CRC Press is an imprint of the
Taylor & Francis Group, an **informa** business

First edition published 2023
by CRC Press
6000 Broken Sound Parkway NW, Suite 300, Boca Raton, FL 33487-2742

and by CRC Press
4 Park Square, Milton Park, Abingdon, Oxon, OX14 4RN

CRC Press is an imprint of Taylor & Francis Group, LLC

© 2023 Taylor & Francis Group, LLC

ISBN: 978-0-367-50254-6 (hbk)
ISBN: 978-0-367-50326-0 (pbk)
ISBN: 978-1-003-04927-2 (ebk)

DOI: 10.1201/9781003049272

Typeset in Nimbus Roman
by KnowledgeWorks Global Ltd.

Contents

Authors

Malay Kumar Das received a BE degree in mechanical engineering (1989) from Bengal Engineering College, Shibpur, India; MTech. in mechanical engineering (2003) from IIT Kanpur, India and PhD in mechanical engineering (2008) from the Pennsylvania State University, USA. He has been a faculty member in IIT Kanpur since 2008. He has authored approximately 60 technical papers, 3 patents and 1 edited book.

Pradipta K. Panigrahi received BTech. degree in mechanical engineering (1987) from UCE Burla, Sambalpur, India; MTech. in mechanical engineering (1992), MS in system (computer) science (1997) and PhD in mechanical engineering (1997) from Louisiana State University, Baton Rouge, USA. He has been a faculty member in IIT Kanpur since 1998. He was Head of the Laser Technology Program from 2010–2013 and Head of Mechanical Engineering Department from 2014–2017. Presently, he is the Head of Photonics Science and Engineering Program, Center for Lasers and Photonics at IIT Kanpur. He has authored more than 200 technical papers, 3 patents, 1 text book, 2 edited books and 2 research monographs.

1 Introduction

1.1 DEFINITION AND IMPORTANCE

Design is defined as a *creative process* by which new methods, devices, processes and techniques are developed to solve new or existing problems. Nowadays, most industries are interested in producing new and high-quality products at minimal cost while satisfying the increasing concerns about environmental impact and safety. It is no longer adequate just to develop a product that performs the desired task due to increasing worldwide competition. It is important to optimize the process or system such that a chosen quantity, known as the objective function, is maximized or minimized. For a given system, the profit, productivity, product quality etc. may be maximized or the cost per item, investment, energy input etc. may be minimized.

Several classical industries e.g. the steel industry have become less important in recent years due to the advent of many new materials, such as composites and ceramics, and new manufacturing processes. Therefore, it is important to keep abreast of new developments and use new techniques for product improvement and cost reduction. The design and optimization of new processes/systems and optimization of existing ones are closely linked to the prosperity of a given company. Design and optimization methods have been traditionally applied to mechanical systems such as those involved with transmission, vibration, control and robotics. In recent years, there has been a tremendous growth in development and use of thermal systems in which fluid flow and transport of energy play a dominant role. Some examples of thermal systems are:

(a) Manufacturing Systems: Continuous casting, plastic screw extrusion, optical fiber drawing, hot rolling
(b) Power Systems: Solar energy, nuclear energy, coal power plant, gas-fired power plant, wind power plant
(c) Cooling Systems: Electronic equipments, gas turbine blades, gas turbine combustion chamber, nuclear power plants
(d) Transportation Systems: Aircraft propulsion (turbojet engine), IC engines
(e) Fluid Distribution Systems: Industrial system, residential system

This book focuses on the design and optimization methods related to thermal systems.

1.1.1 DESIGN VERSUS ANALYSIS

We are quite familiar with the **analysis** of engineering problems using information derived in the basic areas i.e. statics, dynamics, thermodynamics, fluid mechanics and heat transfer. There is little interaction between different disciplines during analysis. All inputs needed for the problem are usually given, and the results are generally

DOI: 10.1201/9781003049272-1

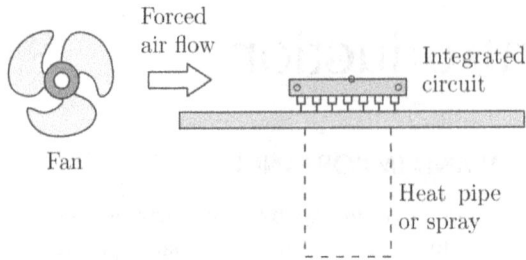

Figure 1.1   An arrangement for electronic cooling applications.

unique and *well defined* during analysis. The solution to a given problem can be carried out to completion during analysis. Such problems are termed as *close-ended*. Some examples of problems for analysis are:

1. Parabolic fully developed flow through pipe/channels
2. One-dimensional conduction through a flat wall
3. Fluid flow and heat transfer over a flat plate
4. Isentropic work done during a compression process

The **design** process is *open-ended*; i.e. the results are not well known or well defined at the onset. The *inputs* may also be vague or incomplete for design. Therefore, it is necessary to seek additional information or to employ approximation and assumptions. There is usually considerable *interaction* between various disciplines i.e. technical areas and those concerned with cost, safety and the environment during design. A *unique solution* is not generally obtained by the design. One may have to choose from a range of acceptable solutions. A solution that satisfies all the requirements may not be obtained by the design. Therefore, the designer may have to relax some of the requirements to obtain an acceptable solution. *Trade-off* generally forms a necessary part of the design, since certain characteristics of the system may have to be given up in order to achieve other goals such as lower cost or lower environmental impact.

A design solution is not unique. It depends on a variety of factors and many of them are non technical. The following electronics cooling problem can be used to demonstrate the difference between design and analysis. The electronics cooling problem in Figure 1.1 can be either *close-ended* or *open-ended* depending on the problem statement. It is termed as close-ended when the information on energy dissipated by the integrated chip, geometry, material properties of the circuit board and forced flow conditions are available. We have to solve for the temperature distribution in the component and circuit board, which is an example of close-ended problem. The solution methods for this problem may be either analytical or numerical. In this case, the solution is well defined and unique. There are no trade-offs, and no additional considerations are required.

This problem is termed as an *open-ended problem* when the problem statement is to find the appropriate materials, geometry and dimensions so that the temperature

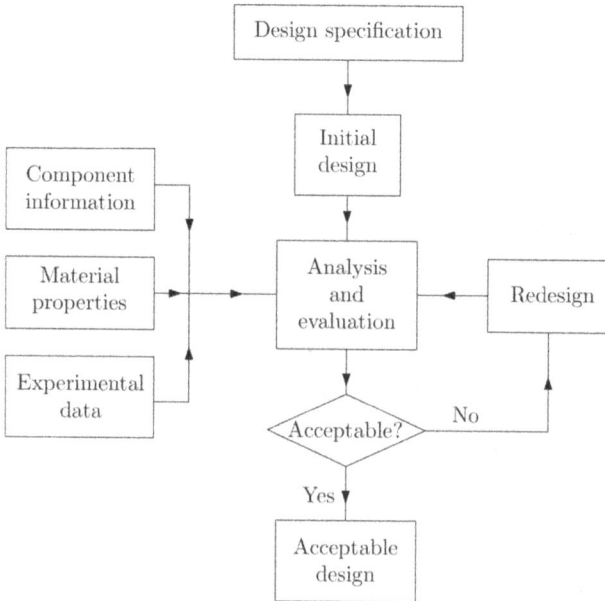

Figure 1.2 Schematic of a typical design procedure.

(T_c) of integrated chip is maintained below a certain value (T_{max}). There is no unique answer for this problem because many combinations of materials, dimensions, geometry, fan capacity etc. may satisfy the $T_c < T_{max}$ requirement. It is also possible that a satisfactory solution cannot be found for the given conditions. An additional cooling method such as heat pipe or spray may have to be included in the system. Different approaches, often known as conceptual designs, may be considered for satisfying the given requirements.

The design is often much more involved than analysis because of the open-ended nature of design problems. Extensive information on analysis of various thermal processes is available in literature. However, the corresponding design problems have received much less attention. The design and analysis are very different in their objectives and goals. Analysis usually forms the basis for design process.

1.1.2 SYNTHESIS FOR DESIGN

Several components and their corresponding analyses are brought together in design to yield the characteristics of the overall system. This is known as synthesis. The designer cannot consider only the heat transfer aspects while ignoring the strength of materials and manufacturing aspects. The design process incorporates information and outputs from different types of models, including experimental and numerical results from existing systems. Figure 1.2 shows the flow chart of a design process. Here, the details of the component information, material characteristics and experimental data are used. The cost, properties and characteristics of materials are also provided as input during the design process. Additional effects such as safety, legal, regulatory and environment considerations are also synthesized in order to obtain satisfactory design.

1.1.3 SELECTION VERSUS DESIGN

Selection and design are frequently employed together in the development of a system. Selection involves determining the specifications of items for the given task that are easily available in market over the ranges of interest. A choice is made from the various types of items available with different ratings on the basis of these specifications. Standard items that can be selected from the catalog are valves, control sensors, heaters, flow meters, storage tanks, pumps, compressors, fans, condensers and so on. Design is involved in the development of these components. However, for a given system, the design of these individual components may be avoided in the interest of cost, time and convenience.

1.2 THERMAL SYSTEM DESIGN ASPECTS

Thermal systems involve several issues that are unique compared to other systems. Some key design aspects within the context of thermal systems are discussed in the following sections.

1.2.1 ENVIRONMENTAL ASPECTS

There are several types of pollution sources of a thermal system: (1) air, (2) water, (3) thermal, (4) solid waste and (5) noise pollution. The design engineer has to implement different pollution control strategies based on the types of pollution.

Air pollution treatment equipment falls into two general types: (1) particulate removal by mechanical means, such as cyclones, filters and scrubbers etc. and (2) gas component removal by absorption, condensation and incineration etc.

Physical, chemical and biological waste treatment measures can be used for *liquid wastes* effluents. It is advisable to consider the recovery of valuable liquid-borne products to avoid costly waste treatment measures prior to waste treatment.

Thermal pollution resulting from the direct discharge of warm water into lakes, rivers and streams can be ameliorated by cooling towers.

Solid waste can be handled by incineration and pyrolysis etc. Coupling waste incineration with steam or hot water generation may provide economic benefit.

It is necessary to understand the individual noise sources, their acoustic properties and how they interact to cause the overall noise pollution for effective *noise control*.

Traditionally, an *end-of-the-pipe* approach that mainly addresses the pollutants emitted from the system is used. In this approach, the procedure for treatment of pollutants is considered. This is not the best approach. Rather the environmental aspects should be considered during the conceptual stage of a design, which is known as design for environment (DFE). In DFE, the environmentally preferred aspects of a system are treated as design objectives not as constraints. Designers are required to anticipate negative environmental aspects throughout the life cycle and engineer them out. Efforts need to be directed for reducing the creation of waste i.e. changing the process technology and/or plant operation and replacing input materials which are the sources of toxic waste with more benign materials. Compliance with environmental regulations should be considered throughout the design process. It should not be deferred to the end when options might be foreclosed due to earlier decisions. Addressing such regulations early may result in fundamentally better process choices that reduce the size of required clean-up. However, some end-of-the-pipe clean-up might be required to meet the government regulations. Costs to control pollution are generally much higher if left for resolution after the facility has begun operation. Thus, the appropriate pollution control measures i.e. the type of pollutant being controlled and the features of the available control equipment need to be specified by the design team. It should be noted that when reporting a project, it is imperative for the design engineer to submit *environmental impact statement*.

1.2.2 SAFETY ISSUES

The service life of a thermal system will not always be trouble-free. Occasional failure of some piece of equipment is likely to happen. Safety should be one of the quality factors at the beginning of the life cycle-design process. The design team should take note of the following points:

1. The design team is responsible for anticipating unsafe events and designing the system so that a local failure cannot amplify to an *overall system failure* or disaster.
2. One approach for safety analysis is to test the response of each component via computer simulation at *extreme conditions* that are not part of the normal operating plan.
3. Safety studies should be undertaken throughout the design process. It is not wise to defer the safety issues to the end of design process because decisions made earlier during design process may foreclose effective alternatives.
4. Exposure to toxic materials should be prevented or minimized during the operation of a thermal system. Machinery must be guarded with some protective devices. It is important for the design engineer to consider safety aspects. However, an indiscriminate application of safety factors is not a good practice because it may lead to over-design and the system can become uneconomical.

1.3 RELIABILITY, AVAILABILITY AND MAINTAINABILITY (RAM)

Reliability, availability and maintainability (RAM) are three important design attributes. *Reliability* is defined as probability that a system will successfully perform specified functions for specified environmental conditions over a prescribed operating life. *Maintainability* is defined as probability that the system can be repaired and maintained easily and economically within a specified time period. *Availability* is defined as the measure of how often the system can be available (operational) when needed.

1.4   BACKGROUND INFORMATION AND DATA SOURCES

Design engineers should be aware of current advances in their fields and related fields. They should read technical literature, attend industrial expositions and professional society meetings and develop a network of *professional contacts*. A design engineer should have both private and public sources of information/data to support a design project. Private sources of information consist of guarded *proprietary information* accumulated by individual companies, which is usually not available to outsiders. Public sources of information include open technical literature. There is a rapid growth of literature in recent times. Therefore, it is becoming increasingly difficult to search effectively for specific types of information. *Online databases* have helped to facilitate these searches. *Handbooks and review articles* describing current technology are useful resources. The *Thomas Registrar* also provides an exhaustive listing of manufactured items.

The *patent literature* is one of the useful public sources of information. It can provide ideas that can assist in achieving a design solution. It can also help in avoiding approaches that will not work.

Codes and standards provide information in the form of allowable limits on the performance of various systems. It can be helpful in shortening the design time and reducing uncertainty in performance of the system. It can also improve product quality and reliability. *Court of law* normally considers use of codes and standards as a sign of good engineering practice even if there is no legal compulsion to do so.

1.5   WORKABLE, OPTIMAL AND NEARLY OPTIMAL DESIGNS

The initial step in the design process is to specify the requirements of a system quantitatively i.e. formulate the *design specifications*. A *workable design* is the one that meets all the design specifications. The *optimal design* is the one that is best among several workable designs. Several alternative criteria for specifying the best can be applied depending on the type of application i.e. lowest cost, lowest weight, maximum reliability etc. A true optimum is generally impossible to determine due to system complexity and uncertainties associated with a thermal system. Therefore, the designer often accepts a design that is close to optimal and is known as *nearly optimal design*.

Let us consider Bejan et al.'s [1996] example of a counter-flow heat exchanger design. One of the key design variables is the minimum temperature difference between the two streams $(\Delta T)_{min}$. The temperature difference between the two streams of a heat exchanger is a measure of nonideality (irreversibility) according to thermodynamics. The heat exchangers approach ideality as the temperature difference between two streams approaches zero. This source of nonideality exerts a penalty on the fuel supplied to the overall plant, as a part of the fuel is used to feed the source of irreversibility. Therefore, we would prefer to reduce the stream-to-stream

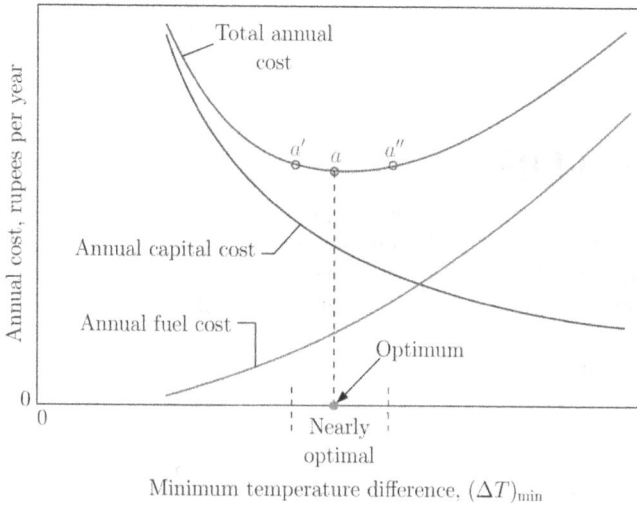

Figure 1.3 An example showing optimal and nearly optimal design of a counter-flow heat exchanger.

temperature difference for reducing operating costs. However, as the temperature difference narrows, more heat transfer area is required for the same rate of energy exchange from heat transfer perspective. More heat transfer area corresponds to higher materials requirement and therefore higher cost of heat exchanger.

Figure 1.3 compares different costs involved in acquisition and operation of a counter-flow heat exchanger. The capital cost includes material cost and fabrication cost, which reduces with increase in temperature difference between the streams. The fuel cost increases with increase in temperature difference due to higher irreversibility. The total cost is summation of fuel cost and capital cost. The point labeled a corresponds to the design with minimum total annual cost. Between location a' and a'', there is no significant change in annual cost. The $(\Delta T)_{min}$ between a' and a'' is termed as nearly optimal design. Optimal design offers more options for selection of final design. Specifying $(\Delta T)_{min}$ outside the nearly optimal range would be a design error.

Notes:

1. The above example of a heat exchanger design involves just one design variable i.e. temperature difference between streams. Several design variables may be considered and optimized simultaneously.
2. Thermal systems typically involve several components that interact with one another in complicated manner. Because of component inter-dependency, a unit of fuel saved by reducing the nonideality of one component (heat exchanger) may lead to waste of one or more units of fuel elsewhere in the system. As a result, there will be an increase in fuel consumption of the overall system. Therefore,

the objective of a design is to optimize the overall system consisting of several components. Optimization of one component does not guarantee that the overall system is working in an optimum condition.

1.6   STAGES OF THE DESIGN PROCESS

The flow chart of a design's life cycle is presented in Figure 1.4. The broad design process is considered and the role of the design team has been discussed. This design process is not the only possible choice, and alternative strategies can be employed. There are three distinct stages of the design as follows:

1. Understanding the problem
2. Concept development
3. Detailed design

The first step is to explore the viability of the overall design process when the requirement of the thermal system is specified. The design project may begin with an idea that something is worth doing. The design teams has to be formed for evaluating the idea and defining the specification of the design. Different concepts for achieving the design objective can be developed subsequently. The detailed design is carried out after finalizing the concept. Figure 1.4 shows the flow chart with details of different design stages. Iteration, concurrent design reviews and pilot plants are important steps at all stages of the design process.

Iteration: The iteration identifies and corrects any problem as early as possible during design stages. Through iteration, more knowledge is obtained and utilized even though iteration involves some cost.

Concurrent Design: The design process involves a wide range of skills and experience. The complete system design is well beyond the capabilities of a single individual and a group effort is essential. All the departments of a company are involved in the concurrent design process. Therefore, decisions can be made earlier with better knowledge, avoiding delay and errors and shortening the design process. Concurrent design combines the efforts of process engineering, manufacturing, maintenance, cost accounting, environmental engineering, production engineering, marketing and so on. Concurrent design overcomes the weakness inherent in traditional design approaches where evolving designs are passed from one department to another with minimal communication. There is little appreciation on the significance of design decisions made earlier by another department during traditional design process. As a result, there is a poor understanding of the impact of subsequent changes on such decisions.

Design Reviews: Design reviews are carried out in the form of presentations to other design team members. This is a key element of concurrent design approach. It provides formal opportunities to share information and solicit inputs for improving the design. The design team members may challenge one another by frequently asking "why" for thinking more deeply about underlying assumptions and the viability of proposed approaches to problems arising during the design process.

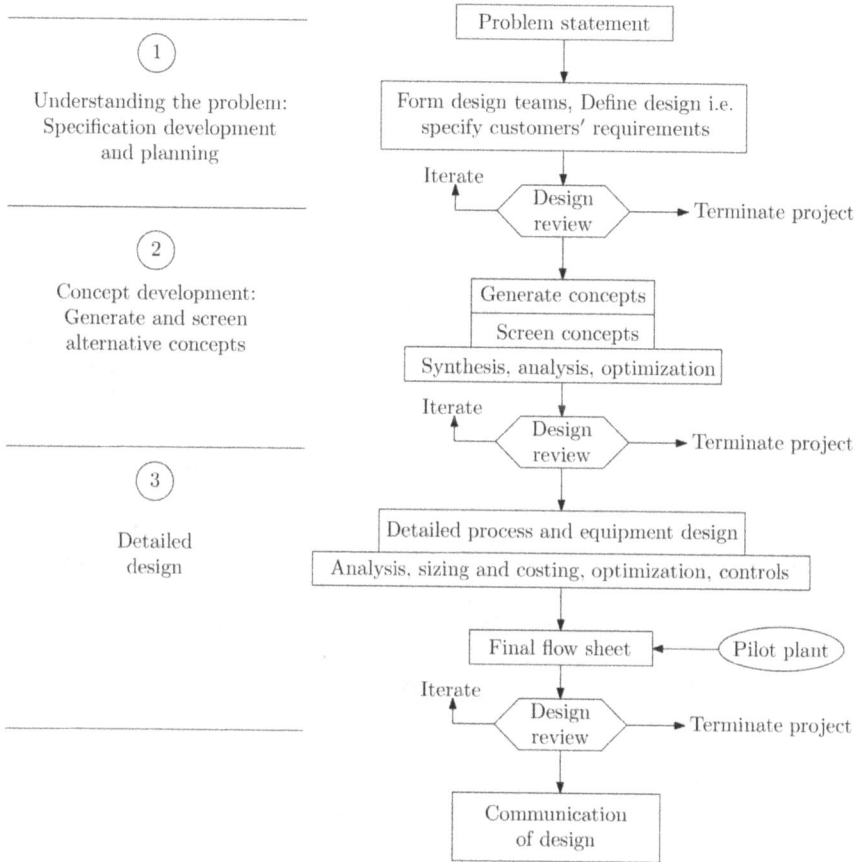

Figure 1.4   Different stages of the design process.

Pilot Plants: Laboratory testing, prototyping or pilot plant tests may be necessary to finalize the design when there is lack of accurate design data.

1.6.1   DFX STRATEGIES

DFX strategies are another concept of the design process where, DF denotes "design for" and X identifies a characteristic to be optimized. For example, DFA stands for "Design for Assembly", DFM stands for "Design for Manufacturing" and DFE stands for "Design for Environment". All the critical factors that influence Xs should be identified at the concept development stage and considered throughout the life cycle of a design. Different stages of designs are elaborated in the following sections.

1.6.2 FORMULATION OF THE DESIGN PROBLEM

The formulation of the design problem involves specification of the following: (1) requirements, (2) given parameters, (3) design variables, (4) limitations and constraints and (5) safety, environmental and other considerations [Jaluria, 2007].

1.6.2.1 Requirements

The requirements form the basis for design and evaluation of different designs. It is necessary to express the design requirements quantitatively and the permitted variation or tolerance level of these quantities. Let a *water flow system* is needed to obtain a specific volume flow rate of a system. There may be variations in the operating conditions leading to change in the flow rate (Q). Therefore, it is essential to determine the possible increase or decrease in the flow rate that can be tolerated by the system. We may have to design the system for delivering the desired flow rate Q_0 with a possible maximum flow variation of ΔQ. This may be expressed quantitatively as $Q_0 - \Delta Q \leq Q \leq Q_0 + \Delta Q$.

Suppose, we have to design a *water cooler*. In this case, the flow rate Q_0 and the desired temperature T_0 at the outflow are the requirements. In addition to the flow rate, Q expressed as before, the temperature is specified as $T_0 - \Delta T \leq T \leq T_0 + \Delta T$, where $\pm \Delta T$ is the acceptable variation in the outflow temperature.

Note: It is critical to determine the *main requirements* of the system and to focus on satisfying these requirements because it is often difficult to meet all the desired characteristics of the system. Therefore, the requirements that are not particularly important for the chosen application may have to be ignored. It is recommended to satisfy the most essential requirements. Subsequently, attempts can be made to satisfy other less important requirements by varying the design within the specified constraints and limitations.

Example: For a refrigeration system, the design should be completed first for the specified temperature and heat removal rate. Subsequent effort can be to find a substitute for the refrigerants R-11 and R-12, or to replace the compressor with one that is more efficient, or to vary the dimensions of the freezer or to improve the temperature control arrangement.

1.6.2.2 Given Parameters

There are parameters which are given or fixed i.e. *materials, dimensions, geometry* and *basic concept* or method during the formulation of the design problem. The other possibility is that some of the materials and dimensions are given and others are to be determined as part of the design. If the basic concept is not fixed or given, different concepts may be considered, resulting in considerable flexibility in design. If most of the parameters are fixed for a particular system, the design problem becomes relatively simple because only a small number of variables are to be determined. For example, for the design of an electronic cooling system, the electronic component size, the geometry and dimensions of the board, the number of electronic components on each board and the distance between two boards may be given. In a solar energy system, sensible heat storage in water may be chosen as the concept, with the dimensions, geometry etc. of the tank as the design variables.

1.6.2.3 Design Variables

The design variables are the quantities that may be varied during design in order to satisfy the given requirements. Attention is focused on these parameters during the design process. The design variables are varied to determine the behavior of thermal system and then chosen so that the system meets the given requirements. It is important to isolate the main design variables of the problem, since the complexity of design procedures is a *strong function* of the number of variables. The variables in the design problem may be classified as (1) hardware, and (2) operating conditions.

Hardware can be components of a system, dimensions, materials, geometrical configuration and other quantities that constitute the structure of system. Varying these parameters generally results in changes in the fabrication process and assembly of system. Changes in the hardware are not easy to implement if existing systems are to be modified for a new design or for optimization.

Operating conditions are the parameters that can be varied relatively easily, over specified ranges, without changing the hardware of a given system, such as temperature, flow rate, pressure, speed, power input etc. The design process yields the ranges for such parameters, with optimization indicating the values at which the performance is optimal.

1.6.2.4 Constraints and Limitations

The design constraints may arise due to material, weight, cost, availability and space limitations. The maximum pressure and temperature that a given component can be subjected to is limited by the properties of its material. For example, the semiconductor devices are very sensitive to temperature. Therefore, the temperatures of electronic equipment are constrained to values about 100 °C. These constraints may be written as $T \leq T_{max}, P \leq P_{max}$. There may be *weight restrictions* in the design of portable computers, airplanes, automobiles and rocket systems. Similarly, *volume constraints* may be important in room air conditioners, household refrigerators etc. Constraints also arise due to *conservation principles*. The energy rejected from a power plant to a cooling pond is $\dot{m}C_p\Delta T$, where $\dot{m}$ is the mass flow rate of cooling water and ΔT is its temperature rise going through the condenser. This energy must be rejected to the environment through heat loss at the water surface. We should have:

$$\dot{m}C_p\Delta T = hA_{surface}(T_{new} - T_{old})$$

where $(T_{new} - T_{old})$ is the rise in the average surface temperature. A limitation on this temperature rise is specified by regulations. As a result, we have restriction on temperature rise in the condensers as well as on the total flow rate.

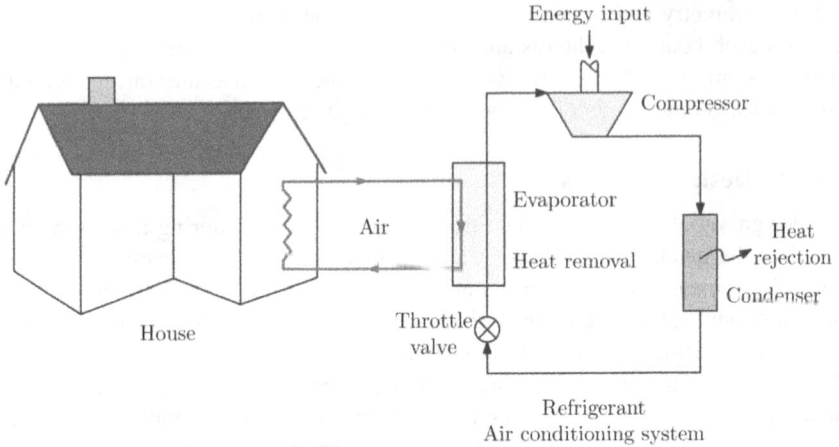

Figure 1.5 An example of the problem statement for air conditioning a house.

1.6.2.5 Safety, Environmental and Other Considerations

Disposal of waste, particularly hazardous waste from chemical plants and radioactive waste from nuclear facilities, is an important consideration that substantially affects the design of a thermal system. For example, safety concerns of *nuclear facilities* demand that adequate safety features be built into the system. To illustrate, the system must shut down if the temperature or heat flux levels exceed safe values. A safety feature would not allow the heater to be turned on if the *fluid level* is too low in a boiler. This will avoid any damage to the heaters and keep the operation safe. Similarly, the *energy source* may be changed from gas to electricity because of safety concerns in an industrial system. An oil furnace may be developed instead of a gas furnace for the same reason. Use of proper refrigerant for an air conditioner may be essential to minimize ozone layer depletion due to environmental considerations.

1.6.2.5.1 Problem Statement of an Air-Conditioning System

Let's consider the design of a building air-conditioning system, where the interior of the building is to be maintained at a temperature of $23 \pm 7\,°C$. The ambient temperature can go as high as $48\,°C$. The rate of heat dissipated from the house is given as 5.0 kW. The geometry and dimension of the building is given. The concept for operation of the building air-conditioning system is shown in Figure 1.5. We can systematically describe the above design problem as follows:

Requirements: Temperature inside the building must be maintained within 16 and $30\,°C$.

Note: In typical cases, the rate of cooling or response time τ_τ is also a requirement. We can write the energy balance relationship for the building as:

$$mC_p\frac{\partial T}{\partial \tau} = Q - Q_r$$

where, mC_p is the thermal capacity of the building air, Q is the thermal load and Q_r is the heat removal rate or loss from the building.

Given: The building geometry, location and dimensions are provided. Maximum ambient temperature is 48 °C. Rate of heat dissipation from the house is 5.0 kW. The time needed (τ_τ) to cool the building to 67% of its initial temperature difference from the ambient is defined here as the characteristic response time.

Constraints: Limitation on size, volume and weight of air conditioner are possible constraints. Maximum air flow rate circulating in the house can also be a possible constraint.

Design Variables: System parts i.e. condenser, evaporator, compressor, valve and the type of refrigerant are possible design variables.

1.7 CONCEPTUAL DESIGNS

The design effort starts with the selection of a conceptual design. The design concept is initially expressed in fairly vague terms as a method that might satisfy the given requirement and constraints. Conceptual design can be either an invention of a new approach not employed before or modifications of the existing systems. *Creativity, originality, experience, knowledge of existing systems* and *information on current technology* play a key role in development of conceptual design.

Innovative Conceptual Design: Brainstorming, where a group of people collectively try to generate a variety of ideas to solve a given problem is one of the techniques for innovative conceptual design. *Design contests* and *awards* to employees with the best ideas can also promote the generation of innovative concepts. Various ideas brought forth must be examined before they are discarded.

Selection from Available Concepts: In an industry, the ideas that have been tried in the past to solve problems similar to the one under consideration should be considered. Existing literature can be used to generate additional information on various concepts and solutions that have been previously employed. The conceptual design for the present problem may then be selected from the list of earlier concepts and modified based on this information. In this case, the basic concept is similar to the earlier concepts and the system design may be quite different. Even those ideas that did not yield satisfactory designs earlier must be considered because of changes in the problem statement and availability of new technology. Different *concepts can also be combined* to yield the conceptual design for a given problem. For example, both forced-air cooling and liquid-immersion cooling may be employed for different parts of an electronic system due to different heat input levels.

Example: Let us consider the task of *transporting iron ore* from the loading dock to the blast furnace of a steel plant. This can be achieved by trucks, trains, conveyor belts and carts. Each of these represents different concepts for the transportation system. The final choice is guided by several factors i.e. the distance over which the material is to be transported, the size and form in which the iron ore is available, and the rate at which the material is to be fed. For small plants, individual carts and

```
                                        ┌ ─ ┤ Redesign ├ ─ ┐
                                        │                  │
┌──────────────┐     ┌──────────┐     ┌────────────┐     ┌──────────────┐
│  Conceptual  │ ──▶ │ Modeling │ ──▶ │ Simulation │ ──▶ │    Design    │
│design details│     │          │ ◀─ ┤│            │     │  evaluation  │
└──────────────┘     └──────────┘     └────────────┘     └──────────────┘
                                                                  │
                                                                  ▼
┌──────────────┐     ┌──────────┐     ┌────────────┐     ┌──────────────┐
│Communication │ ◀── │Automation│ ◀── │  Optimal   │ ◀── │  Acceptable  │
│  of design   │     │and control│    │   design   │     │   designs    │
└──────────────┘     └──────────┘     └────────────┘     └──────────────┘
```

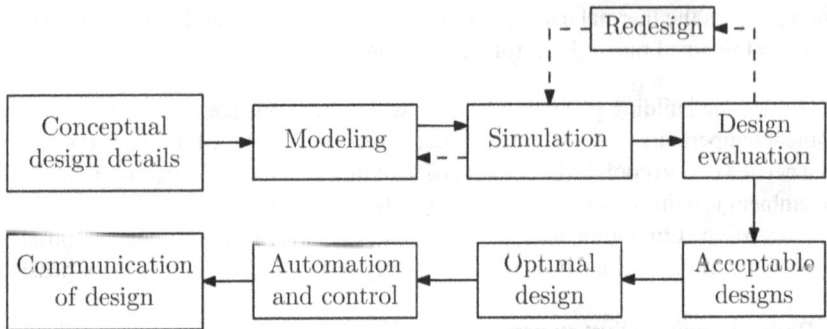

Figure 1.6   Different steps of the detailed design process.

trucks driven by the workers may be adequate. Trains may be the most appropriate method for large plants.

1.7.1   MODIFICATION IN THE DESIGN OF EXISTING SYSTEMS

The simplest approach for obtaining a conceptual design for a given problem is the modification in the design of existing systems. The conceptual design provides the design of existing system along with the possible modifications needed to meet the requirements of a given problem. The overall configuration of the system is kept largely unchanged, and only a few relevant components or subsystems are varied. Many thermal systems in use today have evolved through modifications of existing system over the years.

Example: Rankine cycle is the basic thermodynamic cycle used for steam power plants. Some of the conceptual designs that have been modified are super-heating of vapour leaving the boiler, reheating of steam passing through the turbine, and regenerative heating of the working fluid using stored energy from an earlier process in the system.

1.7.2   STEPS IN THE DESIGN PROCESS

In this section, we present the detailed steps of a design process. The flow chart showing different steps during the design process is shown in Figure 1.6. The main steps of the detailed design process are (1) initial physical system, (2) modeling of system, (3) simulation of system, (4) evaluation of different designs, (5) optimal design, (6) safety features, automation and control and (7) communicating the design.

1.7.2.1   Physical Systems

The starting point in the design process is the details of physical system obtained from conceptual design. The physical system is well defined in terms of the following details: (1) overall geometry and configuration of the system, (2) different components or subsystems that constitute the system, (3) interaction between various components, (4) given or fixed quantities of the system and (5) initial values of the design variables.

1.7.2.2 Modeling

Most practical thermal systems are fairly complex. Therefore, it is necessary to focus on the dominant aspects of the system and *neglect* relatively small effects. The objective is to simplify a given problem and make it possible to investigate its characteristics and behavior for a variety of operating conditions. *Idealization* and *approximation* of processes that govern the system are used to simplify analysis. This is known as modeling.

Modeling of a thermal system yields a set of algebraic, differential or integral equations that describes the behavior of the actual system. The system is described as

$$F_i(x_1, x_2, x_3, \ldots, x_n) = 0, \text{ for } i = 1, 2, 3, \ldots, n.$$

where, x_i and F_i represent, respectively, the physical variables and the equations that govern the system. In thermal systems, the nonlinear ordinary and partial differential equations are often encountered. Discretized equations are derived on the basis of numerical techniques such as finite difference and finite element methods giving rise to a numerical model. The numerical/analytical results must be validated preferably by comparisons with available experimental data.

Modeling of a thermal system also makes use of *dimensional analysis*. The governing dimensionless groups are determined for a thermal system using the principle of dimensional analysis. This simplifies the analysis of experimental and simulation results by reducing the number of parameter that needs to be varied for characterization of a given process/system. Modeling not only simplifies the problem but also eliminates relatively minor effects that only serve to confuse the main issues. It also allows satisfactory inclusion of experimental results into the overall model.

Modeling is generally first applied to *individual components*, parts or subsystems of a thermal system under consideration. The individual models are subsequently combined considering the interactions between various components. Different sub-models are linked to each other through boundary conditions and the flow of mass, momentum and energy. When these individual models are coupled with each other, the overall model for the thermal system is obtained. The main subsystems of an air-conditioning system are shown in Figure 1.7. The model of the subsystems are coupled through the fluid and energy transport i.e. the outlet from the evaporator is the inlet to the compressor whose outlet is inlet to the condenser.

1.7.2.3 Simulations

Simulation is the process of subjecting the model of a thermal system to various inputs. The behavior of the system at various operating conditions helps in predicting

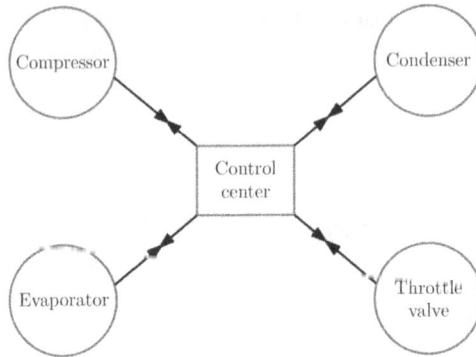

Figure 1.7   The main subsystems that combine to make an air-conditioning system.

the characteristics of the actual physical system. Both hardware and operating conditions are varied to study the system characteristics. A very important question that must be answered in any numerical simulation is how accurately it represents the actual or real world system. This involves ascertaining the validity of the various approximations used during modeling as well as estimating the accuracy of numerical algorithm. Sometimes, simpler or similar systems for which experimental results are available may be simulated to validate the models.

A common approach for simulating thermal system is to fix the hardware and vary the operating conditions over the desired ranges. The hardware is then changed to give a different design and the process is repeated. The simulation of the system is carried out with different design variables until an acceptable design or a range of acceptable designs is obtained.

Example: The output of the heat exchanger can be expressed as

$$Q = F(D_1, D_2, L, t_1, t_2, \dot{m}_1, \dot{m}_2, T_{1i}, T_{2i})$$

$$T_{20} = G(D_1, D_2, L, t_1, t_2, \dot{m}_1, \dot{m}_2, T_{1i}, T_{2i})$$

Here, the heat transfer rate, Q and outlet temperature of the outer fluid, T_{20} are the outputs from the model of heat exchanger. Mass flow rate $(\dot{m})$, length (L), thickness (t) and temperature (T) are the inputs to the model, where subscript 1 and 2 correspond to the inner and outer fluid stream, respectively. Subscript i corresponds to the inner fluid.

The diameters and the length may be chosen so that the constraints due to size or space limitations are not violated. Tube diameter (D_1, D_2) and thickness (t_1, t_2) choices may be restricted by the availability with the manufacturer to reduce costs. Each combination of the three design variables D_1, D_2 and L represents a system design, which can be obtained by keeping two of them fixed while varying the other one. Each system design is subjected to different flow rates and temperatures, which represent the operating conditions.

1.7.2.4 Acceptable Design Evaluations

An acceptable design is one that satisfies the given requirements for the system without violating the given constraints (safety, environmental regulation and financial constraints). In almost all practical cases, there are many possible solutions to a given design problem. The acceptable design obtained is by no means unique. The best design may be chosen on the basis of a given criterion such as minimum cost or highest efficiency. For the counter-flow heat exchanger, the requirements and constraints may be written as:

Requirements: $Q = Q_0 \pm \Delta Q$, $T_{20} = T_0 \pm \Delta T$
Constraints: $(D_1)_{min} < D_1 < D_2 - 2t_2$
$D_2 < (D_2)_{max}$, $L < L_{max}$
Operating conditions: $\dot{m}_1$, T_{1i}
Fixed quantities: $\dot{m}_2 = (\dot{m}_2)_0 \pm \Delta \dot{m}_2$, $T_{2i} = (T_{2i})_0 \pm \Delta T_{2i}$

1.7.2.5 Optimal Designs

Optimization is a systematic approach used to minimize a chosen quantity or function applied to a number of acceptable designs. The optimization problem can be expressed in the following manner:

Objective function: $U(x_1, x_2, x_3, \ldots, x_n) \to$ minimum/maximum
Equality constraint: $G_i(x_1, x_2, x_3, \ldots, x_n) = 0$
Inequality constraint: $H_j(x_1, x_2, x_3, \ldots, x_m) \le C_j$

where, U is the objective function, $x_1, x_2, \ldots, x_n$ are the design variables, G_i is the equality constraint, H_j is the inequality constraint and $j = 1, 2, 3, \ldots, m$ are the number of inequality constraints.

Example: Let us consider the example of a heat exchanger. The objective is to minimize the cost of equipment. Minimizing the total material used for manufacturing will help in reducing the total cost. Therefore, the cost can be represented by the volume of the material, which can be written in terms of design variables as

$$V = \pi D_1 L t_1 + \pi D_2 L t_2$$

Design variables D_1, L, t_1, D_2, t_2 are to be varied in the domain of acceptable designs in order to minimize objective function, V. The market availability of different tube sizes may also be included in this process to employ dimensions that are easily obtainable without significantly affecting the optimum. Once the optimal design is obtained, the operating conditions m_1 and T_{1i} may also be varied to determine if the costs could be further minimized by adjusting the flow rate and inlet temperature at the fluid stream while keeping the other fluid stream fixed. Thus, the overall costs may be minimized.

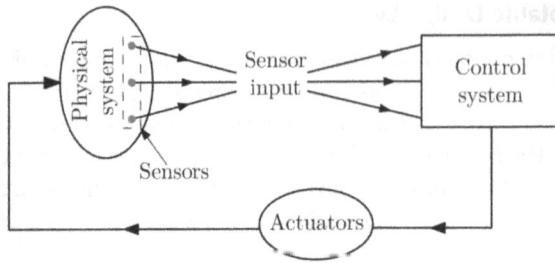

Figure 1.8 A schematic of a safety arrangement in a thermal system. (Jaluria [2007].)

1.7.2.6 Safety Features, Automation and Control

An important ingredient of a thermal system for safe and successful operation is the control scheme, which not only ensures safety of the system and the operator but also maintains the operating conditions within specified limits for satisfactory output. Sensors that monitor the temperatures, pressures, flow rates and other physical quantities in the system are employed to turn off the system or the inflow of material and energy into the system if the safety of people working on it is threatened. Figure 1.8 shows the typical safety arrangement of a thermal system. The possible sensors used for sensing the behavior of the physical system can be thermistors, thermocouples, flow meters, pressure transducers and so on. The actuators reduce the flow rate, increase/decrease the heat input, turn off power supply etc. as a control measure.

1.7.2.7 Communicating the Design

Communicating the final design to the client or customer and to those who will implement the design i.e. fabrication section is an important step for the overall success of a project. It is necessary to highlight the salient features of the design and justify how it meets the design requirements. The basic approach adopted in the development of a design, including information on modeling and simulation must also be presented. The accuracy and validity of the results should be discussed. Design team consists of individuals who have diverse backgrounds to study different aspects of a design problem. Therefore, good communication is very important. The communication depends on the target audience. The mode of communication by the design group depends on the target groups. When communicating with the *fabrication section*, the design group needs to communicate through detailed engineering drawings, parts list and materials list. When communicating with the group involved in *prototype testing*, the computer programs and numerical simulation results need to be supplied by the design group. When communicating with the *customer*, the working models and important results of the system under different operating conditions must be provided. The design team needs to communicate with the *management* through reports and presentations. The design team may also have to communicate with

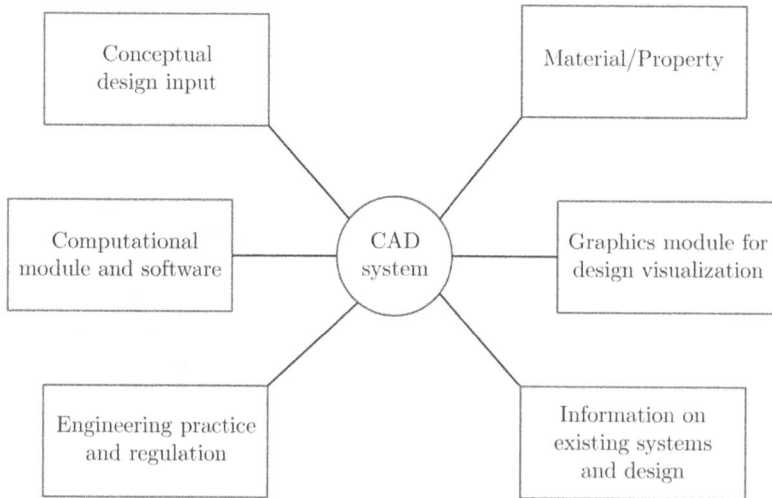

Figure 1.9   Various elements of a typical CAD system. (Jaluria [2007].)

the *patents and copyright group* of the organization for protecting the design from duplication by the competitors. The design team needs to communicate with the sales and marketing team for feedback and improvement of the existing design.

1.7.3   COMPUTER-AIDED DESIGNS

The term computer-aided design largely refers to an independent or stand-alone system such as single or multiple work stations or personal computers. The interactive use of the computer system considers various design options and obtains an acceptable or optimal design. A CAD system involves several aspects that help in the iterative design process. A representative CAD system and its components are shown in Figure 1.9. Some of the important features of a CAD system are

1. Interactive application of the computer system
2. Graphical display of result
3. Graphical input of geometry and variables
4. Available software for analysis
5. Available database for considering different options of material and components
6. Knowledge base from current engineering practice
7. Storage and information from earlier designs

Computer software codes for analysis are often based on finite element or finite volume or finite difference method like Ideas, Nastran, Fluent etc. Additional codes for curve fitting, numerical solutions of ordinary algebraic equations, differential equations, integration, differentiation, matrix inversions like Maple, Mathematica, MathCAD, MATLAB etc. are also used.

Notes:

1. The *modeling aspect* is the most involved part of thermal systems compared to other CAD systems. The remaining aspects are similar to CAD systems for other engineering fields.

2. Most of the CAD systems are devoted to the design of mechanical systems and components such as gears, springs and vibrating devices. It is not easy to develop similar CAD systems for thermal processes due to complexity of thermal systems. Some CAD systems are currently available for simple systems such as heat exchangers, air-conditioners, heating systems and refrigerators. Interactive design is generally not possible for more elaborate thermal systems, since numerical simulation might involve considerable CPU time and memory requirements. *Parallel computer systems* hold promise for the CAD of practical thermal systems.

3. The extent to which CAD of thermal systems can be applied is limited by the availability of *property data* in suitable forms.

4. CAD relies heavily on suitable process equipment design programs. Libraries of programs are available for *designing or rating* one of the most common thermal system components i.e. heat exchangers. Software for other types of equipment, including *piping networks*, are also available.

5. Common software allow the engineer to model the behavior of a system or system components under specified conditions and do thermal analysis, sizing, costing and optimization required for concept development. They are developed along two lines i.e.

   a. Sequential-modular approach
   b. Equation-solving approach

   In the sequential-modular approach, library models associated with the various components are called in sequence by an executive program. Here, the output stream data for each component is the input for the next component. In the equation-solving approach, all the equations representing individual components are assembled as a set of equations for simultaneous solution.

6. Optimization is an important component of CAD systems. Complex thermal systems are described in terms of a large number of equations, including nonlinear equations and non-explicit variable relationships. Therefore, the term optimization implies *improvement* rather than calculation of a global mathematical *optimum*.

   Conventional optimization procedure may be sufficient for relatively simple thermal systems. However, in thermal systems, costs and performance data are seldom in the form required for optimization. The conventional optimization methods can become unwisely time-consuming and costly with increasing system complexity. The method of *thermo-economics* is a better approach as it can assist in improving system efficiency and reducing the product costs by pinpointing the required changes in structure and parameter values of a thermal system.

PROBLEMS

1. You have to design a water cooler for drinking water. The water intake on a summer day is at 30 °C and the cooler must supply drinking water in the range of 12–24 °C at a maximum flow rate of 4.5×10^{-3} m^3/min. Give the requirements for the design. Develop an appropriate conceptual design and suggest the relevant design variables and constraints.

2. You have to design a cooling tower for heat rejection from a power plant. The rate of heat rejection to a single tower is given as 100 MW. Ambient air at temperature of 25 °C and relative humidity of 0.8 is to be used for removal of heat from the hot water coming from the condensers of the power plant. The temperature of the hot water is 15 °C above the ambient temperature. Give the formulation of the design problem in terms of the fixed quantities, requirements, constraints and the design variables.

3. The condensers of a 300 MW power plant operating at a thermal efficiency of 35% are to be cooled by the water from a nearby pond. The intake water is available at 22 °C. The temperature of the water discharged back into the lake must be less than 30 °C. Quantify the design problem for the cooling system.

4. You have to design a thermal system for the storage of thermal energy using an underground tank of water. The tank is buried with its top surface at a depth of 2.5 m. It is a cube of 1.5 m on each side. The water in the tank is heated by circulating water through a solar energy collection system. The heat input to the water is due to the solar energy flux. Characterize the design problem in terms of the fixed quantities and design variables.

5. Formulate the design problem for a hot rolling manufacturing process. The steel plate achieves reduction of thickness from 2.5 cm to 1.5 cm at a feed rate of 1.5 m/s during hot rolling.

6. Formulate the design problem of a water supply system to a water treatment plant. The water needs to be supplied from a river to the water treatment plant at 0.2 m^3/min flow rate and a pressure of 4 atm.

7. Discuss the nature, type and possible locations of sensors that need to be used for safety as well as control of a thermal system, which heat short metal rods in a gas furnace and then bend them into desired shapes in a metal forming process.

8. Select the ideal liquid for immersion cooling of an electronic system.

9. You are designing the tank of a water cooler. You have to decide on the location of inlet and outlet port of water from the tank. What is the ideal location for the water tap? Discuss with reasoning.

10. As an engineer at Tata Motors, you are asked to design an engine cooling system. The system should be capable of removing 20 kW of energy from the engine of a car at a speed of 75 km/h and ambient temperature of 35 °C. The system consists of radiator, fan and flow arrangement. The dimensions of the engine are given. The distance between the engine of the car and the radiator must not exceed 2.0 m and the dimensions of the radiator must not exceed $0.45 \times 0.45 \times 0.15$ m^3.

REFERENCES

A. Bejan, G. Tsatsaronis, and M. Moran. *Thermal Design and Optimization*. John Wiley & Sons Inc, 1996.

Y. Jaluria. *Design and Optimization of Thermal Systems*. CRC Press, 2007.

2 Modeling and Simulation Basics

2.1 INTRODUCTION

Modeling is one of the crucial steps in design of thermal systems. It is defined as the process of simplifying a given problem, so that it can be represented in terms of a *system of equations* for analysis or a physical *arrangement for experimentation*. Simulation is the process of obtaining information on the *behavior* and *characteristics* of the real system by analyzing, studying or observing a model of the system. In this chapter, various issues related to modeling and simulation of a thermal system are discussed.

2.2 TYPES OF MODELS

There are two types of models: (1) *descriptive* and (2) *predictive* (Jaluria 2007). Descriptive model is a *working model* of compressor, turbine or IC engines, which may be used to explain how the device works. Model made of plastics may have a cutaway section to show the internal mechanism. Predictive models can be used to predict the performance of a given system. For example, the equation describing the cooling of a hot iron rod immersed in a cold water pond represents a predictive model. It allows one to obtain the temperature variation with time, and also to determine the dependence of physical variables such as initial temperature of the rod, water temperature and material properties. The predictive model can also be classified as: (1) analog model, (2) mathematical model, (3) physical model and (4) numerical model.

2.2.1 ANALOG MODELS

An analog model is based on the analogy or similarity between different physical phenomena [Jaluria, 2007]. This type of model helps to use solution and results from a familiar problem to obtain the solutions for a different unsolved problem.

Example: The conduction heat transfer through a composite wall of a cold storage wall can be solved as an analogous electrical circuit. The following analogy is applicable based on the principle of heat transfer.

- Thermal resistance $\equiv$ Electrical resistance
- Heat flux $\equiv$ Electrical current
- Temperature difference $\equiv$ Electrical voltage

It may be noted that the Ohm's law can be employed to compute the total thermal resistance. The heat flux for a given temperature difference can be calculated using

DOI: 10.1201/9781003049272-2

the total thermal resistance. Another example of analog model is the use of Reynolds analogy,

$$\frac{C_f}{2} = St Pr^{\frac{2}{3}}$$

$$\text{where,} \quad St = \frac{h}{\rho V C_p} \quad C_f = \frac{\tau_w}{0.5 \rho V^2}$$

Here, St is the Stanton number, C_f is the skin friction coefficient, h is the heat transfer coefficient, ρ is the density, V is the velocity and τ_w is the wall shear stress. This equation is valid for Prandtl number (Pr) between 0.6 and 60. The skin friction data can be used for calculation/estimation of heat transfer while using the Reynolds analogy.

2.2.2 MATHEMATICAL MODELS

A mathematical model is one that represents the performance and behavior of a given system in terms of mathematical equations. There are two types of mathematical models: (1) theoretical and (2) empirical. *Theoretical models* use physical principles such as conservation laws, e.g. conservation of mass, momentum and energy, to derive the governing equations. *Empirical models* use curve fitting of experimental or simulation data to obtain mathematical representations of the thermal system behavior.

2.2.3 PHYSICAL MODELS

A physical model is one that resembles an actual system and is generally used to obtain experimental results on the behavior of systems.

Example: A scaled-down model of a truck positioned inside a wind tunnel can be used to study the drag force acting on the truck. Similarly, water channels are used to investigate the forces acting on ships and submarines. The physical model may be a scaled-down version of the actual system or a full-scale experimental model to study the basic characteristics of the system or a prototype, which is essentially the first complete system to be checked in detail before the start of production.

2.2.4 NUMERICAL MODELS

A numerical model facilitates the solution of mathematical model using computational platform. This type of model helps to obtain quantitative results on the behavior of a system for different operating conditions and design parameters using a computer. This model involves selecting the appropriate method for solution e.g. finite difference or finite element method. The mathematical equations are discretized to put them in a form suitable for digital computation, appropriate numerical parameters such as grid size, time step etc. are chosen and numerical solution is obtained by executing the numerical code.

2.3 MATHEMATICAL MODELING

A general step-by-step procedure is outlined for mathematical modeling of thermal systems on the basis of different considerations that arise in these systems. Generally, there is no unique model for a typical thermal system. Some guidelines that may be useful for developing an appropriate model are discussed here.

2.3.1 TRANSIENT/STEADY STATE

Time brings in an additional independent variable of a mathematical model, which increases the complexity of problem. Therefore, it is important to determine whether time variation effects can be neglected. There are two main *characteristic time scales* that need to be considered. The first time scale, τ_r, refers to the *response time* of the material or body under consideration. The second time scale, τ_c, refers to the *characteristic time* of variation in ambient operating conditions.

Example: The response time, τ_r, for a lumped body at uniform temperature subjected to a step change in ambient temperature with convective cooling or heating is given by the expression

$$\tau_r = \frac{\rho C V}{h A}$$

where, ρ is the density, C is the specific heat, V is the volume of the body, A is the surface area and h is the convective heat transfer coefficient.

Let's consider different probable cases based on the relative magnitude of τ_r and τ_c.

2.3.1.1 Case 1: $\tau_c \to \infty$ (Large τ_c)

In this case, the environmental conditions don't change with time and the system may be treated as steady state. At the start of the process, the conditions change sharply over a short time and transient effects are important. Steady-state conditions are attained with increase in time.

Example: Let us consider an *unheated electronic chip*, which is heated by an electric current. The chip temperature increases with time and after some overshoot in temperature, it attains steady state due to the balance between heat loss to the environment and the heat input (see Figure 2.1).

2.3.1.2 Case 2: $\tau_c \ll \tau_r$

In this case, the operating conditions change very rapidly as compared to the response of the material. Here, the system may be approximated as steady with the operating conditions are taken at their mean values.

Example: For a *deep lake*, the surface temperature may reflect the effect of ambient fluctuations. However, the bulk fluid may not show effect of temperature fluctuations because the response time of lake is much higher than the ambient temperature fluctuation due to large mass of lake water.

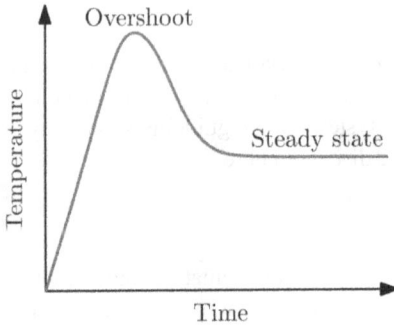

Figure 2.1 Dynamic response i.e. temperature of an electronic chip heated by electric current.

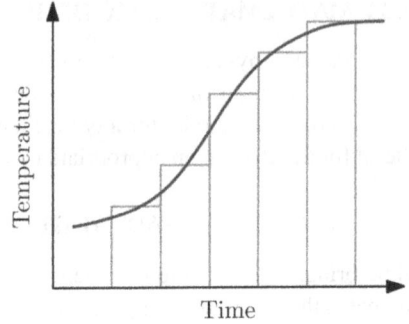

Figure 2.2 Example of quasi-steady modeling showing the variation of ambient temperature.

2.3.1.3 Case 3: $\tau_c \gg \tau_r$

This refers to the case where the material or body responds very quickly but the operating or boundary conditions change very slowly.

Example: The solar flux has a slow variation with time on a sunny day while the *solar collector* has a rapid response. In such cases, the system may be modeled as *quasi-steady*, with the steady problem being used at different time instants (see Figure 2.2). This implies that the system goes through a sequence of steady states, each characterized by constant operating or environmental conditions. However, as time elapses, the steady-state results do vary because of the changes in environmental conditions. This situation is modeled as quasi-steady modeling, where the ambient temperature variations are replaced with finite number of steps, with the temperature being constant over each step (Figure 2.2).

2.3.1.4 Case 4: Periodic Processes

The behavior of a thermal system repeats over a given time period, τ_p, in a periodic process. The main advantage of modeling a system as periodic is that steady-state results need to be obtained only over the time period of the cycle (see Figure 2.3) leading to considerable saving in computation time.

2.3.1.5 Case 5: Transient

When none of the above approximations is applicable, the system has to be modeled as a general time-dependent problem, with the transient terms i.e. time varying terms included in the model. This is the most complicated circumstance for simulating the system behavior.

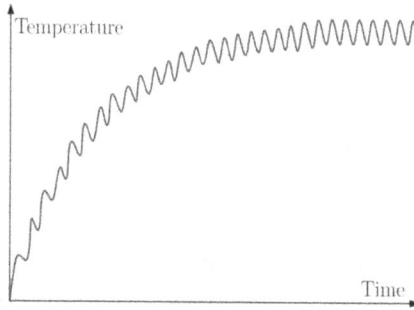

Figure 2.3   An example of the periodic process, where the body is subjected suddenly to a periodic variation in the heat input.

2.3.2   NUMBER OF SPATIAL DIMENSIONS

The spatial dimension required for modeling a thermal system refers to the number of spatial dimensions needed to describe a given system. Though all practical systems are three-dimensional (3-D), they can often be approximated as two-dimensional (2-D) or one-dimensional (1-D). 3-D modeling is generally avoided unless absolutely essential because of additional complexity in obtaining a solution of the governing equations. The results from a 3-D model are also not easy to interpret. The simplification from a 3-D model to a 1-D/2-D model is based on the geometry of the system under consideration and on the boundary conditions.

Example: Let us consider a 3-D conduction problem of a solid block of dimension $(L \times H \times W)$ in x, y and z directions, respectively. For constant thermal conductivity (k) and no heat source in the material, the governing equation is expressed as

$$\frac{\partial^2 T}{\partial x^2} + \frac{\partial^2 T}{\partial y^2} + \frac{\partial^2 T}{\partial z^2} = 0$$

Using dimensionless variables, $X = \frac{x}{L}, Y = \frac{y}{H}, Z = \frac{z}{W}, \phi = \frac{T}{T_{ref}}$
We have

$$\frac{\partial^2 \phi}{\partial X^2} + \frac{L^2}{H^2} \frac{\partial^2 \phi}{\partial Y^2} + \frac{L^2}{W^2} \frac{\partial^2 \phi}{\partial Z^2} = 0$$

Since X, Y and Z all vary from 0 to 1, all the second derivative terms in this equation can be of the same order of magnitude i.e. from 0 to 1. Hence, the overall magnitude of each term in this equation is determined by magnitude of the coefficients i.e.

if $\dfrac{L^2}{W^2} \ll 1$ (Last term is neglected i.e. no z-variation is assumed)

if $\dfrac{L^2}{H^2} \ll 1$ (Second term is neglected i.e. no y-variation is assumed)

When both the above conditions are satisfied, the problem becomes 1-D as both the second and third are negligible.

Note: If the heater elements are mounted in the Z-direction on the surface of the block, the spatial variation in the Z-direction cannot be neglected due to the boundary condition effect even though $L^2/W^2 << 1$.

2.3.3 LUMPED MASS APPROXIMATION

The temperature, species concentration or any other transport variables are assumed to be uniform within the domain of interest for lumped mass approximation. The variables are lumped and no spatial variation within the region is considered. The variables change only with time for transient problems, resulting in ordinary differential equation (ODE) instead of partial differential equation (PDE).

Example: A heated body (V is the volume and A is the surface area) at an initial temperature of T_o cools in an ambient medium at temperature T_a by convection, with h being the heat transfer coefficient.

The energy equation for the heated body is

$$\rho C V \frac{dT}{dt} = -hA(T - T_a)$$

After non-dimensionalization, we have

$$\rho C V \frac{d\theta}{dt} = -hA\theta$$

where, $\theta = T - T_a$, $\theta_o = T_o - T_a$ and $\tau = \frac{\rho C V}{hA}$ is the time constant.
For the solution of this equation, we have

$$\theta = \theta_o \exp\left(\frac{-hA\tau}{\rho C V}\right) = \theta_o e^{-t/\tau}$$

The applicability of the lumped body approximation is based on the ratio of the conductive resistance (L/kA) to the convective resistance $(1/hA)$. If this ratio is much smaller than 1.0, the convective resistance dominates and the temperature variation in the material is negligible compared to that inside the fluid. This ratio is expressed in terms of Biot number $(Bi = hL/k)$. If $Bi << 1$, the lumped analysis is valid.

2.3.4 SIMPLIFICATION OF BOUNDARY CONDITIONS

Most practical systems and thermal processes involve complicated, nonuniform boundary conditions. Considerable simplification can be obtained without significant loss of accuracy by approximating the boundary as *smooth* with *simpler geometry* and *uniform conditions*.

First Example: A large cylinder (with diameter D) is itself approximated as a flat surface for convective heat transfer calculation if the thickness of the boundary layer δ adjacent to the surface is much smaller than the diameter D of the cylinder i.e. $\delta/D << 1$.

Second Example: A given temperature distribution over a boundary may be replaced by the average value if the amplitude of the variation in temperature, ΔT, is small compared to the mean temperature, T_{avg}, i.e. $\Delta T / T_{avg} \ll 1$.

2.3.5 NEGLIGIBLE EFFECTS

Major simplifications in the mathematical modeling of thermal systems can be achieved by neglecting effects that are relatively small. Estimation of the relevant quantities for the problem are used to eliminate considerations that are of minor consequence. For example, estimates of convective and radiative loss from a heated surface may be used to determine if the radiation effects are important and should be included in the model. The convective heat transfer from a plate of surface area, A, with respect to ambient temperature, T_a, and heat transfer coefficient, h, can be estimated from

$$Q_c = hA(T - T_a)$$

The radiative heat transfer from the plate of emissivity, ε, to a surrounding at temperature, T_{surr}, can be estimated from

$$Q_r = \varepsilon \sigma A(T^4 - T_{surr}^4)$$

Here, the approximated or expected values of the surface temperature (T) may be employed to estimate the relative magnitudes of these transport rates i.e. Q_c and Q_r. At relatively low temperatures, the radiative transfer may be neglected, and at high temperatures, it may be the dominant mechanism.

2.3.6 IDEALIZATIONS

Practical systems and processes have undesirable energy losses, friction losses and many more losses that affect the system behavior. However, idealizations are made to simplify the model, focus on the main consideration and avoid aspects that are often difficult to characterize such as *frictional effects*, *leakage* and *contact resistance*. Actual systems are evaluated assuming the ideal behavior and the resulting performance of the system are specified in terms of an *efficiency* or *effectiveness*.

First Example: Frictional losses are neglected to simplify the models for many systems with moving parts, using a performance-related factor to characterize an actual system.

Second Example: Thermodynamic devices such as turbines, compressors, pumps and nozzles are analyzed as ideal and the efficiency of the device is used to model actual systems.

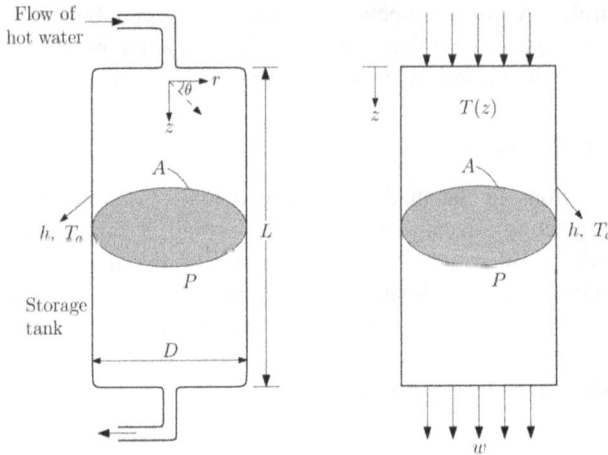

Figure 2.4 A schematic of a storage tank of a solar collector. (Jaluria [2007].)

2.3.7 MATERIAL PROPERTIES

It is important to employ accurate material property data for satisfactory mathematical modeling of any thermal system. The material properties are usually dependent on physical variables such as temperature, pressure and species concentration. In non-Newtonian fluid like molten polymer, the viscosity of the fluid depends on the shear rate and thus on the flow field. Properties may be taken as constant even when the properties vary with temperature and other variables. For example, thermal conductivity is assumed constant if the change in the properties, say thermal conductivity, Δk, is small compared to the average value, k_{avg}, i.e. $\frac{\Delta k}{k_{avg}} \ll 1$. Here, the change in properties is evaluated over the anticipated range of variables that affect the property. *Curve fitting* is often used to represent the variation of the relevant properties when constant property approximations cannot be made because of large ranges of these variables.

2.4 SAMPLE MATHEMATICAL MODELING EXAMPLES

Sample examples related to the mathematical modeling of thermal systems are presented below i.e. the storage tank of a solar collector and an electric heat treatment furnace.

2.4.1 STORAGE TANK OF SOLAR COLLECTOR

Let us consider the modeling of a storage tank used in a solar collector (Figure 2.4). The storage tank is assumed cylindrical in cross section with $L/D = 8$, $D = 40$ cm and it is 5 mm thick. The tank is made of stainless steel. The temperature range in the system is 20 °C–90 °C. A mathematical model for the storage tank needs to be developed for determining the temperature distribution of water. Appropriate non-dimensionalization of the governing equation should be carried out to obtain the relevant governing parameters for the system. Different steps for developing the mathematical model are presented here.

2.4.1.1 Lumped Mass Approximation

In the present case, $k_{stainless}/k_{water} = 23.59$ and steel wall is thin i.e. ratio of wall thickness (5 mm) to tank diameter (40 cm) is small. Therefore, the energy storage and the temperature drop in the wall may be neglected.

2.4.1.2 Material Properties

In the temperature range of 20 oC–90 °C, $\Delta \rho / \rho_{avg}$ and $\Delta K / k_{avg}$ are $<< 1$. Therefore, the variation in material properties of the medium may be neglected.

2.4.1.3 Spatial Dimensions

The original problem can be specified in cylindrical coordinate as: (T(r, θ, z)) i.e. 3-D distribution of temperature. Because of *axisymmetry*, we can assume no change in temperature in θ direction. Thus, the temperature becomes function of r and z, say $T(r,z)$. Hot water is discharged at the top. Therefore, the water in the tank is stably stratified, with warmer and thus lighter fluid lying above colder and thus denser fluid. This curbs recirculating flow in the tank and promotes horizontal temperature uniformity. Therefore, we can assume no variation in temperature in radial direction, and the temperature is expressed as $T(z)$.

2.4.1.4 Simplifications

The vertical velocity (w) in the tank can be assumed as uniform across each cross-section by employing the average value. This approximation is attributed to absence of recirculating flow in the tank. The problem is now substantially simplified as the uniform vertical downward velocity can be easily obtained from the flow rate. The coupled convective flow has to be determined, without this simplification, making the problem far more involved.

2.4.1.5 Governing Equation

Assuming incompressible flow, constant fluid properties and negligible viscous dissipation, the general energy equation is

$$\rho c_p \left[\frac{\partial T}{\partial t} + u \frac{\partial T}{\partial x} + v \frac{\partial T}{\partial y} + w \frac{\partial T}{\partial z} \right] = k \left[\frac{\partial^2 T}{\partial x^2} + \frac{\partial^2 T}{\partial y^2} + \frac{\partial^2 T}{\partial z^2} \right] + Q'$$

where, Q' is the heat loss per unit volume to the surrounding. This equation can be simplified for the present case as

$$\rho c_p \left[\frac{\partial T}{\partial t} + w \frac{\partial T}{\partial z} \right] = k \frac{\partial^2 T}{\partial z^2} - hP(T - T_\infty)$$

2.4.1.5.1 Initial Condition

$$\text{At} \quad t = 0, \quad T(z) = T_a$$

2.4.1.5.2 Boundary Condition 1

$$\text{At} \quad t > 0, \quad \left.\frac{\partial T}{\partial z}\right|_{z=L} = 0 \quad \text{(insulated tank at the bottom)}$$

2.4.1.5.3 Boundary Condition 2

$T = T_o$ at $z = 0$,   (discharge temperature of hot water from solar collector)

2.4.1.6 Dimensionless Parameters

$$\theta = \frac{T - T_a}{T_o - T_a}, \quad \tau' = \frac{\alpha t}{L^2}, \quad Z = \frac{z}{L}$$

Here, L^2/α is taken as the reference time scale, τ_r. Then, the dimensionless governing equation is obtained as

$$\frac{\partial \theta}{\partial \tau'} + W\frac{\partial \theta}{\partial Z} = \frac{\partial^2 \theta}{\partial Z^2} - H\theta$$

where, the non-dimensional parameters are $W = wL/\alpha$ and $H = hPL^2/AK$.

2.4.1.7 Dimensionless Initial/Boundary Conditions

At $\tau' = 0$, we have $\theta(\tau') = 0$.
For $\tau' > 0$, $\partial\theta/\partial Z = 0$ at $Z = 1$ and $\theta = 1$ at $Z = 0$.
At steady state, the governing equation is

$$W\frac{\partial \theta}{\partial Z} = \frac{\partial^2 \theta}{\partial Z^2} - H\theta$$

This differential equation can be solved analytically.

2.4.2 AN ELECTRIC HEAT TREATMENT FURNACE

Let's consider the design of an electric heat treatment furnace (Figure 2.5). The heat treatment furnace and the material to be heat treated are initially at room temperature, T_r. The material is raised to a desired temperature level, followed by gradual cooling by controlling the energy input to the heaters. We have to develop a mathematical model for this system.

    Let's assume:

1. For the wall and insulation, the thickness is much smaller than the corresponding height and width.

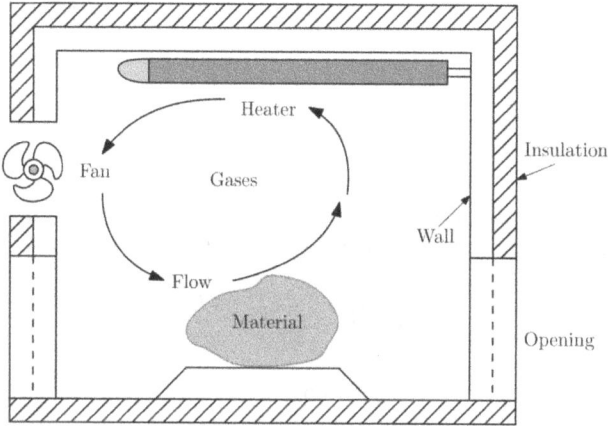

Figure 2.5   The sketch of an electric heat treatment furnace. (Jaluria [2007].)

2. The material to be heat treated is relatively small compared to the dimension of the furnace.

The constituents of the thermal system are (1) material (m), (2) heater (h), (3) gas (g), (4) walls (w) and (5) insulation (i). Let's develop the mathematical model for each component of the heat furnace.

(a) *Material:* Let the material being heat treated is made of metal. Since it's size is given as small, the Biot number is expected to be small. Therefore, the material may be modeled as *lumped mass*.

Therefore, the Governing Equation (GE) for the metal block (using enclosure radiation analysis) is

$$(\rho CV)_m \frac{dT_m}{d\tau} = h_m A_m (T_g - T_m) + \varepsilon_m \sigma A_m (F_{mh} T_h^4 + F_{mw} T_w^4 - T_m^4) \qquad (2.1)$$

where, F_{mh} is the geometrical view factor between the material and the heater and F_{mw} is the geometrical view factor between the material and the wall. The density, specific heat, volume and area of the material are denoted as ρ, c_p, V and A_m, respectively. The subscripts for material, gas, wall and air are denoted as m, g, w and h, respectively. The convective heat transfer coefficient from the material is denoted as h_m. In equation 2.1, the wall and the heater are assumed as black, and the energy reflected at the material surface is assumed to be negligible.

(b) *Heater:* The heater can be treated as lumped mass because it is a thin metal strip. The governing equation for the heater can be expressed as

$$(\rho CV)_h \frac{dT_h}{d\tau} = Q(t) + h_h A_h (T_g - T_h) + \sigma A_h (T_h^4 - F_{hw} T_w^4 - F_{hm} \varepsilon_m T_m^4) \qquad (2.2)$$

where, Q is the heater input.

(c) *Gas:* The gases are driven by fan and a well mixed condition is expected to arise. Therefore, a uniform temperature is assumed in the gases.
The governing equation for gas is

$$(\rho CV)_g \frac{dT_g}{dt} = h_h A_h (T_h - T_g) - h_m A_m (T_g - T_m) - h_w A_w (T_g - T_w) \qquad (2.3)$$

(d) *Wall:* Since the thickness of the walls is much smaller than the other two dimensions, the conduction transport in the walls may be approximated as 1-D.

$$(\rho C)_w \frac{\partial T_w}{\partial t} = \frac{\partial}{\partial x} \left(k_w \frac{\partial T_w}{\partial x} \right) \qquad (2.4)$$

(e) *Insulation:* Similarly, for the insulation, we can write

$$(\rho C)_i \frac{\partial T_i}{\partial t} = \frac{\partial}{\partial x} \left(k_i \frac{\partial T_i}{\partial x} \right) \qquad (2.5)$$

2.4.2.1   Initial and Boundary Conditions

- *For Equation 2.1 (Material):* The only condition needed is the initial condition as

$$T_m = T_r \text{ at } \quad t = 0$$

where, T_r is the room temperature.

- *For Equation 2.2 (Heater):* The initial condition is

$$T_h = T_r \quad \text{at} \quad t = 0$$

- *For Equation 2.3 (Gas):* The initial condition is

$$T_g = T_r \quad \text{at} \quad t = 0$$

- *For Equation 2.4 (Wall):* The initial condition is

$$T_w = T_r \quad \text{at} \quad t = 0$$

The boundary condition is

$$\text{For} \quad t > 0, \quad \text{at} \quad x = 0: \qquad -k_w \frac{\partial T_w}{\partial x} = h_w (T_g - T_w)$$

$$\text{At} \quad x = d_1, \quad T_w = T_i$$

where, d_1 is the wall thickness. In this equation, x corresponds to the distance along the wall thickness.

- *For Equation 2.5 (Insulation):* The initial condition is

$$T_i = T_r, \quad \text{at} \quad t = 0$$

The boundary condition is

$$\text{At} \quad x = d_1 + d_2, \qquad -k_i \frac{\partial T_i}{\partial x} = h_e (T_i - T_\infty)$$

$$\text{At} \quad x = d_1, \qquad -k_w \frac{\partial T_w}{\partial x} = -k_i \frac{\partial T_i}{\partial x}$$

where, d_2 is the thickness of insulation. The temperature of the ambient medium is denoted as T_∞. The preceding system of equations with corresponding boundary conditions represents the mathematical model for the heat treatment furnace.

Notes:

1. In the above equations [(a) to (e)], the various convective heat transfer coefficients are assumed to be known. Heat transfer correlations available in the heat transfer literature may be used for the purpose. In actual practice, greater accuracy is obtained by solving the convective flow in the gases for the present geometrical configuration using computational fluid dynamics (CFD) approach.
2. If the last two components i.e. wall and insulation are also assumed as lumped, then a system of ordinary differential equations is obtained instead of the PDE.

$$(\rho C V)_w \frac{dT_w}{dt} = h_w A_w (T_g - T_w)$$

$$(\rho C V)_i \frac{dT_i}{dt} = h_e A_i (T_i - T_\infty)$$

2.5 DIMENSIONAL ANALYSIS

Dimensional analysis refers to the derivation of dimensionless governing parameters that determine the behavior of a thermal system. This is carried out largely to reduce the number of independent variables required for generalizing the results so that these dimensionless parameters can be used for a wide range of conditions. This helps in describing the behavior of a system. Dimensional analysis is also helpful in the development of empirical model from experiments. There are two main approaches for deriving the dimensionless parameters of a given problem:

1. Combination of variables (Buckingham pi theorem and Rayleigh theorem)
2. Non-dimensionalization of governing equations

The combination of variable approaches has been extensively discussed in regular Fluid Mechanics books. For the governing equation approach, the governing equation is first written in terms of physical variables i.e. time, space and so on. Characteristic or reference quantities are then chosen on the basis of experience and physical nature of the system. These reference quantities are used to non-dimensionalize the variables that are present in the governing equations. The non-dimensionalization is

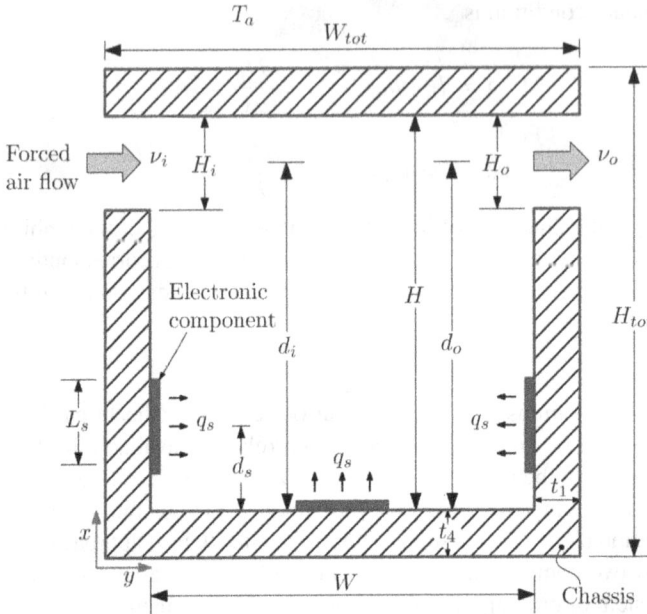

Figure 2.6 The electronic device, with three electrical chips as heat sources and forced air cooling. (Jaluria [2007].)

then applied to each variable of the governing equations. The governing equation is transformed to dimensionless equations, with all the dimensionless groups appearing as coefficients in the equation. Similarly, the boundary conditions are also non-dimensionalized to yield any additional dimensionless parameters that might arise.

2.5.1 EXAMPLE OF AN ELECTRONIC DEVICE

An example showing the determination of important non-dimensional parameters for an electronic device (see Figure 2.6) is discussed here. The electronic cooling system adopts forced air cooling with three electronic components mounted inside the chassis. The following assumption is used:

1. The dimension in the third direction is large and the problem may be treated as 2-D.
2. The velocities are uniform over the inlet and outlet with magnitudes V_i and V_o, respectively.
3. The temperature at the inlet is T_i and fully developed conditions i.e. $\frac{\partial T}{\partial y} = 0$ may be assumed at the exit of inlet and outlet. The outer wall of the system loses energy to ambient air at temperature T_a with a given convective heat transfer coefficient h.
4. The flow is laminar and has constant properties.

We need to write down the governing equations and boundary conditions for calculating the temperature distributions of the system. Subsequently, we have to non-dimensionalize the governing equation to obtain the governing dimensionless parameters of the system.

The continuity equation of the fluid medium with assumption of incompressibility and constant density can be written as

$$\overline{\nabla}.\overline{V} = 0 \tag{2.6}$$

For the momentum equation, the Boussinesq approximation and constant viscosity assumption is used. The inlet temperature is equal to the ambient temperature, and β is the coefficient of volumetric expansion. The momentum equation is

$$\rho\left(\frac{\partial\overline{V}}{\partial t} + \overline{V}.\overline{\nabla}\overline{V}\right) = -\overline{\nabla}p + \overline{g}\beta(T - T_i) + \mu\nabla^2\overline{V} \tag{2.7}$$

The energy equation assuming constant thermal conductivity and neglecting the pressure work and dissipation terms is

$$\rho C_p\left(\frac{\partial T}{\partial t} + \overline{V}.(\overline{\nabla}T)\right) = k\nabla^2 T \tag{2.8}$$

The boundary conditions for the system are as follows:

Velocity: No slip conditions i.e. zero velocity is assumed at the solid boundaries and the velocities at inlet and outlet are equal to V_i and V_o, respectively.

Temperature:

(a) *At the Inner Surface of the Enclosure:*

$$T = T_s \qquad \text{(Continuity of the temperature)}$$

$$k_f\left(\frac{\partial T}{\partial n}\right) = k_s\left(\frac{\partial T}{\partial n}\right)_s \qquad \text{(Continuity of the heat flux)}$$

Here, k_f is the thermal conductivity of fluid, the subscript s denotes solid material properties and n is the coordinate normal to the surface.

(b) *At the Left Source:* Energy dissipated by the source per unit width

$$q_s = L_s\left(-k_f\frac{\partial T}{\partial y} + k_s\frac{\partial T}{\partial y}\right)$$

(c) *At the Outer Surface of the Enclosure Walls:*

$$-k_s\frac{\partial T}{\partial n} = h(T - T_i)$$

(d) *At the Inlet:* The temperature is uniform at T_i.

(e) *At the Outlet:* The temperature profile is developed.

$$\frac{\partial T}{\partial y} = 0$$

2.5.1.1   Non-Dimensionalization

The energy input governs the heat transfer processes, and the inlet conditions determine the forced air flow in the enclosure. Therefore, the characteristic physical quantities can be assumed as V_i, H_i, T_i and q_s. Thus, the geometric dimension of the problem are non-dimensionalized by H_i, the velocity is non-dimensionalized as $\overline{V}$ by V_i and time, t is non-dimensinalized by H_i/V_i to give dimensionless time $\tau' = t(V_i/H_i)$. The non-dimensional temperature θ is defined as

$$\theta = \frac{T - T_i}{\Delta T} \qquad \text{where, } \Delta T = q_s/k$$

Now the dimensionless equations i.e. the continuity, momentum and energy, respectively, for the convective flow are

$$\overline{\nabla^*}.\overline{V^*} = 0 \tag{2.9}$$

$$\frac{\partial \overline{V^*}}{\partial \tau} + \overline{V^*}.(\overline{\nabla^*}\overline{V^*}) = -\overline{\nabla^*}p^* + \left(\frac{GR}{Re^2}\right)\theta\overline{e} + \frac{1}{Re}(\nabla^*)^2\overline{V^*} \tag{2.10}$$

$$\frac{\partial \theta}{\partial \tau} + \overline{V^*}.(\overline{\nabla^*}\theta) = \frac{1}{RePr}(\nabla^*)^2\theta \tag{2.11}$$

Here, the * denotes the dimensionless quantity, $p^* = p/\rho V_i^2$ and α is the thermal diffusivity. Unit vector along g is denoted by $\overline{e}$. The dimensionless parameters that arise are

$$\text{Reynolds no., } Re = \frac{V_i H_i}{\gamma} \; ; \quad \text{Grashoff no., } Gr = \frac{g\beta H_i^3 \delta T}{\gamma^2} \; ; \quad \text{Prandtl no., } Pr = \frac{\gamma}{\alpha}$$

$$\left(\frac{\partial \theta}{\partial n^*}\right)_f = \frac{k_s}{k_f}\left(\frac{\partial \theta}{\partial n^*}\right)_s$$

$$\frac{\partial \theta}{\partial n^*} = \left(\frac{hH_i}{k_s}\right)$$

Here, we get additional dimensionless numbers, k_s/k_f i.e. thermal conductivity ratio and hH_i/k_s i.e. Biot number due to non-dimensionalization of the boundary condition.

2.6   CURVE FITTING

Material properties are usually available as discrete data at various values of the independent variable i.e. temperature or pressure. For example, mass diffusivity may be a function of temperature. Curve fitting is employed to obtain appropriate empirical relation. Curve fitting can also be used to represent *numerical results* in a compact and convenient form. These curve-fitted equations can be used for optimization of a thermal system. Results from numerical computation and *experimentation* are represented by means of smooth curve. The equation of the curve can be used to obtain values at intermediate points where data are not available and to model the characteristic of the system.

Two main approaches for curve fitting are (1) *exact fit* and (2) *best fit*.

Exact Fit: A curve that passes through each given data point is said to be in exact fit. This approach is particularly appropriate for data that are very accurate such as calibration results, material properties data and if only a *small number of data points* are available. In this case, the number of parameters in the approximating curve must be equal to the number of data points. The determination of a large number of parameters in the curve becomes very involved if extensive data are available. In that case, the curve obtained is not very convenient to use. There are several coefficients with very small value, which may lead to inaccurate representation of data. Therefore, there is no reason to ensure that the curve passes through each and every data point if extensive data are available. Some of the methods available for exact fit are

1. General form of a polynomial
2. Lagrange interpolation
3. Newton's divided difference polynomial
4. Splines

Best Fit: A best-fit curve represents general trend of data, without necessarily passing through every given point. The most commonly used approach for a best fit is the method of least squares in which the sum of the squares of the errors is minimized.

$$S = \sum_{i=1}^{n}(e_i)^2 = \sum_{i=1}^{n}(y_i - f(x_i))^2$$

where, n is the number of data points, e_i is the error, y_i is the actual data and $f(x_i)$ is the curve-fitted data. We can have (1) linear regression, (2) polynomial best fit, (3) non polynomial forms and (4) more than one independent variables.

2.6.1 LEAST SQUARE METHOD

The least square method is often needed for situations in which the data scatter is significant and in which no prior smoothing has been applied to the data and in which the functional form is not evident and a general polynomial is used. The least square procedure for an mth order polynomial generates curve-fitting coefficients $a_0, a_1, \ldots, a_m$ such that the sum of the squares of the differences between the curve produced and the n-data points ($n > m + 1$) is minimized.

The sum of the residuals, S_r, for n-data points (x_i, y_i), where $y_i = f(x_i)$, is the curve fit equation is written as

$$S_r = \sum_{i=1}^{n}(y_i - a_0 - a_1 x_i - a_2 x_i^2 - \cdots - a_m x_i^m)^2$$

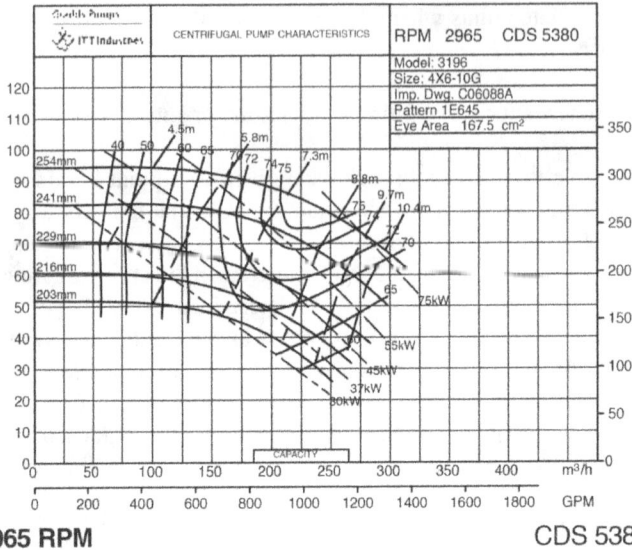

Figure 2.7   A typical pump characteristic, $H = g(Q,N)$. (Goulds 3196, Pump Performance Curves.)

For minimization of this residual, the expression for derivatives, $\frac{\partial S_r}{\partial a_0} = \frac{\partial S_r}{\partial a_1} = \cdots =$ $\frac{\partial S_r}{\partial a_m} = 0$. We have $m+1$ equations for the unknowns $a_0, a_1, \ldots, a_m$ i.e.

$$\sum y_i = na_0 + a_1 \sum x_i + a_2 \sum x_i^2 + \cdots + a_m x_i^m$$
$$\sum x_i y_i = a_0 \sum x_i + a_1 \sum x_i^2 + a_2 \sum x_i^3 + \cdots + a_m x_i^{m+1}$$
$$\vdots$$
$$\sum x_i^m y_i = a_0 \sum x_i^m + a_1 \sum x_i^{m+1} + a_2 \sum x_i^{m+2} + \cdots + a_m x_i^{2m}$$

where, all summations are from 1 to n. This set of equations constitutes a linear system of equations for the unknowns $a_0, a_1, \ldots, a_m$. The solution of this set of equations gives the coefficients, $a_0, a_1, \ldots, a_m$ of the curve fit equation.

2.6.2   TWO INDEPENDENT VARIABLE CASES

There are many situations, where it is desirable to generate a curve fit for two independent variables i.e. $y = f(x,z)$. Figure 2.7 shows a typical pump performance curve indicating the relationship between (H), speed (N) and flow rate (Q). One satisfactory polynomial representation for the pump curve is

$$H = a_0(N) + a_1(N)Q + a_2(N)Q^2$$

The dependence of speed is contained in the coefficients $a_o(N), a_1(N)$ and $a_2(N)$. This dependence on speed can be presented as

$$a_i(N) = b_{i0} + b_{i1}N + b_{i2}N^2$$

Therefore, we can write

$$H = (b_{00} + b_{01}N + b_{02}N^2) + (b_{10} + b_{11}N + b_{12}N^2)Q + (b_{20} + b_{21}N + b_{22}N^2)Q^2$$

2.6.2.1 Curve-Fitting Procedure

We generate data points between the head (H) and the flow rate (Q) at three or more speeds.

$$H_1 = a_{10} + a_{11}Q + a_{12}Q^2, \quad \text{at speed } N_1 \text{ perform first curve fit.}$$

$$H_2 = a_{20} + a_{21}Q + a_{22}Q^2, \quad \text{at speed } N_2 \text{ perform second curve fit.}$$

$$H_3 = a_{30} + a_{31}Q + a_{32}Q^2, \quad \text{at speed } N_3 \text{ perform third curve fit.}$$

- Use (a_{10}, N_1), (a_{20}, N_2) and (a_{30}, N_3) to evaluate b_{00}, b_{10}, b_{20} (perform fourth curve fit).

$$a_0 = b_{00} + b_{10}N + b_{20}N^2$$

- Use $(a_{11}, N_1), (a_{21}, N_2)$ and (a_{31}, N_3) to evaluate b_{01}, b_{11}, b_{21} (perform fifth curve fit).

$$a_1 = b_{01} + b_{11}N + b_{21}N^2$$

- Use $(a_{12}, N_1), (a_{22}, N_2)$ and (a_{32}, N_3) to evaluate b_{02}, b_{12}, b_{22} (perform sixth curve fit).

$$a_2 = b_{02} + b_{12}N + b_{22}N^2$$

Note: If, instead of three speeds, we use four speeds (N_1, N_2, N_3, N_4), the number of the curve fit will be seven instead of six.

2.7 NUMERICAL MODELING

The mathematical equations of thermal systems are mostly coupled in nature. These equations can be nonlinear algebraic, ordinary differential equations and PDE. Analytical solutions of these equations are rarely possible, and we have to resort to numerical techniques to obtain the desired results. The solution strategy of the thermal systems behavior depends on the nature of the governing equations as follows:

1. *Set of Linear Algebraic Equations:* There are two general approaches for solution of these problems. They are the direct method and iterative method.
   (a) Direct Method: The following techniques can be classified as direct method:
       (A) Gaussian elimination
       (B) Gauss–Jordan elimination
       (C) Matrix decomposition methods
       (D) Matrix inversion methods

    (b) Iterative Method: The following techniques can be classified as iterative method.

        (A) Gauss–Seidel method

2. *Single Nonlinear Algebraic Equations:* The following techniques can be used to solve single nonlinear algebraic equation:

    (a) Search method

    (b) Bisection method

    (c) Newton–Raphson method

3. *System of Nonlinear Algebraic Equations:* The following techniques can be used for solving set of nonlinear algebraic equations:

    (a) Newton's method

    (b) Successive substitution method

4. *Ordinary Differential Equations:* The initial value and boundary value ordinary differential equations can be solved by the following methods:

    (a) Initial Value Problem

        (A) Runge–Kutta methods

        (B) Predictor-corrector methods

    (b) Boundary Value Problems

        (A) Shooting methods

5. *Partial Differential Equations:* The partial differential equations can be solved by the following techniques:

    (a) Finite difference method

    (b) Finite element method

    (c) Control volume method

    (d) Boundary element method

    (e) Spectral methods

The details of these numerical techniques can be found in any general purpose numerical technique textbooks. The sample codes are now available in numerical recipes, libraries (IMSL, NAG etc.) and so on. There are also numerous software e.g. MATLAB, MATHEMATICA and many more for easy implementation of these techniques. A design engineer is expected to use the available software for avoiding delay in the design process. In a special circumstance, the design team may have to implement it own code. In that situation, the design team has to refer the relevant literature in detail.

The numerical model for a complete thermal system consists of several parts or subsystems. It may contain programs that have been *developed by the user*, available in the *public domain*, standard programs available on the computer and *commercially* available general purpose programs (Ansys, Comsol, Phoenics, Fluent etc.). The numerical model may also be linked with the available information on material properties, heat transfer correlations and other relevant information. It is important to be conversant with the algorithm used in the software and be aware of its *applicability, accuracy and limitations*.

The first step in the mathematical and numerical modeling of a thermal system is isolation of various parts or components of the system. Differences in *geometry,*

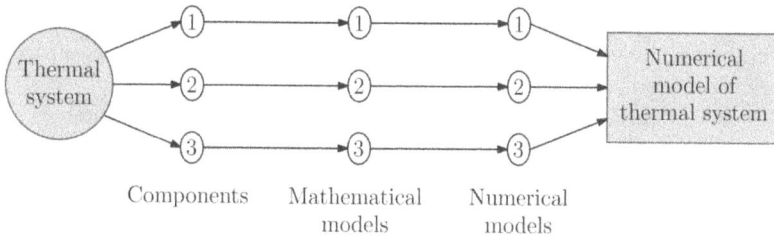

Figure 2.8 Schematic explaining the general approach to developing an overall model for a thermal system.

material, function, thermodynamic state, location and other such characteristics can be used to isolate the components. The mathematical model for each component is developed separately. Subsequently, the corresponding numerical modes are developed. The individual numerical models are combined to represent the numerical model of thermal system. Figure 2.8 shows a flow chart for developing the complete model of a thermal system.

Example: The walls and the ceiling of a room may be treated as separate components because of the difference in heat transfer mechanism. The walls and the outside insulation in a furnace may be treated as different components because of the difference in material.

2.7.1 ACCURACY AND VALIDATION

Numerical models for individual parts of the system should be tested and validated. Subsequently, these models can be merged to yield the overall numerical model of the thermal system. The main considerations that form the basis for validation are:

1. Results should be independent of arbitrary *numerical parameters* (grid, size, time step etc.).
2. The numerical model is subjected to a range of operating conditions and the results obtained are examined for physical consistency.
3. Simulation results should be obtained for a few highly idealized situations and compared with analytical solutions. Similarly, simulation data may be obtained for a few simple geometries and conditions for which experimental data are available.
4. Simulation results should be compared with the prototype behavior.

2.8 IMPORTANCE OF SIMULATION

Simulation can be useful in the following situations:

1. *Evaluation of Different Designs for Selection of an Acceptable Design:* The cost will be prohibitive if each of the designs is fabricated and tested for acceptability.

System simulation is employed to investigate each design and to determine if the given requirements and constraints are satisfied, thus yielding an acceptable or workable design.

2. *Study of System Behavior under Off-Design Conditions:* During actual operation of a thermal system, there is a deviation from design conditions because of variation in energy input, differences in raw materials fed into system, changes in characteristics of components with time and changes in environment conditions. The results obtained from simulation under off-design conditions thus indicate *versatility* and *robustness* of the system. These simulation results help to build safety features i.e. possible shutdown of the system or warning lights to indicate possible damage to the system.

3. *Study of Design Variables Effect for Optimization of System:* The result obtained from simulation may often be curve fitted to yield algebraic equations, which greatly facilitate the optimization of a thermal system design.

4. *Improvement or Modification of Existing Systems:* The simulation results can be used to correct a problem in the existing system. It can also be used to modify a system and improve its performance. Considerable savings may be obtained by using simulation in this manner.

5. *Investigation of Sensitivity of the Thermal System to Different Variables:* System simulation is used to determine sensitivity of the system performance to design variables. This helps the designer to decide if slight alterations can be made by using already available pipes, blowers, pumps and fans in the interest of reducing cost without significant sacrifice in system characteristics.

2.9   DIFFERENT CLASSES OF NUMERICAL SIMULATION

The numerical simulation can be classified based on the implementation procedure of the numerical scheme i.e. the number of *variables* (time), the *numerical approach* and the *type of input* to the system as described here.

2.9.1   DYNAMIC OR STEADY STATE

When the simulation is carried out with time as a variable, we classify it as dynamic simulation, otherwise it is a steady simulation. We may have a thermal system whose response (temperature) is steady most of the time except during startup and shutdown. As the system is steady most of the time, the simulation may be classified as steady.

2.9.2   CONTINUOUS OR DISCRETE

When we carry out the simulation of fluid flow in a refrigeration system, power plant etc., the flow of the fluid is taken as continuous with no finite gaps in between. This type of simulation is termed as continuous simulation. The example of discrete simulation is the heat treatment of glass of *television screen* in an oven, where the television screen elements travel in a conveyor belt inside the oven. In this situation,

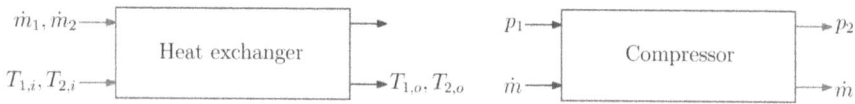

Figure 2.9　Block representation of a heat exchanger and compressor.

the mass, momentum and energy balances for each item (television screen) are considered separately to determine its temperature distribution as a function of time. An aggregate or average behavior of all screens may be obtained at the end to quantify the process if needed.

2.9.3　DETERMINISTIC OR STOCHASTIC

Simulations can be termed as deterministic or stochastic depending on the type of input. The design variables and inputs are assumed to be specified with precision in deterministic simulation. The input conditions are not known precisely in stochastic simulation, and a probability distribution with dominant frequency, an average and an amplitude of variation is provided. For example, when dealing with *consumer demands* for power, hot water supplies etc., probabilistic descriptions are often employed. The corresponding simulation is known as stochastic simulation. The turbulent flow and flow control applications also require specification of the turbulent flow i.e. frequency and amplitude of fluctuation which can also be specified using probability distribution.

2.10　FLOW OF INFORMATION

The strategy for system simulation is often guided by the nature and characteristics of information flow (temperature, velocity, flow rate and pressure) between different parts of the system. Different types of information flow arrangements arise depending on the nature of thermal system. The information flow takes place as inputs and outputs for different blocks in the information flow diagram. Different possible blocks representing the components of a thermal system are illustrated in Figure 2.9. The mass flow rate of two streams $(\dot{m}_1, \dot{m}_2)$ and temperature of respective streams $(T_{1,i}, T_{2,i})$ are the inputs embedded into a heat exchanger. The mass flow rate of the respective streams and the corresponding temperature at the exit of the heat exchanger are the output information received from the heat exchanger. Similarly, pressure (p_1) and mass flow rate $(\dot{m})$ are the input information embedded to a compressor, and pressure (p_2) and mass flow rate $(\dot{m})$ are the output information received from the compressor.

2.11　BLOCK REPRESENTATION

A thermal system is presented using block representation, where each component is represented as block with characteristic inputs and outputs. The equations that

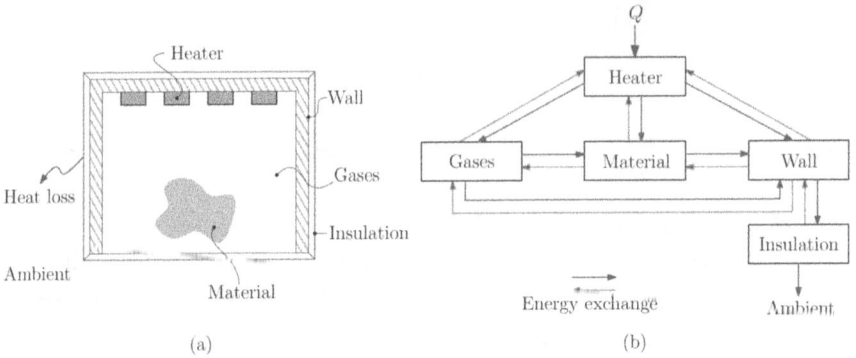

Figure 2.10 (a) Schematic of an electric furnace and (b) the information flow diagram of different blocks in the furnace.

express the relationship between the variables of a component or block are written within the block. The governing equations for different components of a thermal system may be written as

$$f_1(x_1, x_2, x_3, \ldots, x_n) = 0$$
$$f_2(x_1, x_2, x_3, \ldots, x_n) = 0$$
$$f_3(x_1, x_2, x_3, \ldots, x_n) = 0$$
$$\vdots \quad \vdots \quad \vdots \quad \vdots \quad \vdots$$
$$f_n(x_1, x_2, x_3, \ldots, x_n) = 0$$

Function f_1 may correspond to the compressor, and function f_2 may correspond to the heat exchanger (Figure 2.9).

2.11.1   INFORMATION FLOW DIAGRAM

The information flow diagram helps in setting up the simulation platform of a thermal system. Energy transfer occurs between different parts of the system. For example, in an electric furnace (see Figure 2.10), heater exchanges thermal energy with the walls, the gas and the material simultaneously. Similarly, the material undergoing heat treatment exchanges energy with the heater, walls and gases. A part inside the furnace is simultaneously coupled with several others through the energy exchange mechanisms. Therefore, simultaneous solution of the mathematical model corresponding to each component is carried out for the simulation of electric furnace.

In the vapor compression system (see Figure 2.11), the output from one component feeds to the next as an input. The overall arrangement is sequential i.e. one part depends only on the preceding one. Therefore, *sequential calculation procedure* is appropriate for the simulation of vapor compression system.

If these governing equations are linear, they may be solved by direct or iterative methods. Direct methods are used for small sets of equations and for tri-diagonal

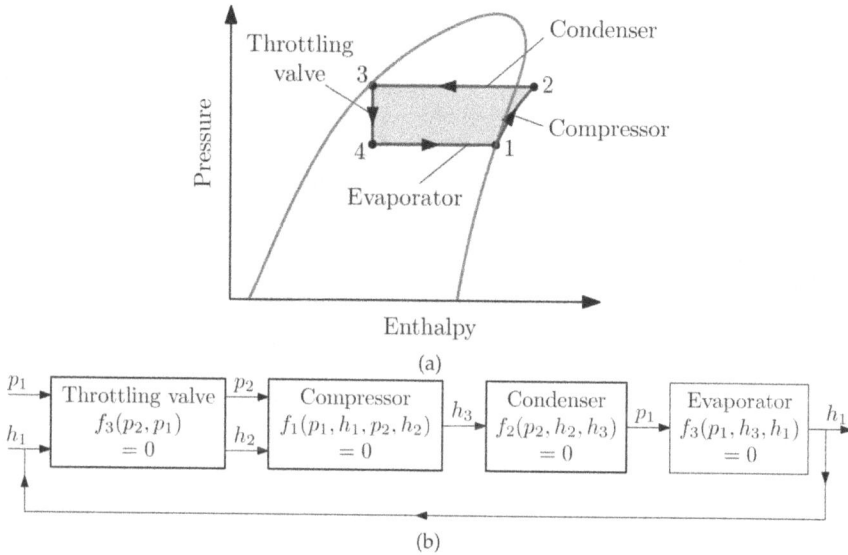

Figure 2.11 (a) The vapor compression cycle and (b) information flow diagram of a vapor compression cooling system.

system. Iterative methods are used for large sets of equations. If these equations are nonlinear, either successive substitution or the Newton–Raphson method is used. The successive substitution method is particularly well-suited for *sequential information flow diagrams*. The Newton–Raphson method is appropriate for an information flow diagram in which a strong interdependence arises between different parts of the system. *The convergence* characteristic is generally better in Newton's method than the successive substitution method. However, Newton's method is more complicated.

2.12 INITIAL DESIGN

The initial design is decided after the formulation of problem and a decision on the conceptual design. The analysis of system through modeling and simulation is used for the evaluation of an initial design for its acceptability. The initial starting design affects the convergence of iterative design process and often influences the final acceptable or optimal design. Therefore, the development of the initial design is a critical step in the design process and considerable care and effort must be exerted. Some of the commonly used methods for obtaining an initial design are as follows:

1. *Selection of Components to Meet Given Requirements and Constraints:* The design of a thermal system may involve selection of different components that meet the given requirements or specifications. For example, a *reverse Brayton cycle* is commonly used in aircrafts to cool the cabin. The detailed design may include selection of the components that satisfy constraint on pressures and temperatures.

The design process is carried out by including the efficiency of components of the thermal system.

2. *Existing System Designs:* Suppose an engineer has to design a forced air furnace for *continuous heat treatment* of silicon wafers. Similar systems that are being used for other processes such as baking of circuit boards and curing of plastic components may be employed to obtain initial estimates of the heater characteristics, wall material and conveyor design.

3. *Library of Previous Designs:* Similar designs used previously can be chosen as initial design of a thermal system. Some of these designs that were previously discarded might be satisfactory for different design requirements. For example, an air compressor may have been discarded earlier because the pressure or the flow rate is too low.

4. *Expert Knowledge:* Several ideas developed over the years form the basis of expert knowledge. The expert knowledge can play a major role in determining what is feasible and what is not. This is because many aspects of the thermal system are very complex to analyze. For example, contact resistance between surfaces, surface roughness, fouling in heat exchangers, losses due to friction and so on cannot be quantified for simulation of a thermal system.

2.13 ITERATIVE REDESIGN FOR CONVERGENCE

There are several variables in a design problem. Therefore, there are many criteria for convergence. Two approaches can be used for deciding criteria for convergence.

1. We should focus on quantity or condition that is of particular significance to the problem at hand.

2. Combination of different conditions may be selected as the convergence criteria i.e. several different requirements may be combined as convergence criterion that is used to be followed in an iterative process and to characterize its convergence characteristics.

Let $x_1, x_2, \ldots, x_n$ are the quantities of interest in a thermal system and d_i is the respective requirement, which may be specified as

$$x_i = d_i, x_i \leq d_i \text{ or } x_i \geq d_i$$

This can also be expressed as

$$(x_i - d_i) = 0, (x_i - d_i) \leq 0, (x_i - d_i) \geq 0$$

Possible combination of these requirements include:

$$Y = (x_1 - d_1) + (x_2 - d_2) + (x_3 - d_3)$$

or

$$Y = |x_1 - d_1| + |x_2 - d_2| + |x_3 - d_3|$$

or

$$Y = (x_1 - d_1)^2 + (x_2 - d_2)^2 + (x_3 - d_3)^2$$

For example, in a heat exchanger, the design parameter may be

$$Y = \left[\left(\frac{T_o - T_r}{T_r} \right)^2 + \left(\frac{Q - Q_r}{Q_r} \right)^2 \right]^{1/2}$$

where, T_o is the cold fluid outlet temperature and Q is the heat transfer rate. The subscript r refers to the required value. The iteration is carried out by varying the design variables and the design parameter (Y) is evaluated at the end of each iteration. When Y reaches the desired value Y_r, the iteration is terminated, which can be expresssed as

$$\|Y - Y_r\| \leq \in$$

Weighting factors (w) may be used to stress the importance of certain requirements over the others. Accordingly, the expression for convergence of the design can be modified as

$$Y = \left[\left(\frac{T_o - T_r}{T_r} \right)^2 + w \left(\frac{Q - Q_r}{Q_r} \right)^2 \right]^{1/2}$$

where, w is the weighing factor for the heat transfer requirement.

2.14 SAMPLE THERMAL SYSTEM DESIGN EXAMPLES

In the following sections, we discuss implementation issues related to the simulation of two thermal systems: (1) a piping network and (2) a gas turbine.

2.14.1 A PIPING NETWORK PROBLEM

Figure 2.12 shows the schematic of the piping network with a pump. Two pumps in parallel are pumping water between two reservoirs with an elevation difference $h = 91.5$ m. The pipes have diameter $D = 30.5$ cm and length $L = 3048$ m. The minor loss coefficient $K = 543$ and $\varepsilon/D = 0.001$. The pump-1 uses 33.02 cm rotor and pump-2 uses 30.48 cm rotor. The pump characteristic is shown in Figure 2.13.

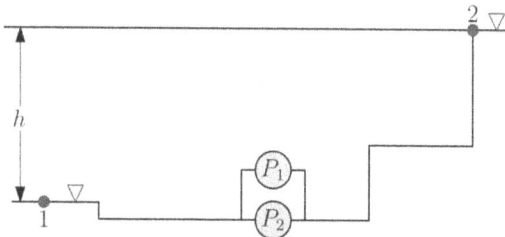

Figure 2.12   Schematic of a piping network.

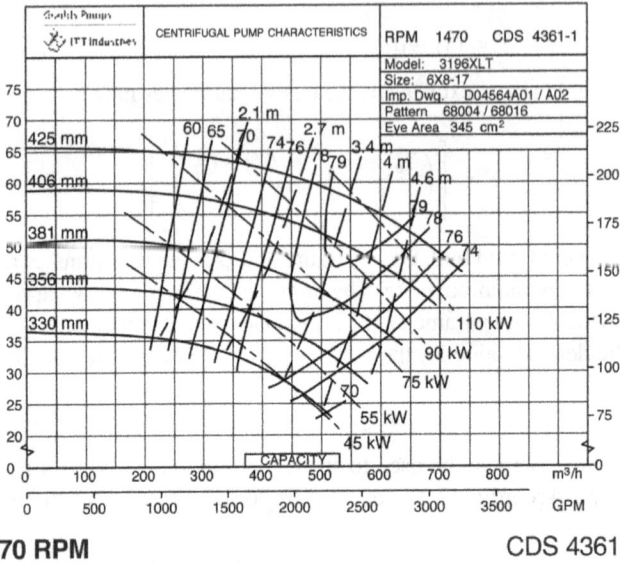

Figure 2.13 The pump performance curve for the pump used in Figure 2.12. (Goulds 3196, Pump Performance Curves.)

It is required to find the flow rate if water at 21 °C is the pumped fluid. The design parameter e.g. piping diameter and material, or the type of pump can be changed according to the flow rate requirement.

Solution

The system of equations for the piping network problem can be obtained by using pump curve and energy equation. Using pump curve, the $(H - Q)$ relation for pump 1 and pump 2 can be curve fitted using the data for the respective rotor diameter as

$$\Delta H_1 = a_{01} + a_{11}Q_1 + a_{21}Q_1^2$$

$$\Delta H_2 = a_{02} + a_{12}Q_2 + a_{22}Q_2^2$$

Since pumps are in parallel, we have

$$\Delta H_1 = \Delta H_2$$

Applying the energy equation between location -1 and location -2, we have

$$\Delta H = h + \left(\frac{fL}{D} + K\right)\frac{V^2}{2g} = h + \left(\frac{fL}{D} + K\right)\frac{Q_T^2}{2gA^2}$$

where, $Q_T = Q_1 + Q_2$. Here, K is the minor loss coefficient. The friction factor is a function of pipe material/roughness and Reynolds number. The empirical relationship discussed in Section 6.2 can be used for the friction factor calculation. Thus, we have three equations and three unknowns.

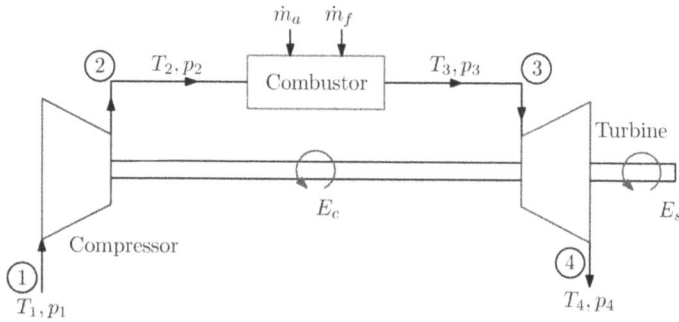

Figure 2.14 Schematic of a gas turbine system.

Alternatively, we can assume f to start with and then verify it once converged solution is obtained. In that situation, we will have two equations and two unknowns.

2.14.2 A GAS TURBINE PROBLEM

Figure 2.14 shows a gas turbine system, and the compressor and turbine characteristics are shown in Figure 2.15. Air is used as the working fluid. Inlet temperature to the compressor $T_1 = 38\,°C$ and 795 l/hr of JP-4 (heat content 302380 kJ/kg) fuel is used. Assume $P_2 = P_3$. The steady-state value of all variables needs to be reported. Write down the necessary governing equations required to simulate the gas turbine system. The mass flow rates can be varied as the design parameters till the required design objective i.e. shaft power is obtained.

Solution

Compressor: Using the data from the discharge pressure (P_2) versus flow rate $(\dot{m}_a)$ and applying the curve-fitting procedure, we have

$$P_2 = a_{po} + a_{p1}\dot{m}_a + a_{p2}\dot{m}_a^2$$

where, a_{p0}, a_{p1} and a_{p2} are constants.

Using compressor power (E_c) vs. discharge pressure (P_2) data in curve fitting, we have

$$E_c = a_{eo} + a_{e1}P_2 + a_{e2}P_2^2$$

The compressor power can also be expressed as the temperature difference between inlet and outlet as

$$E_c = \dot{m}_a C_p (T_2 - T_1)$$

Combustor: Assuming that all added fuel contributes in increasing the temperature, an energy balance yields,

$$\dot{q} \approx (\dot{m}_a + \dot{m}_f) C_p (T_3 - T_2)$$

where, $\dot{q}$ is known from the fuel enthalpy value and mass flow rate of fuel, $\dot{m}_f$.

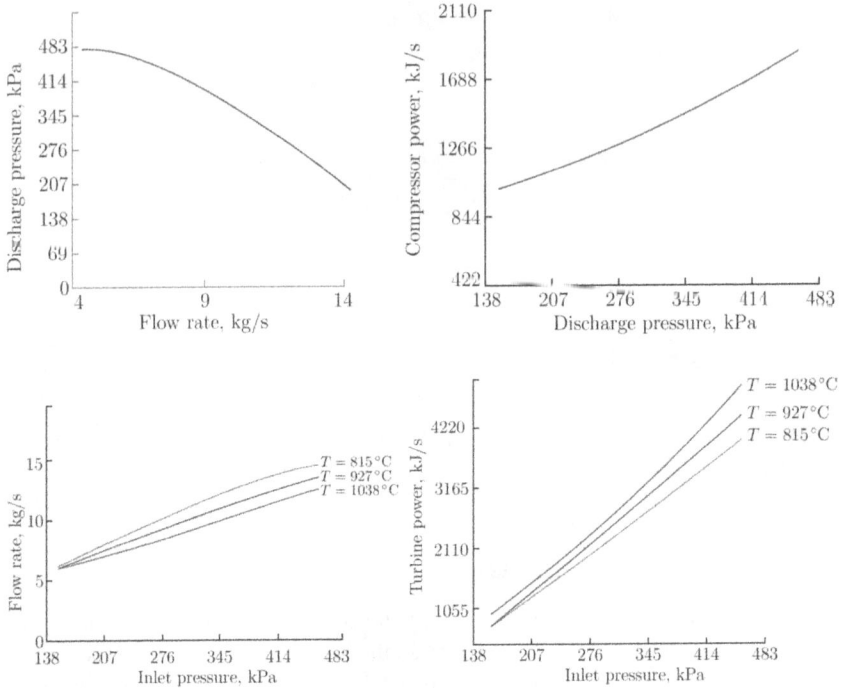

Figure 2.15 The compressor (top) and turbine (bottom) characteristics for the gas turbine system in Figure 2.14.

Turbine: From the turbine characteristic curve between flow rate $(\dot{m}_T)$ and inlet pressure (P_3) at different turbine inlet temperature (T_3), the curve-fitting equation can be written as

$$\dot{m}_T = a_{mo}(T_3) + a_{m1}(T_3)P_3 + a_{m2}(T_3)P_3^2$$

where, a_{m0}, a_{m1} and a_{m2} are constants. Similarly, from the turbine power (E_t) vs. inlet pressure (P_3) at different turbine inlet temperatures (T_3), the curve-fitting equation can be expressed as

$$E_t = a_{to}(T_3) + a_{t1}(T_3)P_3 + a_{t2}(T_3)P_3^2$$

The total turbine power, which is a combination of compressor power and shaft power, can be expressed as

$$E_t = E_c + E_s$$

The constants $a_{mo}, a_{m1}, a_{m2}, a_{to}, a_{t1}$ and a_{t2} are the functions of turbine inlet temperature (T_3). The unknowns are $P_2, \dot{m}_a, E_c, T_2, T_3, E_t$ and E_s (seven unknowns). So we have seven unknowns and seven equations. The Newton–Raphson method can be used to obtain the steady-state solution of the problem.

2.14.3 FIN DESIGN

Fin geometry is commonly seen in a variety of thermal systems. The detailed heat transfer analysis of fin is available in regular heat transfer books. Here, we outline the design procedure of fins. The fundamentals related to fin geometry are briefly discussed to maintain continuity.

Let's look at composite wall geometry of a plane wall and cylinder (see Figure 2.16). The problem can be analyzed as analogous electric circuit with the thermal resistance due to combination of conduction by the composite wall of different thermal conductivities (k_1, k_2 and k_3) and convection at the inner (hot) and outer (cold) regions (h_h and h_c). The effect of deposition of a film or scale on the surface can greatly increase the resistance to heat transfer between the fluids. The effect of this additional resistance called fouling can be estimated by adding additional resistance called *fouling resistance* (R_f).

For the plane wall, with composite wall thicknesses L_1, L_2 and L_3, the heat transfer is expressed as (Figure 2.16)

$$q_k = \frac{T_h - T_c}{\frac{1}{h_h A} + \frac{L_1}{K_1 A} + \frac{L_2}{K_2 A} + \frac{L_3}{K_3 A} + \frac{1}{h_c A} + \frac{R_f}{A}}$$

Here, R_f is the fouling resistance due to scale deposit formation on the surface. Thus, the overall heat transfer rate is dependent on the various conductive, convective and fouling resistances. The heat transfer rate can only be increased by decreasing the sum of resistances. One effective technique to enhance the heat transfer rate is by adding extended surfaces or fins.

The heat transfer book (Incropera et al., [2018]) reports analytical solutions for several single fin geometries. Assuming a single fin with circular cross section (P: perimeter, A: cross-sectional area, k: thermal conductivity), the following assumptions are used for the estimation of heat transfer by the fin:

1. Constant cross-sectional area
2. 1-D condition in x-direction
3. Constant material thermal conductivity
4. Steady state
5. Constant value of heat transfer coefficient ($\overline{h_c}$)

The heat transfer q_{fin} at the fin's base for different cases are

1. *Infinitely Long Fin (as $x \rightarrow \infty, T \rightarrow T_\infty$):*

$$q_{fin} = \sqrt{\overline{h_c} P k A} (T_s - T_\infty)$$

2. *Finite Fin with an Insulated End:*

$$q_{fin} = \sqrt{\overline{h_c} P k A} (T_s - T_\infty) \tanh(mL)$$

(a) Plane composite wall

(b) Cylindrical composite wall

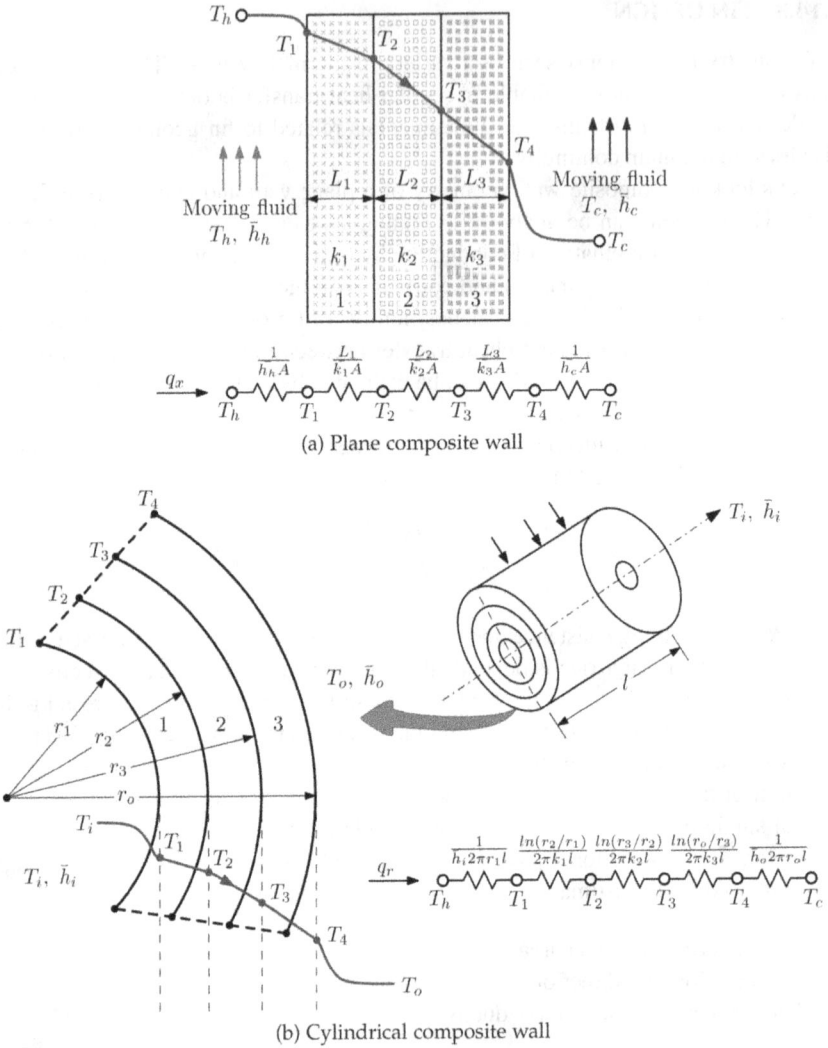

Figure 2.16 Schematic of (a) plane composite wall and (b) cylindrical composite wall along with the electrical resistance analogy.

3. *Finite Fin with End Loss by Convection:*

$$q_{fin} = \sqrt{\overline{h_c}PkA}(T_s - T_\infty)\dfrac{\sinh(mL) + \left(\dfrac{\overline{h_L}}{mk}\right)\cosh(mL)}{\cosh(mL) + \left(\dfrac{\overline{h_L}}{mk}\right)\sinh(mL)}$$

where, $\overline{h_L}$ is the average convective heat transfer co-efficient at the exposed end of fin and $m^2 = \overline{h_c}P/kA$. The above expressions are awkward to use in situations where

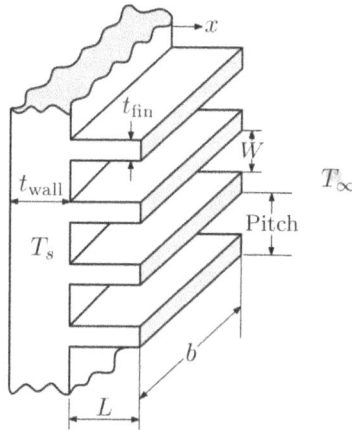

Figure 2.17 Schematic with nomenclature of a multi-fin geometry. (Hodge and Taylor [1999].)

multiple fins are present and where the temperature at the base of the fin, T_s, is not known. The fin efficiency is more useful in these situations.

$$\text{Fin efficiency, } \eta_f = q_{fin}/(q \text{ if entire fin is at } T_s)$$

For a circular pin fin of diameter D and length L with an insulated end, the fin efficiency can be derived as

$$\eta_f = \frac{tanh\sqrt{4L^2\overline{h_c}/kD}}{\sqrt{4L^2\overline{h_c}/kD}}$$

Similar expressions for other fin geometries i.e. parabolic, triangular, hyperbolic and rectangular geometries are available. Once the fin efficiency is known, the heat transfer of a fin can easily be evaluated.

2.14.3.1 Multiple Fins

Let's outline the design procedure of a multiple fin geometry. A typical multiple fin geometry is shown in Figure 2.17. We can obtain the following expressions from the geometry of the multiple fin.

$$A = \text{Total heat transfer area} = A_f + b(\text{Pitch} - t)$$

where, A_f = Total exposed fin area = $b(2L+t)$, η_f is fin efficiency and η_t is total surface effectiveness.

The total heat transfer is due to the combined effect of the fin and no-fin area i.e.

$$q_{total} = q_{fin} + q_{no-fin}$$

Using the total surface effectiveness (η_t),

$$q_{total} = A\eta_t \overline{h_c}(T_s - T_\infty) = \frac{T_s - T_\infty}{1/(A\eta_t \overline{h_c})}$$

We can write,

$$A\eta_t \overline{h_c}(T_s - T_\infty) = A_f \eta_f \overline{h_c}(T_s - T_\infty) + (A - A_f)\overline{h_c}(T_s - T_\infty)$$

Using this expression, we can write for the plane composite wall with multiple fins at both inner and outer wall as

$$q_{total} = \frac{T_h - T_\infty}{\frac{1}{\eta_{ti}h_h A} + \frac{R_{fi}}{A} + \sum_{j=1}^{J}\frac{L_j}{K_j A} + \frac{1}{\eta_{to}A\overline{h_c}}}$$

where, η_{ti} is total surface effectiveness of the finned hot surface, J is number of composite walls and η_{to} is total surface effectiveness of the finned cold surface.

Note: The addition of fins increases the surface area. But at the same time, it also introduces a conductive resistance over that portion of the original surface at which the fins are attached. The addition of fins will therefore *not always increase the rate of heat transfer.* If given a choice of whether to place the fins on the high $\overline{h_c}$ side or the low $\overline{h_c}$ side, the choice is to place fins on the low $\overline{h_c}$ side.

Fins provide a mechanism for reducing the convective resistance of low convective heat transfer situations. The final judgment as to whether or not fins are useful depends on the *convective heat transfer co-efficient, fin material thermal conductivity* and *fin geometry.* The ratio $\frac{h}{k}$ is an important indicator of fin effectiveness for a given geometry; the larger the ratio, the less effective is the fin. The *fin thickness* is also important, since if all other factors are the same, the thicker the fin, the lesser is the conductive resistance of the fin and larger is the fin effectiveness.

2.14.3.2 An Example of a Fin Problem

A mild steel ($K = 45$ W/m-°C) wall with 6.35 mm thickness separates a cold fluid from a hot fluid. A fin of thickness 25.4 mm and center to center spacing of 7.6 mm is placed on the wall.

(a) Estimate the fin performance (compare the value of U for no fin, fins in the hot side, fins in cold side and fins in both sides) for the following cases, when using rectangular straight fin (see Table 2.1).

(b) For part (a), examine the overall U-value as a function of fin length if the fin volume and pitch remain constant and if fins are placed only on the hot side.

PROBLEMS

1. A heat treatment plant needs hot water at temperature $T_c \pm \Delta T_c$ for the heat treatment process. A storage tank of volume V and surface area A is employed for this

Table 2.1

Different Cases for the Fin Evaluation

Case No.	K (W/m-°C)	$\bar{h}_{hot}$ (W/m²-°C)	$\bar{h}_{cold}$ (W/m²-°C)
1	45	56.8	2839
2	8.7	14196.5	14196.5
3	391	56.8	2839

purpose. Whenever hot water is withdrawn from the tank, cold water at temperature T_c flows into the tank. A heater supplying energy at the rate of Q turns on whenever the temperature reaches $T_c - \Delta T_c$ and turns off when it reaches $T_c + \Delta T_c$. The heater is submerged inside water contained in the tank. Assume uniform temperature in the tank and a convective heat loss to the environment at the surface, with a heat transfer coefficient h. Obtain a mathematical model for this system.

2. The bolts are placed on a conveyor belt during the heat treatment of steel bolts, which passes through a long furnace at speed V, as shown in the figure below. In the first section, the bolts are heated at a constant heat flux q. In the second and third sections, the bolts lose energy by convection to the air at temperature T_a at connective heat transfer coefficients h_1 and h_2 in the two sections, respectively. Assuming validity of lumped mass analysis, obtain the governing equations for the three sections and outline the mathematical model.

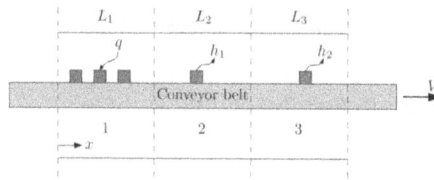

3. The displacement (x) of a particle in a flow is measured as a function of time (t). The data obtained are:

t (s)	0.0	1.0	2.0	3.0	4.0	5.0
x (m)	0.0	2.0	8.0	20.0	40.0	62.0

Obtain a linear best fit of the data. From this fit, calculate the values at t equal to 2.0 and 4.0. Compare these values with the given data and comment on the difference.

4. A water cooler is to be designed to supply cold drinking water with a given time-dependent mass flow rate m. Assume a cubical tank of cold water surrounded by insulation of uniform thickness. Water at the ambient temperature flows into the tank to make up the cold water outflow. The refrigeration unit turns on if the water

temperature reaches a value T_{max} and turns off when it drops to T_{min}. Develop a simple mathematical model for this system.

5. Consider a counter-flow heat exchanger, where the heat loss to the environment is to be included in the mathematical model. The hot fluid flows outside and the cold fluid flows inside, as shown in the figure below. Sketch qualitatively the change that the inclusion of this consideration will have on the temperature distribution in the heat exchanger. Also give the energy equation, taking this loss into account.

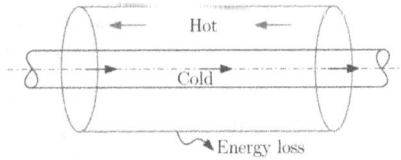

6. A hot surface of a chemical reactor at $90\,^{\circ}C$ is to be cooled by attaching 2.5 cm long, 0.3 cm diameter aluminum pin fins ($k = 237$ W/m-$^{\circ}$C) to it. There are total 28000 number of fins. The temperature of the surrounding medium is $25\,^{\circ}C$, and the heat transfer coefficient on the surface is equal to 30 W/m-$^{\circ}$C. Determine the overall effectiveness of the finned surface. Assume the fin end to be insulated.

7. A fan is used to supply air through the duct. The relation between the volume flow rate, Q (m^3/s.) and pressure difference, P (Pa) is given below:

$$Q = 15 - 75 \times 10^{-6} \times P2; \qquad P = 80 + 10.5 \times Q^{5/3}$$

Calculate the value of Q and P using the Newton–Raphson method. Carry out one iteration and use initial value of $P = 330$ Pa and $Q = 6.5$ m^3/s.

8. Consider a cylindrical rod of diameter d undergoing thermal processing and moving at a speed u as shown below:

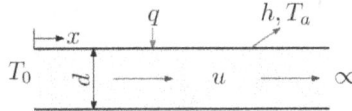

The rod may be assumed to be infinite in the direction of motion. Energy transfer occurs at the outer surface, with a constant heat flux input q and convective loss to the ambient at temperature T_a and heat transfer coefficient h. Assuming 2-D and steady transport, the governing equation and boundary conditions for the problem are given as

$$\rho C U A \frac{\partial T}{\partial x} = kA \left(\frac{1}{r} \frac{\partial}{\partial r} \left(r \frac{\partial T}{\partial r} \right) + \frac{\partial^2 T}{\partial x^2} \right)$$

The boundary conditions are

at $x = 0, T = T_{\infty}$;   at $r = 0, \frac{\partial T}{\partial r} = 0$;   at $r = d/2, k\frac{\partial T}{\partial r} = q - h(T - T_a)$

Use the non-dimensional variable as

$$X = \frac{x}{d}, \quad R = \frac{r}{d}, \quad \theta = \frac{T - T_a}{T_0 - T_a}$$

Find out the relevant non-dimensional parameters using non-dimensionalization of governing equation and boundary condition.

REFERENCES

B. Hodge and R. Taylor. *Analysis and Design of Energy Systems*. Prentice Hall, 1998.

F. P. Incorpera, D. P. DeWitt, T. L. Bergman, and A. S. Lavine. *Fundamentals of Heat and Mass Transfer*. John Wiley, New York, 2018.

Gould pump manual. *Seneca Falls*. NY: Gould Pump Inc. (http://www.gouldspumps.com) 2010.

Y. Jaluria. *Design and Optimization of Thermal Systems*, Second Edition. CRC Press, 2008.

3 Exergy for Design

3.1 INTRODUCTION

The first law concept is "energy cannot be destroyed". The idea that something can be destroyed is useful in the design and analysis of thermal systems, which is not applicable to energy but to exergy (availability). Exergy is a second law concept. It is exergy and not energy that properly gauges the quality (usefulness) of say 1 kJ of electricity generated by a power plant versus 1 kJ of energy in the plant cooling water stream. The exergy analysis should enable more effective use of energy resource because it enables the determination of location, cause and true magnitude of waste and loss. Such information can be used for the design of energy-efficient systems. For example, on the basis of first law reasoning alone, it may be ascertained that the condenser of a power plant is responsible for a plant's seemingly low overall efficiency. However, the exergy analysis reveals that the steam generator is the principal site of thermodynamic inefficiency owing to irreversibility within it, and the condenser is relatively unimportant.

3.1.1 DEFINITION OF EXERGY

An opportunity for doing useful work exists whenever there is communication between two systems in different states. Work can be developed when two systems are allowed to reach equilibrium. When one of the two systems is an idealized system called as environment and the other is some system of interest, *exergy* is the maximum useful theoretical work obtainable as the system interacts to achieve equilibrium. Exergy is a measure of the departure of the state of the system from that of the environment. It is therefore an attribute of both the system and the environment.

3.1.2 ENVIRONMENT

Any system whether a component in a larger system such as a gas turbine in a power plant or a larger system (power plant) itself operates within surroundings of some kind. It is important to distinguish between the *environment* and the system's *surroundings*. The term surroundings refers to everything not included in the system. The term environment applies to some portion of the surroundings. The intensive properties of each phase of the environment are uniform and do not change significantly as a result of any process under consideration. The environment is regarded as free of irreversibilities. All significant irreversibilities are located within the system and its immediate surroundings. The substances of the environment are in stable forms. There is no possibility of developing work from physical or chemical interactions between parts of the environment.

DOI: 10.1201/9781003049272-3

The environment is usually modeled as a simple compressible system, larger in the extent and uniform in pressure, $P_0 = 1$ atm and temperature, $T_0 = 25\,°C$. The enviromental condition may be defined as the average ambient temperature and pressure for the location at which the system under consideration operates.

3.1.3 EXERGY COMPONENTS

In the absence of nuclear, magnetic, electrical and surface tension effects, the total exergy of a system is divided into four components (physical + kinetic + potential + chemical):

$$e = e^{PH} + e^{KN} + e^{PT} + e^{CH} \tag{3.1}$$

When evaluated relative to the environment, the kinetic and the potential energies of a system are, in principle, fully convertible to work as the system is brought to rest relative to the environment. Therefore, kinetic and potential energy correspond to kinetic and potential exergies, respectively. Accordingly, the corresponding exergy components can be written as

$$e^{KN} = 1/2mv^2 \tag{3.2}$$

$$e^{PT} = mgz \tag{3.3}$$

3.1.3.1 Physical Exergy

Physical exergy is defined as the maximum theoretical value of the work of the system as it comes to equilibrium with the environment. Physical exergy for a closed system at a specified state is given by the expression:

$$e^{PH} = (U - U_0) + P_0(V - V_0) - T_0(S - S_0) \tag{3.4}$$

where, U is internal energy, V is volume and S is entropy of the system. The subscript, o refers to when the system is at the restricted dead state. The above expression can be derived by using the energy balance and the entropy balance to the combined system of the closed system and the environment which can be found in a regular thermodynamics book.

3.1.3.2 Dead States

The dead state corresponds to the conditions of mechanical, thermal and chemical equilibrium between the system and environment. There is no possibility of a spontaneous change within the system or the environment, nor can there be an interaction between them under these conditions. Only the conditions of mechanical and thermal equilibrium must be satisfied at the *restricted dead state*. At the restricted dead state, the fixed quantity of matter under consideration is imagined to be sealed in an envelope impervious to mass flow. It is also at zero velocity and elevation relative to the coordinates of the environment and at the temperature T_0 and pressure P_0.

3.1.3.3 Chemical Exergy

Chemical exergy is the exergy component associated with the departure of the chemical composition of a system from that of the environment. The substance comprising the system must be referred to the properties of a suitably selected set of environmental substances. Standard chemical exergy is based on standard environmental temperature T_0 (298.1 K) and pressure p_0 (1 atm). The standard environment is regarded as consisting of a set of reference substances with standard concentrations reflecting as closely as possible to the chemical makeup of the natural environment. The natural environment consists of standard gas phases representing air that includes N_2, O_2, CO_2, $H_2O(g)$ and other gases. Chemical exergy calculated with reference to the alternative specification of the environment is in general good agreement with each other.

3.2 EXERGY BALANCE EQUATION

Similar to mass, energy and entropy, exergy balance equations can be developed for different energy systems and system components. In the following section, exergy balance equations are developed for closed and open systems.

3.2.1 CLOSED SYSTEM

The exergy balance equation can be developed by combining the energy balance and entropy balance equation.

3.2.1.1 Energy Balance of the Closed System

Based on the first law of thermodynamics, the energy balance equation of the closed system between state 1 and state 2 can be written as

$$(U_2 - U_1) + (KE_2 - KE_1) + (PE_2 - PE_1) = \int_1^2 \delta Q - W \qquad (3.5)$$

3.2.1.2 Entropy Balance of the Closed System

Similarly, the entropy change between the two states of the closed system is

$$S_2 - S_1 = \int_1^2 \left(\frac{\delta Q}{T} \right)_b + S_{gen} \qquad (3.6)$$

where, S_{gen} is the entropy generation and T_b is the temperature on the boundary where energy transfer by heat occurs.

Multiplying the entropy balance equation by T_0 and subtracting from the energy balance equation gives

$$(U_2 - U_1) + (KE_2 - KE_1) + (PE_2 - PE_1) - T_0(S_2 - S_1)$$
$$= \int_1^2 \delta Q - T_0 \int_1^2 \left(\frac{\delta Q}{T} \right)_b - W - T_0 S_{gen} \qquad (3.7)$$

The exergy change between the two states of the closed system can be written from the definition of exergy as

$$(e_2 - e_1) = (U_2 - U_1) + (KE_2 - KE_1) + (PE_2 - PE_1) - T_0(S_2 - S_1) + P_0(V_2 - V_1) \tag{3.8}$$

Substituting equation 3.8 in equation 3.7, we get

$$(e_2 - e_1) = \int_1^2 \left(1 - \frac{T_0}{T_b}\right) \delta Q - (W - P_0(V_2 - V_1)) - T_0 S_{gen} \tag{3.9}$$

In equation 3.9, the first term $(e_2 - e_1)$ is the *exergy change*; the second term $\left(1 - \frac{T_0}{T_b}\right) \delta Q$ is the *exergy transfer* associated with the transfer of energy by heat, the third term $(W - P_0(V_2 - V_1))$ is the *exergy transfer* associated with net useful work, which is equal to the work of the system W less the work that would be required to displace the environment whose pressure is P_0 and the last term $(T_0 S_{gen})$ is *exergy destruction* due to irreversibilities within the system.

Note: As $(1 - \frac{T_0}{T_b})$ is the Carnot efficiency, the first quantity in the RHS of equation 3.9 can be interpreted as the work that could be generated by a reversible power cycle receiving energy by heat transfer δQ at the temperature T_b and discharging energy by heat transfer to the environment at temperature T_0.

We can also write the above exergy balance equation in a rate form as

$$\frac{de}{dt} = \sum_j \left(1 - \frac{T_0}{T_j}\right) \dot{Q}_j - \left(\dot{W} - P_0 \frac{dV}{dt}\right) - \dot{E}_D \tag{3.10}$$

where, $\frac{dV}{dt}$ is the rate of change in system volume, $\dot{E}_D$ is the rate of exergy destruction, $\dot{W}$ is the rate of work transfer and $\dot{Q}$ is the rate of heat transfer.

3.2.2   OPEN SYSTEM

A control volume shown in Figure 3.1 represents an open system with exergy transfer at the inlet and outlet. Exergy can be transferred into or out of a control volume. Thus, adding the terms accounting for exergy transfers in equation 3.10, we get

$$\frac{de_{CV}}{dt} = \sum_j \left(1 - \frac{T_0}{T_j}\right) \dot{Q}_j - \left(\dot{W}_{CV} - P_0 \frac{dCV}{dt}\right) + \sum_i \dot{m}_i e_i - \sum_e \dot{m}_e e_e - \dot{E}_D \tag{3.11}$$

where, $\frac{de_{CV}}{dt}$ is the rate of exergy change, $\sum_j \left(1 - \frac{T_0}{T_j}\right) \dot{Q}_j - \left(\dot{W}_{cv} - P_0 \frac{dCV}{dt}\right) + \sum_i \dot{m}_i e_i - \sum_e \dot{m}_e e_e$ is the rate of exergy transfer and $\dot{E}_D$ is the rate of exergy destruction.

At the steady state, we have

$$0 = \sum_j (1 - \frac{T_0}{T_j}) \dot{Q}_j - \dot{W}_{CV} + \sum_i \dot{m}_i e_i - \sum_e \dot{m}_e e_e - \dot{E}_D$$

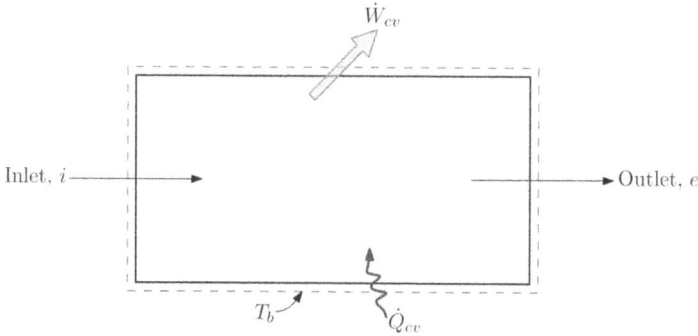

Figure 3.1 The control volume at steady state.

3.2.2.1 Exergy Transfers at Inlets and Outlets

It is essential to evaluate the exergy transfers at inlets and outlets e_i and e_e, respectively, for representing the exergy balance equation. The exergy associated with a stream of matter entering (or exiting) a control volume is the maximum theoretical work that should be obtained when the stream is brought to a dead state and heat transfer takes place with the environment only. This work can be evaluated in two steps. In the first step, the stream is brought to a restricted dead state. In the second step, it is brought from the restricted dead state to the dead state. The contribution of the first step is the physical exergy e^{PH}. The contribution from the second step to the work developed is the chemical exergy e^{CH}. The energy balance for the above control volume (Figure 3.1) can be written as

$$0 = \dot{Q}_{CV} - \dot{W}_{CV} + \dot{m}((h_i - h_e) + 0.5(V_i^2 - V_e^2) + g(Z_i - Z_e)) \tag{3.12}$$

The entropy balance equation is

$$0 = \frac{\dot{Q}_{CV}}{T_b} + \dot{m}(s_i - s_e) + \dot{S}_{gen} \tag{3.13}$$

Combining the above energy rate balance and entropy rate balance of the control volume by eliminating the heat transfer term from these expressions, we have

$$\frac{\dot{W}_{CV}}{m} = \left[(h_i - h_e) - T_b(S_i - S_e) + 0.5(V_i^2 - V_e^2) + g(Z_i - Z_e)\right] - T_b \frac{\dot{S}_{gen}}{\dot{m}} \tag{3.14}$$

For an internal reversible system, $\dot{S}_{gen}$ is equal to zero and the above equation reduces to:

$$\left(\frac{\dot{W}_{cv}}{\dot{m}}\right)_{Int\text{-}rev} = [(h_i - h_e) - T_b(S_i - S_e) + 0.5(V_i^2 - V_e^2) + g(Z_i - Z_e) \tag{3.15}$$

The properties of the stream at the inlet are h, S, V and Z, and at the outlet, the corresponding properties are $h_0, S_0, V_0 = 0$ and $Z_0 = 0$, where h_0 and S_0 are specific

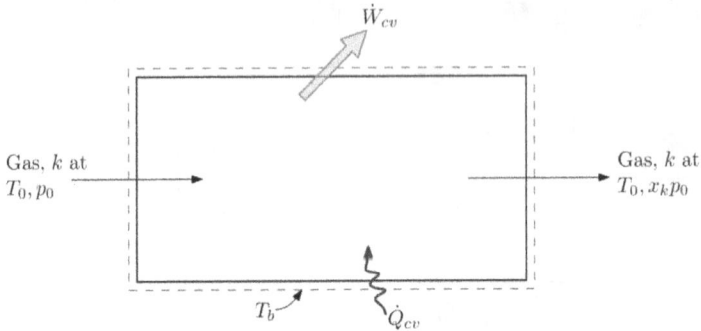

Figure 3.2   Device for evaluating the chemical exergy of a gas and gas mixtures.

enthalpy and specific entropy at the restricted dead state, respectively. The work developed per unit mass is

$$\left(\frac{\dot{W}_{cv}}{\dot{m}}\right)_{\text{Int-rev}} = [(h-h_0) - T_0(S-S_0) + 0.5V^2 + gZ] \qquad (3.16)$$

Note that as the heat transfer occurs with the environment only, the temperature T_b at which heat transfer occurs corresponds to T_0. The total exergy transfer is

$$e = [(h-h_0) - T_0(S-S_0) + 0.5V^2 + gZ + e^{CH}] \qquad (3.17)$$

where physical exergy is

$$e^{PH} = (h-h_0) - T_0(S-S_0) \qquad (3.18)$$

Physical exergy is associated with the temperature and pressure of the stream of matter.

3.2.3   STANDARD CHEMICAL EXERGY OF GASES AND GAS MIXTURES

A device showing the entry and exit of gas through a system (see Figure 3.2) can be used to evaluate the chemical exergy of a gas stream. Let us assume that the gas k enters the device in the gas phase at temperature T_0 and pressure P_0. It expands isothermally with heat transfer only with the environment and exits to the environment at temperature T_0 and partial pressure $P_k^e = x_k^e P_0$, where e denotes the environment and x_k^e is the mole fraction of gas k in the environmental gas phase. The maximum theoretical work per mole of gas k would be developed when the expansion occurs without any irreversibilities. Using the control volume energy equation 3.15, we have

$$e_K^{CH} = \left(\frac{W_{CV}}{m}\right)_{\text{Int-rev}} = (h_i - h_e) - T_0(S_i - S_e) + 0.5(V_i^2 - V_e^2) + g(Z_i - Z_e)$$

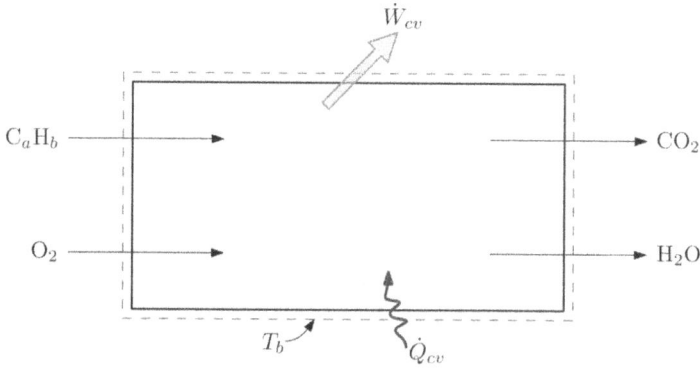

Figure 3.3 Device for evaluating the chemical exergy of hydrocarbon fuel. All substances enter and exit unmixed at pressure P_0 and temperature, T_0. Heat transfer is at temperature T_0 only.

Assuming $V_i = V_e$, $Z_i = Z_e$, we have

$$e_k^{CH} = (h_i - h_e) - T_0(S_i - S_e)$$

Using the relationship

$$dh = Tds - vdp$$

$$e_k^{CH} = -vdp = \int_{P_i}^{P_e} -vdp$$

Using ideal gas relationship $v = \frac{\bar{R}T}{P}$, we have

$$\bar{e}_k^{CH} = -\bar{R}T_0 \ln \frac{P_e}{P_i} = -\bar{R}T_0 \ln \frac{x_k^e P_0}{x_k P_0} = -\bar{R}T_0 ln \frac{x_k^e}{x_k}$$

If a gas mixture at temperature T_0 and pressure P_0 with mole fraction x_k enters, whose mole fraction in the environment is x_k^e, the chemical exergy per mole of the mixture can be obtained by summation as

$$\bar{e}_k^{CH} = -\bar{R}T_0 \Sigma x_k \ln \frac{x_k^e}{x_k}$$

3.2.4 STANDARD CHEMICAL EXERGY OF FUELS

The standard chemical exergy of the substance not present in the environment can be evaluated by considering an idealized reaction of the substance with other substances for which the chemical exergies are known. Figure 3.3 shows a device for evaluating the chemical exergy of a hydrocarbon fuel. All substances enter and exit unmixed at pressure P_0 and temperature T_0. Heat transfer takes place at temperature T_0 only. Let's assume the following reaction equation of the hydrocarbon inside the device.

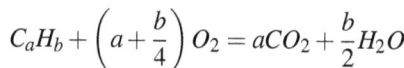

$$C_a H_b + \left(a + \frac{b}{4}\right) O_2 = aCO_2 + \frac{b}{2} H_2O$$

Using the exergy balance of the device shown in Figure 3.3 for steady state, we have

$$0 = \sum_j \left(1 - \frac{T_0}{T_j}\right) \dot{Q}_j - \dot{W}_{CV} + \sum_i \dot{m}_i e_i - \sum_e \dot{m}_e e_e - \dot{E}_D$$

As the temperature at heat transfer is equal to T_0, the exergy change due to heat transfer is equal to zero and we have

$$0 = -\left(\frac{\dot{W}_{CV}}{n_F}\right)_{\text{Int.rev}} + \bar{e}_F^{CH} + (a+b/4)\bar{e}_{O_2}^{CH} - a\bar{e}_{CO_2}^{CH} - (b/2)\bar{e}_{H_2O}^{CH}$$

Thus, we have

$$\bar{e}_F^{CH} = \left(\frac{\dot{W}_{CV}}{n_F}\right)_{\text{Int-rev}} + a\bar{e}_{CO_2}^{CH} + (b/2)\bar{e}_{H_2O}^{CH} - (a+b/4)\bar{e}_{O_2}^{CH} \qquad (3.19)$$

To develop the expression of $\bar{e}_F^{CH}$, we need to find out the expression of $\frac{\dot{W}_{CV}}{n_F}$. The derivation of $\frac{\dot{W}_{CV}}{n_F}$ is shown below. For steady-state operation, the energy equation is expressed as

$$\dot{Q}_{cv} - \dot{W}_{cv} + \dot{m}(h_i - h_e) + 1/2(v_i^2 - v_e^2) + g(Z_i - Z_e) = 0$$

The energy rate balance for the system on per mole of fuel basis reduces to

$$\frac{\dot{W}_{CV}}{\dot{n}_F} = \frac{\dot{Q}_{CV}}{n_F} + \bar{h}_F + (a+b/4)h_{O_2} - ah_{CO_2} - (b/2)h_{H_2O} \qquad (3.20)$$

We can write the entropy balance for the control volume as

$$\frac{\dot{Q}_{CV}}{T_0} + \dot{m}(S_i - S_e) + \dot{S}_{gen} = 0$$

For the present system (Figure 3.3), this equation can be written on per mole basis as

$$0 = \frac{\dot{Q}_{CV}/\dot{n}_F}{T_0} + \bar{S}_F + (a+b/4)\bar{S}_{O_2} - a\bar{S}_{CO_2} - (b/2)\bar{S}_{H_2O} + \frac{\dot{S}_{gen}}{n_F} \qquad (3.21)$$

Eliminating the heat transfer from equations (3.20) and (3.21), the work developed per mole of fuel is

$$\frac{\dot{W}_{CV}}{\dot{n}_F} = [\bar{h}_F + (a+b/4)\bar{h}_{O_2} - a\bar{h}_{CO_2} - (b/2)\bar{h}_{H_2O}]$$

$$-T_0[\bar{S}_F + (a+b/4)\bar{S}_{O_2} - a\bar{S}_{CO_2} - (b/2)\bar{S}_{H_2O}] - T_0\frac{\dot{S}_{gen}}{\dot{n}_F} \qquad (3.22)$$

For the calculation of enthalpy, we can use the following:

$$\bar{h}(T,\rho) = \bar{h}_f^0 + [\bar{h}(T,\rho) - \bar{h}(T_{\text{ref}}, P_{\text{ref}})]$$

where $\bar{h}$ is enthalpy at a state other than the standard state and h_f^0 is the enthalpy of formation.

For ideal gas, the entropy can be written as

$$\bar{S}(T,\rho) = \bar{S}(T,\rho_{ref}) + [\bar{S}(T,\rho) - \bar{S}(T,\rho_{ref})] = \bar{S}^0(T) - \bar{R}\ln\left(\frac{P}{P_{ref}}\right)$$

Using $\dot{S}_{gen} = 0$ for an internally reversible process, we can write

$$\left(\frac{W_{CV}}{\dot{n}_F}\right)_{Int\text{-}rev} = [(\bar{h}_F + (a+b/4)\bar{h}_{O_2} - a\bar{h}_{CO_2} - (b/2)\bar{h}_{H_2O})] \\ - T_0(\bar{S}_F + (a+b/4)\bar{S}_{O_2} - a\bar{S}_{CO_2} - (b/2)\bar{S}_{H_2O}) \tag{3.23}$$

Note that the terms inside the first bracket in the RHS of this equation are known as the higher heating value (HHV) when water is liquid and the lower heating value (LHV) when water is in vapor form. Hence, substituting the work term in the exergy balance equation (3.19), we have

$$\bar{e}_F^{CH} = HHV(T_0, P_0) - T_0\left[\bar{S}_F + (a+b/4)\bar{S}_{O_2} - a\bar{S}_{CO_2} - (b/2)\bar{S}_{H_2O(l)}\right](T_0, P_0) \\ + [a\bar{e}_{CO_2}^{CH} + (b/2)\bar{e}_{H_2O(l)}^{CH} - (a+b/4)\bar{e}_{O_2}^{CH}] \tag{3.24}$$

Alternatively, using Gibbs function $\bar{g} = \bar{h} - T\bar{S}$, we can write

$$\left(\frac{W_{CV}}{n_F}\right)_{Int.rev} = [\bar{g}_F + (a+b/4)\bar{g}_{O_2} - a\bar{g}_{CO_2} - (b/2)\bar{g}_{H_2O}]_{T_0, P_0}$$

Thus,

$$\bar{e}_F^{CH} = \bar{g}_F + (a+b/4)\bar{g}_{O_2} - a\bar{g}_{CO_2} - (b/2)\bar{g}_{H_2O} \\ + a\bar{e}_{CO_2}^{CH} + (b/2)\bar{e}_{H_2O(l)}^{CH} - (a+b/4)\bar{e}_{O_2}^{CH} \tag{3.25}$$

Equations 3.24 and 3.25 can be used for the calculation of fuel exergy depending on the availability of data for the calculation.

3.3   EXERGY DESTRUCTION AND EXERGY LOSS

In the exergy equation, exergy destruction is one of the important components that needs to be evaluated for completeness. Exergy analysis often involves the calculation of performance parameters i.e. exergy destruction ratio, exergy loss ratio and exergetic efficiency. We discuss these terms and their calculation procedure in the following section.

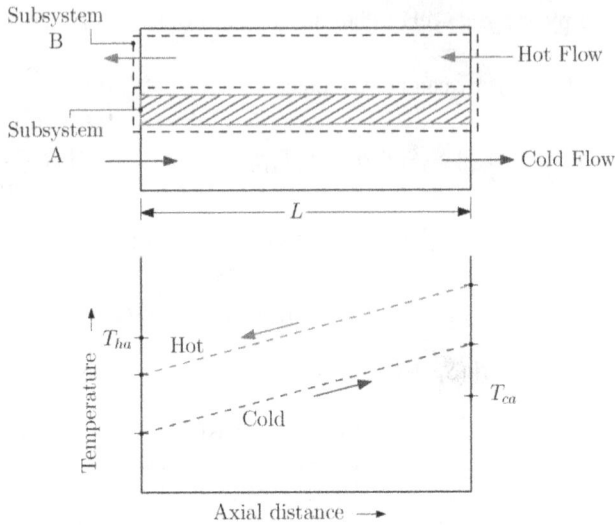

Figure 3.4 Schematic of a counter-flow heat exchanger for calculation of exergy destruction and loss.

3.3.1 EXERGY DESTRUCTION THROUGH HEAT TRANSFER AND FRICTION

Exergy destruction can take place through heat transfer, friction and chemical reaction. In this section, we use a heat exchanger surface example to discuss its calculation procedure and parameters involved. Figure 3.4 shows the hot and cold fluid streams of a heat exchanger and the respective temperature distribution. Subsystem A in Figure 3.4 is the closed system consisting of a wall separating the hot and cold streams, and subsystem B is another subsystem consisting of the fluid flow region of the hot fluid. We define T_{ha} and T_{ca} as the thermodynamic average temperature for the hot and cold streams, respectively.

3.3.1.1 Thermodynamic Average Temperature

For heat transfer experienced by one inlet and one outlet control volume, the heat transfer can be regarded to occur at the thermodynamic average temperature T_a for simplicity i.e.

$$\left(\frac{Q_{cv}}{m}\right)_{int.rev} = T_a(S_e - S_i) = \int_i^e T\,ds$$

where,

$$T_a = \frac{\int_i^e T\,ds}{S_e - S_i} = \frac{(h_e - h_i) - \int_i^e v\,dp}{S_e - S_i}$$

For constant pressure, we have

$$T_a = \frac{h_e - h_i}{S_e - S_i}$$

3.3.1.1.1 Subsystem A

Subsystem A constitutes exergy destruction due to heat transfer only. Using closed system exergy rate equation 3.10,

$$\frac{de}{dt} = \sum_j \left(1 - \frac{T_0}{T_j}\right)\dot{Q}_j - \left(\dot{W} - P_0\frac{dV}{dt}\right) - \dot{E}_D$$

we have

$$0 = \left(1 - \frac{T_0}{T_{ha}}\right)\dot{Q} - \left(1 - \frac{T_0}{T_{ca}}\right)\dot{Q} - \dot{E}_D \tag{3.26}$$

Hence, we can write

$$\dot{E}_D = T_0\dot{Q}\frac{T_{ha} - T_{ca}}{T_{ha}T_{ca}}$$

Heat transfer rate is proportional to the temperature difference between the streams i.e.

$$Q \propto (T_{ha} - T_{ca})$$

Thus, we have

$$\dot{E}_D \propto \frac{T_0(T_{ha} - T_{ca})^2}{T_{ha}T_{ca}} \tag{3.27}$$

The rate of exergy destruction associated with heat transfer varies quadratically with the stream-to-stream temperature difference and inversely with the product of temperature levels.

3.3.1.1.2 Subsystem B

Subsystem B is the control volume encompassing the channel through which hot stream flows. Friction is the only irreversibility. Using the control volume exergy rate equation 3.11,

$$0 = \sum_j \left(1 - \frac{T_0}{T_j}\right)\dot{Q}_j - \dot{W}_{cv} + \sum_i \dot{m}e_i - \sum_e \dot{m}_e e_e - \dot{E}_D$$

where,

$$e = (h - h_0) - T_0(S - S_0) + \frac{1}{2}V^2 + gZ + e^{CH}$$

For subsystem, B, we have

$$\dot{E}_D = \left(1 - \frac{T_0}{T_{ha}}\right)\dot{Q} + \dot{m}[(h_i - h_e) - T_0(S_i - S_e)] \tag{3.28}$$

Note that the potential and kinetic energy change is assumed to be zero and e^{CH} cancels out.

From the energy balance,

$$\dot{Q} = \dot{m}(h_e - h_i)$$

Substituting $\dot{Q}$, we have

$$\dot{E}_D = T_0 \dot{m} \left[(S_e - S_i) - \frac{h_e - h_i}{T_{ha}} \right]$$

Using the definition of T_{ha},

$$\dot{E}_D = -\frac{T_0 \dot{m} \int_i^e v d p}{T_{ha}} - \frac{T_0 \dot{m} h_l}{T_{ha}} = \frac{T_0 \dot{m} \left(\frac{4L}{D} \right) \left(\frac{V^2}{2} \right) f}{T_{ha}} \qquad (3.29)$$

Here, f is the fanning friction factor and h_l is the head loss.

3.3.1.2   Overview

1. It may be noted from equation 3.29 that the head loss and exergy destruction are not synonymous. The head loss accounts for the effect of friction and other resistances in converting mechanical energy to internal energy. Energy is conserved in all such situations. Exergy destruction, on the other hand, accounts for the irreversible reduction in the magnitude of total exergy owing to friction and other resistances. The exergy destruction and head loss differ by the temperature ratio T_0/T_{ha}.
2. The magnitude of exergy destruction is less than that of the head loss for components separating at temperatures above T_0. The magnitude of exergy destruction is greater than the head loss for components operating below T_0. Thus, the effect of friction is especially significant for the system operating at a temperature below that of the environmental temperature i.e. cryogenic systems.
3. The exergy destruction associated with a given value of the head loss is higher at lower working fluid temperatures. Therefore, relatively, greater attention should be paid to the design of lower temperature stages of turbines and compressors compared to higher stages in a power plant.
   There is another situation that requires more attention to the lower stages of compressors and turbines. Suppose "1" and "2" denote the inlet and outlet, respectively, of the compressor, and turbine and the "2s" denotes the isentropic outlet state. For a turbine, the exit temperature after the actual expansion (T_2) exceeds the exit temperature for an ideal expansion without friction (T_{2s}). Since the temperature at state 2 is higher than at state 2s, the specific exergy at state 2 exceeds that at state 2s. Therefore, there is no need to optimize the irreversibility for higher turbine stages. Consequently, the capacity to develop work in the next turbine state is enhanced. For the compressor, $T_2 > T_{2s}$. But higher T_2 here is unfavorable. Because higher temperature requires the next compressor stage to accommodate a greater volumetric flow rate.

3.3.2   EXERGY DESTRUCTION AND EXERGY LOSS RATIOS

The values of the rates of exergy destruction, $\dot{E}_D$, and exergy loss, $\dot{E}_L$, provide thermodynamic measures of system inefficiencies. The exergy destruction ratios y_D and

y_D^* compare the exergy destruction with the exergy rate of the fuel provided to the overall system$(\dot{E}_{F,\text{total}})$ and the total exergy destruction rate within the system $(\dot{E}_{D,\text{total}})$, respectively. The y_D and y_D^* are useful for comparisons among various components of the same system.

$$y_D = \frac{\dot{E}_D}{\dot{E}_{F,total}} \tag{3.30}$$

$$y_D^* = \frac{\dot{E}_D}{\dot{E}_{D,total}} \tag{3.31}$$

The exergy loss ratio is similarly defined as

$$y_L = \frac{\dot{E}_L}{\dot{E}_{F,total}} \tag{3.32}$$

3.3.3 EXERGETIC EFFICIENCY

Exergetic efficiency provides a true performance measure of an energy system from the thermodynamic viewpoint. In defining exergetic efficiency, it is necessary to identify both a product and a fuel for the thermodynamic system being analyzed. The product represents the desired result provided by the system. The fuel represents the resources expended to generate the product. The fuel may not be necessarily restricted to being an actual fuel such as natural gas, oil and coal. Both the product and fuel are expressed in terms of exergy.

An exergy rate balance for a system can be written as

$$\dot{E}_F = \dot{E}_P + \dot{E}_D + \dot{E}_L$$

where $\dot{E}_F$ is exergy of fuel, $\dot{E}_p$ is exergy of product, $\dot{E}_D$ is exergy destruction and $\dot{E}_L$ is exergy losses.

Exergetic efficiency (ε) is defined as

$$\varepsilon = \frac{\dot{E}_P}{\dot{E}_F} = 1 - \frac{\dot{E}_D + \dot{E}_L}{\dot{E}_F} \tag{3.33}$$

The exergetic efficiency can also be written as

$$\varepsilon = 1 - \sum y_D - \sum y_L$$

An important use of exergetic efficiency is to assess the thermodynamic performance of a component, plant or industries relative to the performance of similar components, plants or industries. The performance of a gas turbine can be gauged relative to the typical performance level of other gas turbines. A comparison of exergetic efficiencies for dissimilar devices i.e. gas turbine with heat exchangers is generally not meaningful.

3.3.3.1   How Do We Distinguish between Fuel and Product?

The calculation of exergetic efficiency requires proper classification of the fuel and product. Table 3.1 schematically shows the inlet and outlet streams of important components in a thermal system and the corresponding definition of the product and fuel.

3.3.3.2   Compressor, Pump or Fan

In a compressor, pump or fan, gas or liquid is caused to flow in the direction of increasing pressure and elevation by means of mechanical or electrical power input. As the exergy of the stream increases, we consider the product to be the exergy increase between inlet and outlet $(\dot{E}_2 - \dot{E}_1)$ (Table 3.1).

$$\dot{E}_P = \dot{E}_2 - \dot{E}_1$$

The exergy of fuel is equal to the power input $(\dot{W})$

$$\dot{E}_F = \dot{W}$$

Exergetic efficiency ε is defined as

$$\varepsilon = \frac{\dot{E}_2 - \dot{E}_1}{\dot{W}}$$

3.3.3.3   Turbine or Expander

The exergetic efficiency of the turbine/expander is given by

$$\varepsilon = \frac{\dot{W}}{\dot{E}_1 - \dot{E}_2} = \frac{\dot{E}_P}{\dot{E}_F}$$

For a turbine with one extraction, we have (see Table 3.1)

$$\varepsilon = \frac{\dot{W}}{\dot{E}_1 - \dot{E}_2 - \dot{E}_3} = \frac{\dot{E}_P}{\dot{E}_F}$$

3.3.3.4   Heat Exchanger

The definition of the fuel and product for a heat exchanger can be classified into two cases. The general sketch of a heat exchanger is shown in Table 3.1.

3.3.3.5   Case 1

Heat transfer occurs at or above T_0, and the purpose of the heat exchanger is to increase the exergy of the cold stream $(T_1 \geq T_0)$. An example of this case is a heat recovery steam generator and air preheater. The exergetic efficiency for this situation is defined as

$$\varepsilon = \frac{\dot{E}_2 - \dot{E}_1}{\dot{E}_3 - \dot{E}_4} = \frac{\dot{E}_P}{\dot{E}_F}$$

Table 3.1

Definition of Product and Fuel for Different Types of Energy Systems (i.e. Compressor, Pump or Fan, Turbine, Mixing Unit, Boiler)

Component	Schematic	Exergy Rate of Product, $\dot{E}_P$	Exergy Rate of Fuel, $\dot{E}_F$
Compressor, pump, or fan		$\dot{E}_2 - \dot{E}_1$	$\dot{W}$
Turbine or expander		$\dot{W}$	$\dot{E}_1 - \dot{E}_2 - \dot{E}_3$
Heat exchanger		$\dot{E}_2 - \dot{E}_1$	$\dot{E}_3 - \dot{E}_4$
Mixing unit		$\dot{E}_3$	$\dot{E}_1 + \dot{E}_2$
Gasifier or combustion chamber		$\dot{E}_3$	$\dot{E}_1 + \dot{E}_2$
Boiler		$(\dot{E}_6 - \dot{E}_5) + (\dot{E}_8 - \dot{E}_7)$	$(\dot{E}_1 + \dot{E}_2) - (\dot{E}_3 + \dot{E}_4)$

3.3.3.6 Case 2

When the purpose is to cool the hot stream with the cold stream (refrigeration applications), exergy is transferred from the cold stream to the hot stream; i.e. exergy is transferred in the direction opposite to the direction of heat transfer. For this case, the product and fuel are defined as

$$\dot{E}_P = \dot{E}_4 - \dot{E}_3$$

and

$$\dot{E}_F = \dot{E}_1 - \dot{E}_2$$

3.3.3.7 Mixing Unit

For the mixing unit, the product and fuel are defined with respect to Table 3.1 as

$$\dot{E}_P = \dot{E}_3$$

$$\dot{E}_F = \dot{E}_1 + \dot{E}_2$$

Here, both the hot stream ($\dot{E}_2$) and cold stream ($\dot{E}_1$) exergies contribute to the exergy of the fuel.

3.3.3.8 Gasifier or Combustion Chamber

For the gasifier or combustion chamber (see Table 3.1), the product and fuel are defined as

$$\dot{E}_P = \dot{E}_3$$

$$\dot{E}_F = \dot{E}_1 + \dot{E}_2$$

Here, the reaction product exergy ($\dot{E}_3$) is the product. The oxidants ($\dot{E}_2$) and fuel ($\dot{E}_1$) are termed as fuel.

3.3.3.9 Boiler

With reference to Figure 3.1, the product and fuel for the boiler are defined as

$$\dot{E}_P = (\dot{E}_6 - \dot{E}_5) + (\dot{E}_8 - \dot{E}_7)$$

$$\dot{E}_F = (\dot{E}_1 + \dot{E}_2) - (\dot{E}_3 + \dot{E}_4)$$

For the boiler, the exergies of the flue gas stream (4) and ash stream (3) are treated as losses.

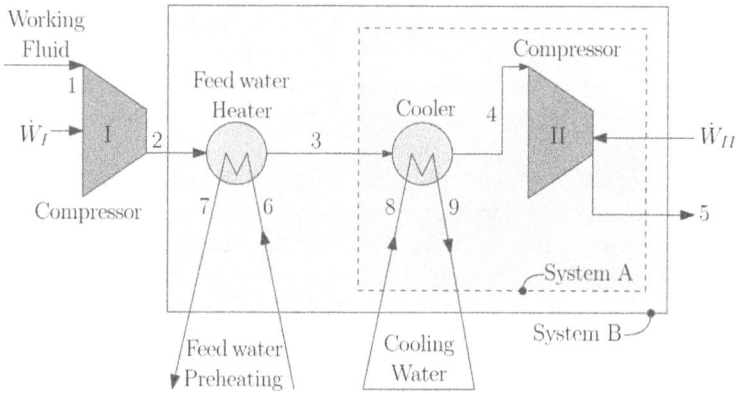

Figure 3.5 Schematic of a compression process with intercooling for the definition of exergetic efficiency.

3.3.3.10 Guidelines for Defining Exergetic Efficiency

The definition of exergetic efficiency may not be obvious in certain situations, where the basic purpose of some components is to improve the performance of another component of the system. Some guidelines for defining the exergetic efficiency are provided below:

1. The definition of exergetic efficiency should be meaningful from both the thermo-dynamic and economic viewpoints.
2. Identifying the fuel as the sum of all exergy inputs and the product as the sum of all exergy outputs can result in misleading conclusions for single plant components.
3. When the purpose of owning and operating a plant component also involves other components, an exergetic efficiency generally should be defined for the enlarged system consisting of the component and other components which it directly affects.

Example: Figure 3.5 shows the schematic of a compression process with inter-cooling, where the heat exchangers (feed water heater and cooler) achieve the cooling of a stream by heating another stream whose exergy gain is discarded. Here, the heat exchanger or cooler exergetic efficiency is defined together with the system component it serves. Two systems are considered (Figure 3.5): (1) system A consists of the cooler and one of the compressors and (2) system B is the combination of system A and the feed water heater.

3.3.3.11 Subsystem A

The exergetic efficiency of subsystem A is defined as

$$\varepsilon_A = \frac{\dot{E}_5 - \dot{E}_3}{\dot{W}_{II}} = \frac{\dot{E}_P}{\dot{E}_F}$$

Note: A cooler serves the compressor by cooling the working fluid. As a result, the power required by the compressor and the compressor size is reduced as it has to handle a lower volume of air. We have assumed that the exergy gain of the cooling water $\dot{E}_9 - \dot{E}_8$ is discarded while writing the above expression for exergetic efficiency. It is counted as exergy loss. The cooler contributes to both the exergy destruction and exergy loss of system A. However, it plays an overall positive role by reducing the cost associated with the compressor, a lower operating cost owing to the reduced power requirement and a lower investment cost owing to the reduction in its size.

3.3.3.12 Subsystem B

The exergetic efficiency of the subsystem B is defined as

$$\varepsilon_B = \frac{(\dot{E}_5 - \dot{E}_2) + (\dot{E}_7 - \dot{E}_6)}{\dot{W}_{II}}$$

Here, the feed water heater provides preheated feed water for use elsewhere and also assists in reducing the power required for compressor II, by cooling the working fluid.

Similarly, a *throttling valve* is a component for which the product is not readily defined when the valve is considered alone. The throttling valve typically serves other components. Accordingly, when formulating an exergetic efficiency, the throttling valve and the components it serves should be considered together.

Note: The exergetic efficiency is generally more meaningful than any other efficiency based on the first law of thermodynamics i.e. thermal efficiency of a power plant, isentropic efficiency of a compressor or a turbine and effectiveness of a heat exchanger.

Example:

1. The thermal efficiency of a cogeneration system is misleading because it treats both the work and heat transfer as having equal thermodynamic value.
2. The isentropic turbine efficiency, which compares the actual process with an isentropic process, does not consider that the working fluid at the outlet of the turbine has a higher temperature in the actual process than in the isentropic process. Consequently, for the actual process, there is higher exergy that can be used in the next component.
3. The heat exchanger effectiveness ($\varepsilon = \frac{q}{q_{max}}$) fails to identify the exergy waste associated with a pressure drop of the heat exchanger working fluids.

3.4   EXERGY ANALYSIS OF A GAS TURBINE-BASED POWER PLANT

A gas turbine-based power plant using cogeneration is shown in Figure 3.6. Table 3.2 presents the state properties of the cogeneration system. The natural gas, methane is used as fuel. Exergy at different points of the cogeneration system can be calculated

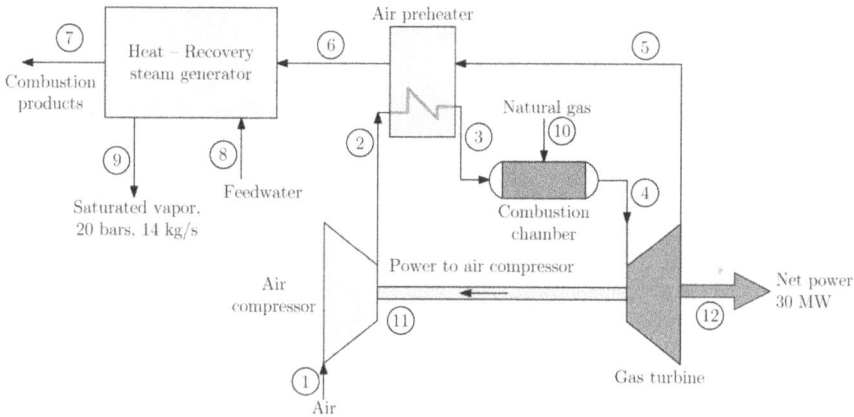

Figure 3.6   A base case of a cogeneration system for exergy analysis.

Table 3.2

State Properties of the Cogeneration System Shown in Figure 3.6

State	Mass Flow Rate (kg/s)	Temperature (K)	Pressure (bar)
1	91.2757	298.15	1.013
2	91.2757	603.738	10.13
3	91.2757	850.000	9.623
4	92.9176	1520.000	9.132
5	92.9176	1006.162	1.099
6	92.9176	779.784	1.066
7	92.9276	426.897	1.013
8	14.0000	298.150	20.000
9	14.000	485.570	20.000
10	1.6419	298.15	12.000

by using the property data for enthalpy and entropy. Table 3.3 shows the exergy values calculated at different locations of the cogeneration system. From Table 3.3, it can be noted that 3.26% ($2.772 \times 100/84.993$) of the fuel exergy is carried out of the system at state 7 and is charged as an exergy loss. Table 3.4 shows that the total summation of exergy destruction of all components is equal to 46.4%. Thus, the overall efficiency caused by exergy destruction and loss is

$$\varepsilon = 1 - \sum y_D - \sum y_L = 1 - 0.4643 - 0.0326 = 0.503$$

Table 3.3

Exergy Values at Different Locations for the Cogeneration System Shown in Figure 3.6

State	E^{PH}	E^{CH}	$E\left(E^{PH}+E^{CH}\right)$
1	0.0	0.0	0.0
2	27.538	0.0	27.53
3	41.938	0.0	41.93
4	101.087	0.3665	101.45
5	38.415	0.3665	38.78
6	21.385	0.3665	21.75
7	2.406	0.3665	2.772
8	0.026	0.0350	0.061
9	12.775	0.0350	12.81
10	0.6271	84.366	84.993

Exergetic efficiency of the components of the cogeneration system

Compressor:

$$\varepsilon = \frac{\dot{E}_2 - \dot{E}_1}{\dot{W}} = \frac{27.53}{29.662} = 92.8\%$$

Turbine:

$$\varepsilon = \frac{\dot{W}}{\dot{E}_4 - \dot{E}_5} = \frac{29.662 + 30}{101.45 - 38.78} = 95.2\%$$

Table 3.4

Exergy Destruction Ratios of Different Components of the Cogeneration System

Component	y_D^* (%)	y_D (%)	Rate (mW)
Combustion chamber	64.56	29.98	25.48
Heat recovery steam generator	15.78	7.33	6.23
Gas turbine	7.63	3.54	3.01
Air preheater	6.66	3.09	2.63
Air compressor	5.37	2.49	2.12
Overall plant	100.00	46.43	39.47

Heat recovery steam generator:

$$\varepsilon = \frac{\dot{E}_9 - \dot{E}_8}{\dot{E}_6 - \dot{E}_7} = \frac{12.81 - 0.061}{21.75 - 2.772} = 67.2\%$$

Air preheater:

$$\varepsilon = \frac{\dot{E}_3 - \dot{E}_2}{\dot{E}_5 - \dot{E}_6} = \frac{41.93 - 27.53}{38.78 - 21.75} = 84.6\%$$

Combustion chamber:

$$\varepsilon = \frac{\dot{E}_4}{\dot{E}_{10} + \dot{E}_3} = \frac{101.45}{84.993 + 41.93} = 79.9\%$$

Notes:

1. The exergy destruction data (Table 3.4) show that the combustion chamber is the major source of thermodynamic inefficiency, followed by a heat recovery steam generator.
2. Exergy destruction is caused by one or more of the three principal irreversibilities associated i.e. chemical reaction, heat transfer and friction. All three irrevesibilities are present in the combustion chamber, where the chemical reaction is the most significant source of exergy destruction. Heat transfer and friction are the sources of exergy destruction for the heat recovery steam generator (HRSG) and air preheater with the most significant irreversibilities being related to the steam-to-steam heat transfer. Exergy destruction in the adiabatic compressor and turbine is caused primarily by friction.
3. Combustion is intrinsically a significant source of irreversibility. A dramatic reduction in its effect on exergy destruction by conventional means can't be expected. Inefficiency in combustion can be reduced by preheating the combustion air. Exergy destruction associated with heat transfer decreases as the temperature difference between the stream is reduced. This is achievable by proper design of the heat exchanger, though there is normally an accompanying increase in exergy destruction by friction. The exergy destruction within the gas turbine and the air compressor decreases as friction is reduced.
4. Measures that improve the thermodynamic performance of one component might adversely affect another leading to no net overall improvement of the system.

3.4.1   GUIDELINES FOR EVALUATING AND IMPROVING THERMODYNAMIC EFFECTIVENESS

The following guidelines from the experience and exergy analysis can be helpful in improving thermodynamic effectiveness.

1. It is important to maximize the use of cogeneration whenever it is feasible for achieving the cost-effective thermal systems.

2. Chemical reaction is a significant source of thermodynamic inefficiency. Therefore, it is a good practice to minimize the use of combustion. As we cannot eliminate the use of combustion, it is advisable to reduce the exergy destruction of the practical combustion system by (a) minimizing the use of excess air and (b) preheating the reactants.
3. Nonidealities associated with heat transfer typically contribute heavily to inefficiency. Thus, a general guideline is that unnecessary or cost-ineffective heat transfer must be avoided.

3.4.2 ADDITIONAL GUIDELINES

1. The lower the temperature level, the greater the need to minimize the steam-to-stream temperature difference.
2. Avoid the use of intermediate heat transfer fluids when exchanging energy between two streams.
3. If there is a significant difference in the heat capacity rates (product of the mass flow rate and the specific heat Cp) of two streams exchanging energy by heat transfer in heat exchanger networks, consider splitting the stream with a larger heat capacity rate.
4. Use the *pinch method* for the heat exchanger network design.

The following guidelines can be applied to friction and mixing irreversibilities:

1. More attention should be paid to the design of the lower temperature stages of turbines and compressors (the last stage of the turbines and the first stage of compressor) compared to the remaining stages of these devices.
2. The use of throttling should be minimized. Explore whether power recovery expanders are cost-effective alternative for pressure reduction.
3. The mixing of streams differing significantly in temperature, pressure or chemical composition should be minimized.
4. The greater the mass flow rate, the greater is the need to use exergy of the stream effectively.
5. The lower the temperature level, the greater is the need to minimize the friction.

3.5 EXERGY ANALYSIS OF A HEAT EXCHANGER

The exergy analysis of a counter-flow heat exchanger geometry is reported in this section (see Figure 3.7). The following assumptions are used:

1. The heat capacity rate of both the streams are equal i.e. $(\dot{m}C_P)_1 = (\dot{m}C_P)_2 = \dot{m}C_P$ (balanced case).
2. Same fluid flows on each side of the heat exchanger.
3. Fluid is modeled as an ideal gas.

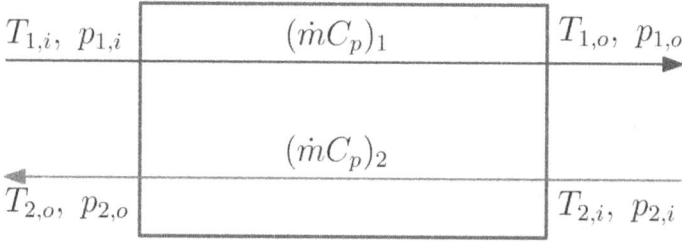

Figure 3.7   A respective geometry of a counter-flow heat exchanger.

Applying the entropy balance, the entropy generation rate of each heat exchanger stream can be written as (Bejan et al. (1995))

$$\dot{S}_{gen} = \dot{m}(S_e - S_i) + \frac{\dot{Q}_{CV}}{T_0}$$

The enthalpy change of each stream can be written as: $dh = Tds - vdp$ and $dh = \dot{m}C_p dT$.

Combining the exergy destruction of two streams, we have

$$\dot{S}_{gen} = (\dot{m}C_P)_1 \, ln \, \frac{T_{1,out}}{T_1} + (\dot{m}C_P)_2 \, ln \, \frac{T_{2,out}}{T_2} - (\dot{m}R)_1 \, ln \, \frac{P_{1,out}}{P_1} - (\dot{m}R)_2 \, ln \, \frac{P_{2,out}}{P_2}$$

$$(3.34)$$

The effectiveness of heat exchanger is expressed as

$$\varepsilon = \frac{T_1 - T_{1,out}}{T_1 - T_2} = \frac{T_2 - T_{2,out}}{T_1 - T_2} \tag{3.35}$$

We can assume the limit of small ΔT & ΔP i.e. high effectiveness of heat exchanger. Thus, we have

$$1 - \varepsilon \ll 1,$$

In other words,

$$\frac{P_1 - P_{1,out}}{P_1} \ll 1, \quad \frac{P_2 - P_{2,out}}{P_2} \ll 1$$

We define entropy generation number as

$$N_S = \frac{\dot{S}_{gen}}{\dot{m}C_P}$$

Substituting the expression of $\dot{S}_{gen}$, we have

$$N_s = (1 - \varepsilon)\frac{(T_2 - T_1)^2}{T_1 T_2} + \frac{R}{C_P}\left[\left(\frac{\Delta P}{P}\right)_1 + \left(\frac{\Delta P}{P}\right)_2\right] \tag{3.36}$$

Here, $(1-\varepsilon)\frac{(T_2-T_1)^2}{T_1 T_2}$ is the irreversibility due to stream-to-stream heat transfer, $\frac{R}{C_p}\left(\frac{\Delta P}{P}\right)_1$ is the irreversibility due to pressure drop along the first stream and $\frac{R}{C_P}\left(\frac{\Delta P}{P}\right)_2$ is the irreversibility due to pressure drop along the second stream.

The total heat transfer irreversibility can be divided into two parts, each describing the contribution made by one side of the heat transfer surface.

Let's assume the thermal resistance of the wall to be negligible. Hence, the overall thermal resistance can be written as

$$\frac{1}{\bar{h}A_1} = \frac{1}{\bar{h}_1 A_1} + \frac{1}{\bar{h}_2 A_2} \tag{3.37}$$

Here, $\frac{1}{\bar{h}A_1}$ is the overall heat thermal resistances based on A_1. Similarly, $\frac{1}{\bar{h}_1 A_1}$ and $\frac{1}{\bar{h}_2 A_2}$ are the thermal resistances based on their respective areas. The above expression can be written as

$$\frac{1}{NTU} = \frac{1}{NTU_1} + \frac{1}{NTU_2} \tag{3.38}$$

Here,

$$NTU = \frac{\bar{h}A_1}{\dot{m}C_p}, \quad NTU_1 = \frac{\bar{h}_1 A_1}{\dot{m}C_p}, \quad NTU_2 = \frac{\bar{h}_2 A_2}{\dot{m}C_p}$$

When $(\dot{m}C_p)_1 = (\dot{m}C_p)_2 = \dot{m}C_p$

$$\varepsilon = \frac{NTU}{1 + NTU} \tag{3.39}$$

for $1 - \varepsilon \ll 1$

$$\varepsilon = 1 - \frac{1}{NTU} \tag{3.40}$$

Combining equations 3.36, 3.38 and 3.40, we can obtain

$$N_S = \underbrace{\frac{\tau^2}{NTU_1} + \frac{R}{C_P}\left(\frac{\Delta P}{P}\right)_1}_{NS_1} + \underbrace{\frac{\tau^2}{NTU_2} + \frac{R}{C_P}\left(\frac{\Delta P}{P}\right)_2}_{NS_2} \tag{3.41}$$

where,

$$\tau = \frac{|T_2 - T_1|}{(T_2 T_1)^{1/2}}$$

One-sided entropy generation numbers $N_{S,1}$ and $N_{S,2}$ have the *same analytical form*. Therefore, the analysis of one of them can be used for the other. Let us rewrite $N_{S,1}$, using $NTU_1 = \left(\frac{4L}{D_h}\right)_1 St_1$

where, $St = \frac{h}{\rho C_P U}$, $D_h = \frac{4A}{P}$ and $A_{surf} = PL$.

Therefore, we can write

$$N_{S,1} = \frac{\tau^2}{St_1}\left(\frac{D_h}{4L}\right)_1 + \frac{R}{C_p}g_1^2 f_1 \left(\frac{4L}{D_h}\right)_1 \qquad (3.42)$$

where, $f_1 = \frac{\rho D_{h,1}}{2G_1^2}\frac{\Delta P_1}{L_1}$ and $St_1 = \frac{\bar{h}_1}{C_p G_1}$.

Let's define dimensional mass velocity $(g_1) = \frac{G_1}{(2\rho P_1)}$ (using $P = \frac{1}{2}\rho V^2, V = \sqrt{\frac{2P}{\rho}}$ and $G = \rho V = \sqrt{2P\rho}$)

We know that $f_1 = f_1(Re_1)$ & $St_1 = St_1(Re_1, Pr)$.

When the mass velocity and Reynolds number are fixed, equation 3.42 shows that the first term of the RHS varies inversely with the ratio $4L/D_h$ and the second term varies directly with the same ratio. Thus, there is a value of the ratio $4L/D_h$ for which $N_{S,1}$ is minimum. The optimum value of $(4L/D_h)$ can be derived as

$$\left(\frac{4L}{D_h}\right)_{1,opt} = \frac{\tau}{g_1[(R/C_P)f_1 St_1]^{1/2}}$$

The corresponding minimum entropy generation number is

$$N_{S,1,min} = 2\tau \left(\frac{R}{C_P}\right)^{1/2} g_1 \left(\frac{f_1}{St_1}\right)^{1/2}$$

Note: In the case of the most common heat exchanger surfaces, the group $(f_1/St_1)^{1/2}$ is only a *weak function of* Re_1. This means that the *mass velocity and minimum rate of entropy generation on one side of the heat exchanger surface are directly proportional to each other.*

Let's look at the entropy generation minimization problem of a heat exchanger surface with two different types of constraints i.e. constraint based on the heat transfer area and the volume of heat exchanger.

3.5.1 AREA CONSTRAINT

From equation 3.42, we see that the one-side irreversibility depends on the parameters fixed by the fluid type and the inlet conditions: τ, R/C_P, Pr, size and geometry of the heat exchanger passage, $(\frac{4L}{D_h})$, Re_1 and g. The degree of freedom out of these three parameters depends on the number of constraints. One important constraint concerns the heat transfer area A_1, where the cost of the heat exchanger surface is a major consideration. The constant-area constraint can be expressed in dimensionless form as

$$a_1 = \frac{A_1}{\dot{m}/\rho V} = \frac{A_1(2\rho P)^{0.5}}{\dot{m}}$$

where, A_1 is the heat transfer area. We can show that

$$a_1 = \frac{(\pi D_h L)_1 (2\rho P)^{1/2}}{\rho A V}$$

$$= (\pi D_h L)_1 \frac{(2P_1\rho)^{1/2}}{\pi D_h^2 V \rho / 4} = \left(\frac{4L}{D_h}\right)_1 \frac{1}{g_1}$$

Thus, we can show that

$$a_1 g_1 = \left(\frac{4L}{D_h}\right)_1$$

Now, equation 3.42 can be written as

$$N_{S,1} = \frac{\tau^2}{a_1 g_1 S t_1} + \frac{R}{C_P} a_1 f_1 g_1^3 \tag{3.43}$$

$$\therefore \qquad N_{S,1} = N_{S,1}(g_1, Re_1) \tag{3.44}$$

For minimizing the entropy generation number subject to a fixed Reynolds number, we have

$$\frac{\delta N_{S,1}}{\delta g_1} = 0$$

$$g_{1,opt} = \left[\frac{\tau^2}{(3R/C_p)a_1^2 f_1 S t_1}\right]^{1/4}$$

$$N_{S,1,min} = \left[\frac{256\tau^6 (R/C_p) f_1}{27 a_1^2 S t_1^3}\right]^{1/4}$$

3.5.2   VOLUME CONSTRAINT

This constraint might be important in the design of heat exchangers for applications *where space is limited.* The volume constraint can be written in dimensionless form as

$$v_1 = V_1 \frac{8P_1}{\gamma \dot{m}}$$

where, V_1 = volume = L_1 * "cross-sectional" area of the passage.

The non dimensionalization is written based on the Poiseuille flow rate expression. We can write

$$v_1 g_1^2 = \left(\frac{4L}{D_h}\right) Re_1$$

Thus, equation 3.42 can be written as

$$N_{S,1} = \frac{\tau^2 Re_1}{v_1 g_1^2 S t_1} + \frac{R}{C_P} \frac{v_1 f_1 g_1^4}{Re_1}$$

For a volume constraint, we can write

$$N_{S,1} = N_{S,1}(g_1, Re_1)$$

If we fix Re_1, the optimal mass flow rate and the corresponding minimum irreversibility are

$$g_{1,opt} = \left[\frac{\tau^2 Re_1^2}{2(R/C_P)v_1^2 f_1 St_1} \right]^{1/6}$$

$$N_{S,1,min} = \left[\frac{27\tau^4 (R/C_P)Re_1 f_1}{4v_1 St_1^2} \right]^{1/3}$$

3.5.3 COMBINED AREA AND VOLUME CONSTRAINT

When both the area A_1 and the volume V_1 of the heat exchanger passage is constrained simultaneously, there is only one degree of freedom left for optimizing the thermodynamic performance of the passage. Combining a_1 (constant area constraint in dimensionless form), and v_1 (dimensionless volume constraint) with equation 3.42 yields

$$N_{S,1} = \frac{\tau^2 v_1}{a_1^2 St_1 Re_1} + \frac{R}{C_P} \frac{a_1^4 f_1 Re_1^3}{v_1^3} \qquad (3.45)$$

In commercial pipes at large Re_1, f_1 & St_1 are relatively *insensitive to changes* in Re_1. Therefore, $N_{S,1}$ is a function of Re_1 only. Then, $N_{S,1}$ is minimum when

$$Re_{1,opt} = \frac{v_1}{a_1^{3/2}} \left[\frac{\tau^2}{3(R/C_P)f_1 St_1} \right]^{1/4} \qquad (3.46)$$

3.5.4 UNBALANCED HEAT EXCHANGER

Let us take limiting cases of unbalanced heat exchangers (for which the capacity rate differs). The unbalance is described by

$$\omega = \frac{(\dot{m}C_P)_1}{(\dot{m}C_P)_2} > 1$$

3.5.5 COUNTER-FLOW HEAT EXCHANGER

When $NTU \to \infty$, $T_{2,out} \to T_1$, using energy balance,

$$(\dot{m}C_P)_1(T_{1,out} - T_1) = (\dot{m}C_P)_2(T_2 - T_{2,out})$$

we have

$$T_{1,out} = T_1 - \frac{1}{\omega}(T_1 - T_2)$$

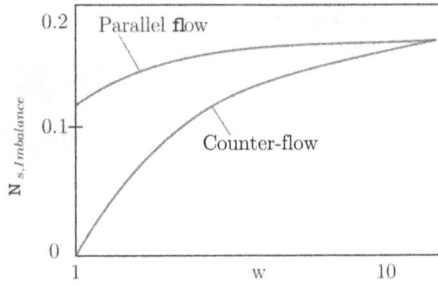

Figure 3.8 Comparison of the entropy generation number $(N_{s,Imbalance})$ as a function of imbalance (w).

Using equations 3.45 and 3.46 with the above relation of outlet temperature T1 and assuming $\Delta P_1 = \Delta P_2 = 0$, we have

$$N_{S,Imbalance} = \frac{\dot{S}_{gen}}{(\dot{m}C_P)_2} \tag{3.47}$$

$$= ln\left\{ \left[1 - \frac{1}{\omega}\left(1 - \frac{T_2}{T_1}\right)\right]^{\omega} \frac{T_1}{T_2} \right\} \tag{3.48}$$

3.5.6 PARALLEL FLOW HEAT EXCHANGER

Similar to the counter-flow heat exchanger, combing equation 3.36 with the effectiveness—NTU relation and assuming $NTU \rightarrow \infty$, $\Delta P_1 = \Delta P_2 = 0$

$$N_{S,Imbalance} = \frac{\dot{S}_{gen}}{(\dot{m}C_P)_2} = ln\left\{ \left(\frac{T_2}{T_1}\right)^{\omega}\left[1 + \left(\frac{T_1}{T_2} - 1\right)\left(\frac{\omega}{1+\omega}\right)\right]^{1+\omega} \right\} \tag{3.49}$$

Figure 3.8 compares the entropy generation of an unbalanced parallel flow and counter-flow heat exchanger with respect to imbalance N_s.

Notes:

1. The irreversibility of a parallel-flow arrangement is considerably greater than the irreversibility of the counter-flow arrangement.
2. For $\omega \rightarrow 1$ (balanced case), the irreversibility of parallel-flow arrangement is finite, indicating that the counter-flow arrangement is superior.
3. Looking at the equation of $N_{S,1}$ & $N_{S,2}$ we can see that the entropy generation vanishes as $NTU \rightarrow \infty$, $\Delta P_1 \rightarrow 0$, $\Delta P_2 \rightarrow 0$ for a balanced heat exchanger. But, for unbalanced heat exchanger, it does not vanish even under such ideal conditions.

3.6 EXERGY ANALYSIS OF A REFRIGERATION SYSTEM

This section describes the exergy analysis of a refrigeration system. Figure 3.9 shows that the results from this analysis reinforce the understanding on the trade-off nature

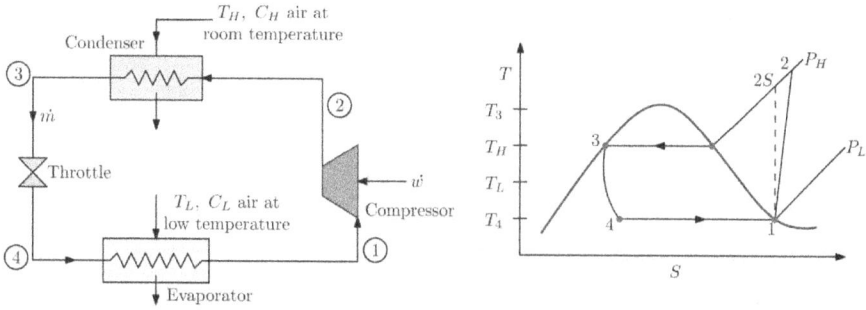

Figure 3.9   Refrigeration plant model with a vapor compression cycle.

of the design, the refrigeration plant model and its T–S diagram. The compressor power is expressed as

$$\dot{W} = \dot{m}(h_2 - h_1) = \frac{\dot{m}(h_{2S} - h_1)}{\eta_C}$$

where, η_C is the isentropic efficiency of the compressor.

The refrigerant is cooled in the condenser from the super heated vapor state ② to the saturated liquid state ③ by contacting with a stream of ambient air at temperature T_H. Although the refrigerant temperature varies from T_2 to T_3, most of the heat transfer between the refrigerant and ambient air occurs during the condensation of the refrigerant i.e. when the refrigerant temperature is at T_3. Thus, we can approximate the condenser heat transfer rate $\dot{Q}_H$ in terms of a heat exchanger's effectiveness based on the condensation temperature.

$$\dot{Q}_H = \dot{m}(h_2 - h_3) = \varepsilon_H (\dot{m}C_P)_H (T_3 - T_H) \tag{3.50}$$

where, the effectiveness of the condenser is expressed as

$$\varepsilon_H = 1 - exp\left(-\frac{U_H A_H}{C_H}\right) \tag{3.51}$$

Similarly, the heat transfer to the refrigerant in the evaporator is

$$\dot{Q}_L = \dot{m}(h_1 - h_4) = \varepsilon_L C_L (T_L - T_4) \tag{3.52}$$

where, T_l is the temperature of the refrigerated space and the evaporator effectiveness is

$$\varepsilon_L = 1 - exp\left(-\frac{U_L A_L}{C_L}\right) \tag{3.53}$$

The coefficient of performance (COP) of the refrigeration cycle is given by

$$COP = \frac{\dot{Q}_L}{\dot{W}} \tag{3.54}$$

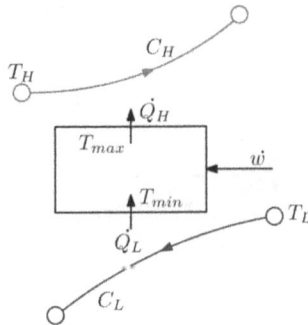

Figure 3.10 Refrigerator model with two heat exchangers and an internally reversible cycle.

where, $\dot{W}$ is the input power to the compressor. We have to decide how to divide the constant available heat transfer area A between A_H (condenser heat transfer area) and A_L (evaporator heat transfer area) to maximize the COP.

Let us look at the process diagram shown in Figure 3.10 for evaluating the role of heat transfer areas A_H and A_L. We can assume the following:

1. The temperature variation of the refrigerant through the condenser is neglected. The constant temperature of the condensing steam is denoted as T_{max}.
2. The evaporator temperature is denoted as T_{min}.
3. Isentropic efficiency of the compressor is equal to 100%.
4. Irreversibility of the expansion valve is neglected.

Figure 3.10 shows the refrigerator model with two heat exchangers and internally reversible cycle.

As the cycle is assumed to be internally reversible. We may write

$$\frac{\dot{Q}_H}{T_{max}} = \frac{\dot{Q}_L}{T_{min}} \tag{3.55}$$

when C_H stream is colder than T_{max}, we have

$$\therefore \quad \dot{Q}_H = \dot{\varepsilon}_H (T_{max} - T_H) \tag{3.56}$$

when the C_L stream is warmer than T_{min}, we have

$$\dot{Q}_L = \varepsilon_L (T_L - T_{min}) \tag{3.57}$$

The energy balance is

$$\dot{W} + \dot{Q}_L = \dot{Q}_H \tag{3.58}$$

Using equations 3.55 to 3.58, we have

$$\frac{T_H}{\dot{Q}_H} = \frac{T_L}{\dot{Q}_L} - \left(\frac{1}{\varepsilon_H C_H} + \frac{1}{\varepsilon_L C_L} \right) \tag{3.59}$$

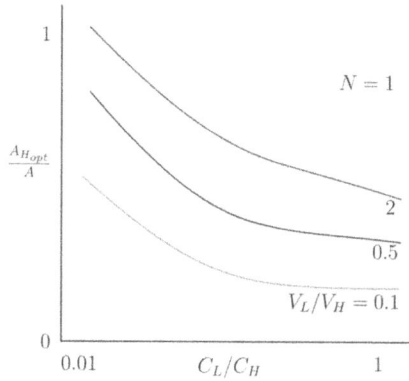

Figure 3.11 Optimal condenser heat transfer area for a refrigeration cycle.

The definition of COP indicates that $\dot{Q}_H$ should be small for a fixed value of $\dot{Q}_L$ to get higher COP ($COP = \frac{\dot{Q}_H}{\dot{Q}_H - \dot{Q}_L}$). This is possible when the LHS of the above equation is maximum i.e. the term in the bracket should be minimum.

When $\dot{Q}_L$ is fixed, maximization of COP is equivalent to minimizing the expression in brackets above.

Using the expressions for ε_H and ε_L as given before, we can write

$$\frac{1}{\varepsilon_H C_H} + \frac{1}{\varepsilon_L C_L} = \frac{(C_H)^{-1}}{1 - exp(-yN)} + \frac{(C_L)^{-1}}{1 - exp[-(1-y)N(U_L/U_H)(C_L/C_H)]} \quad (3.60)$$

where, y = Area of ratio = $\frac{A_H}{A}$, $A = A_H + A_L$ and $N = NTU$ based on total area = $\frac{U_H A}{C_H}$.

Specifying the ratios C_L/C_H, U_L/U_H & N, the RHS of the above equation varies only with respect to y. Thus, minimizing the above expression with respect to y gives the optimal area ratio $\frac{A_{H,opt}}{A}$. The variation of the optimal condenser area as a function of heat capacity ratio is presented in Figure 3.11.

Note: The optimal A_H is generally not equal to $1/2A$ and A_L. The optimal way of dividing A between the two heat exchanger surfaces depends on the relative external characteristics of the two HES i.e. C_L/C_H & U_L/U_H.

3.7 EXERGY STORAGE SYSTEM

Figure 3.12 shows an exergy storage system by sensible heating. The system is charged with exergy when the hot gas entering the system is cooled while flowing through the heat exchanger immersed in the liquid bath. The gas is eventually discharged into the atmosphere. When the system is discharged, the stored exergy can be retrieved at least in part. Let's assume the following:

1. The bath is well mixed thermally, so that at any given time its temperature is uniform $T(t)$.

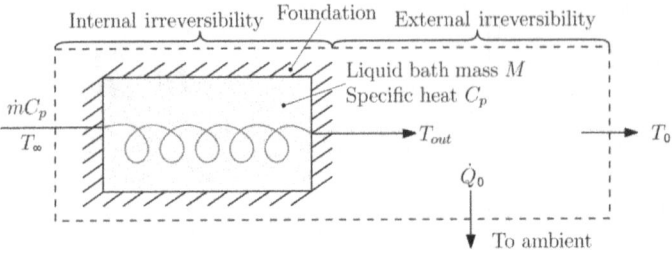

Figure 3.12 A system used for storing exergy by sensible heating.

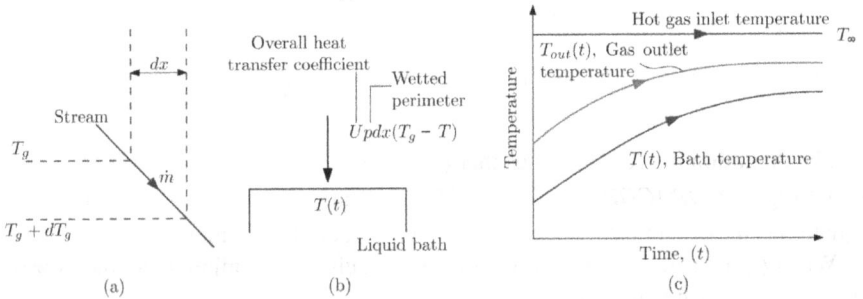

Figure 3.13 (a) A differential element of the hot gas temperature variation, (b) the heat transfer model between the hot gas liquid bath and (c) the transient variation of different components of the storage system.

2. The initial bath temperature equals the environment temperature T_0.
3. The gas obeys the ideal gas model.

   Figure 3.13 (a) and (b) shows the schematic of heat transfer between the hot gas and liquid bath.

   The energy balance between the hot gas and liquid bath can be written as:

$$Updx(T_g - T) + \dot{m}C_P dT_g = 0 \qquad (3.61)$$

where, U is the overall heat transfer coefficient between gas tube and liquid bath, $\dot{m}$ is the gas flow rate and P is the perimeter of the gas tube. Integrating from $x = 0(T_g = T_\infty)$ to $x = L(T_g = T_{out})$, we have

$$\frac{T_{out}(t) - T(t)}{T_\infty - T(t)} = exp(-NTU) \quad \text{where,} \quad NTU = \frac{UPL}{(\dot{m}C_P)} \qquad (3.62)$$

The energy balance for the liquid can be written as

$$MC\frac{dT}{dt} = \dot{m}C_P(T_\infty - T_{out}(t)) \qquad (3.63)$$

where, M is the mass of liquid and C is the specific heat of liquid. Integrating from at $t = 0$ $(T = T_0)$, we get

$$MC(T(t) - T_0) = \dot{m}C_P(T_\infty - T_{out}(t)) \times t \tag{3.64}$$

Combining equations 3.62 and 3.63 yields

$$\frac{T(t) - T_0}{T_\infty - T_0} = 1 - exp(-y\varphi) \tag{3.65}$$

$$\frac{T_{out}(t) - T_0}{T_\infty - T_0} = 1 - yexp(-y\varphi) \tag{3.66}$$

where, $y = 1 - exp(-NTU)$ and $\varphi = \frac{\dot{m}C_P}{MC}t$.

Neglecting the pressure drop in gas-side irreversibilities, the rate of entropy generation for the total system (dashed) can be written as

$$\dot{S}_{gen}(t) = \dot{m}C_P ln\frac{T_0}{T_\infty} + \frac{\dot{Q}_0}{T_0} + \frac{d}{dt}\left(MC\ln\frac{T}{T_0}\right) \tag{3.67}$$

where, $\dot{Q}_0 = \dot{m}C_P(T_{out} - T_0)$.

Integrating equation 3.67 and using $T_{out}(t)$ from equation 3.66, we have

$$\frac{1}{MC}\int_0^t \dot{S}_{gen}dt = \frac{1}{MC}\left(\dot{m}C_P ln\frac{T_0}{T_\infty}t\right) + \int_0^t \frac{\dot{m}C_p(T_{out}(t) - T_0)}{MCT_0}dt + Mcln\frac{T(t)}{T_0}dt \tag{3.68}$$

Substituting equations 3.65 and 3.66, we have

$$\frac{1}{MC}\int_0^t \dot{S}_{gen}dt = \varphi ln\frac{T_0}{T_\infty} + \frac{\dot{m}C_P}{T_0}(T_\infty - T_0)\left(1 - e^{-y\varphi}\right)\frac{t}{MC} + \frac{ln\left(T_0 + (T_\infty - T_0)\left(1 - e^{-y\varphi}\right)\right)}{T_0}$$

$$= \varphi ln\frac{T_0}{T_\infty} + \varphi\frac{T_\infty - T_0}{T_0}\left(1 - ye^{-y\varphi}\right) + ln\left(1 + \frac{T_\infty - T_0}{T_0}\left(1 - e^{-y\varphi}\right)\right)$$

$$= \varphi\left(ln\frac{T_0}{T_\infty} + \frac{T_\infty - T_0}{T_0}\right) + ln\left[1 + \frac{T_\infty - T_0}{T_0}\left(1 - e^{-y\varphi}\right)\right] - \frac{T_\infty - T_0}{T_0}\left(1 - e^{-y\varphi}\right) \tag{3.69}$$

The result is more instructive, if the $T_0\int_0^t \dot{S}_{gen}dt$ is divided by the exergy of the gas drawn from the hot gas supply.

Using $e^{PH} = (h - h_0) - T_0(S - S_0)$ with ideal gas relation, we can write:

$$\dot{m}e_\infty = \dot{m}C_P\left(T_\infty - T_0 - T_0 ln\frac{T_\infty}{T_0}\right)$$

$$= \dot{m}C_p\left(T_\infty - T_0 - T_0\left(ln(1 + \tau)\right)\right)$$

$$= \dot{m}C_pT_0\left(\tau - ln(1 + \tau)\right) \tag{3.70}$$

$$= \frac{\varphi MC}{t}T_0\left(\tau - ln(1 + \tau)\right)$$

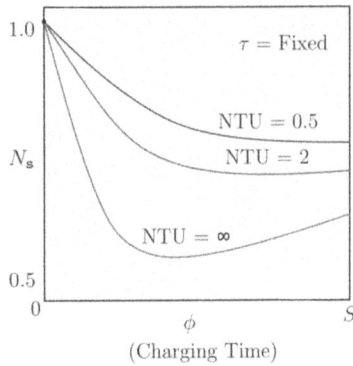

(Charging Time)

Figure 3.14 The entropy generation number as a function of charging time at a fixed dimensionless temperature difference (τ).

where, τ is the dimensionless temperature difference given as

$$\tau = \frac{T_\infty - T_0}{T_0} \tag{3.71}$$

Thus, the entropy generation number,

$$N_S = \frac{T_0}{\dot{m}te_\infty} \int_0^t \dot{S}_{gen}dt = 1 - \frac{\tau\left[\left(1 - e^{-y\varphi}\right) - \ln\left(1 + \tau\left(1 - e^{-y\varphi}\right)\right)\right]}{\varphi\left(\tau - ln(1+\tau)\right)} \tag{3.72}$$

Notes:

1. $N_S = N_S(\varphi, NTU, \tau)$. The entropy generation number is a function of charging time and NTU, as shown in Figure 3.14.
2. There exists a φ_{opt} when N_S reaches the minimum value.
3. In the limit $\varphi \to 0$, the entire exergy content of the hot stream is destroyed by *heat transfer to the liquid bath*.
4. In the limit $\varphi \to \infty$, the gas stream leaves the bath as hot as it enters at ($T_{out} \simeq T_\infty$) and its exergy content is destroyed entirely by direct *heat transfer to the atmosphere*.
5. The smallest N_s occurs when φ is of order 1. This means that we should terminate the heating process when the thermal inertia of the hot gas used ($\dot{m}C_Pt$) is comparable with the thermal inertia of the liquid (MC).

3.8 SOLAR AIR COLLECTOR

The schematic of a solar air collector is shown in Figure 3.15. The relevant parameters associated with the solar air collector are marked in the figure. The energy

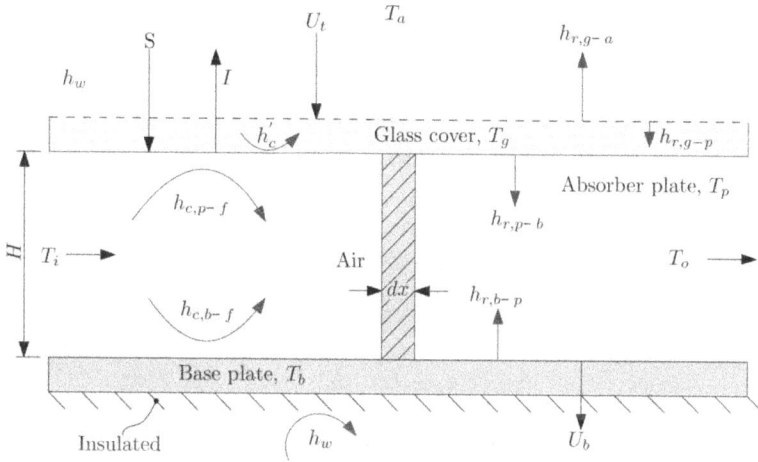

Figure 3.15   Schematic of a solar collector.

balance equations for different components of the solar air collector can be written using the following assumptions:

1. The system operates under steady-state conditions.
2. The heat transfer through glass covers, absorber plate and insulation plate is 1–D in the direction perpendicular to the air flow.

The energy balance equation is written for a differential element with length dx and width dz at a distance x from the inlet. The energy balance equation for a glass cover is expressed as

$$\alpha_g I W dx + (h_{r,g-p} + h'_c)(T_p - T_g)W dx = (h_w + h_{f,g-a})(T_g - T_a)W dx \qquad (3.73)$$

where, α_g is absorptance of glass cover, I is radiation intensity (Wm^{-2}), W is width of collector (m), $h_{r,g-p}$ is radiative heat transfer coefficient from outer glass cover to absorber plate, h'_c is convective heat transfer coefficient from absorber plate to glass cover (WK^{-1}m^{-2}), T_p is absorber plate temperature (K), T_g is glass cover temperature (K), h_w is wind heat transfer coefficient (WK^{-1}m^{-2}), $h^{r,g--a}$ is radiative heat transfer coefficient from glass cover to the sky (WK^{-1}m^{-2}) and T_a is ambient temperature (K).

The energy balance equation for the absorber plate can be expressed as

$$\tau_g \alpha_p I W dx = h_{r,p-b}(T_p - T_b)W dx + h_{c,p-f}(T_p - T_f)W dx + U_t(T_p - T_a)W dx \qquad (3.74)$$

where, τ_g is transmittance of glass cover, α_p is absorptance of absorber plate, $h_{r,p-b}$ is convective heat transfer coefficient between absorber and bottom plate (WK^{-1}m^{-2}), T_b is base plate temperature, $h_{c,p-f}$ is convective heat transfer coefficient between air and absorber plate (WK^{-1}m^{-2}), T_f is air flow temperature (K) and U_t is top heat loss coefficient (WK^{-1}m^{-2}).

The energy balance equation of an insulated base plate can be written as

$$\left[h_{c,b-f}(T_b - T_f) + U_b(T_b - T_a) + h_{r,p-b}(T_b - T_p)\right] W\,dx = 0 \qquad (3.75)$$

where, $h_{c,b-f}$ is convective heat transfer coefficient between air and the bottom plate (WK^{-1}m^{-2}) and U_b is bottom heat loss coefficient (WK^{-1}m^{-2}).

The energy balance equation for air channels can be expressed as

$$\dot{m}c_p dT_f = \left[h_{c,p-f}(T_p - T_f) + h_{c,b-f}(T_b - T_f)\right] W\,dx \qquad (3.76)$$

where, $\dot{m}$ is mass flow rate of air (kg s^{-1}) and c_p is the specific heat capacity of air.

3.8.1 HEAT TRANSFER COEFFICIENT

The heat transfer coefficient used in the energy equation can be estimated using the empirical correlations available in literature as follows:

The top heat loss coefficient U_t in solar air collectors can be evaluated empirically using [Duffie et al., 1974]

$$U_t = \left[\frac{N}{\frac{C}{T_p}\left[\frac{T_p - T_a}{N+f}\right]^e} + \frac{1}{h_w}\right]^{-1} + \frac{\sigma(T_p + T_a)(T_p^2 + T_a^2)}{\left[\varepsilon_p + 0.00591 N h_w\right]^{-1} + \left[\frac{2N+f-1+0.133\varepsilon_p}{\varepsilon_g}\right] - N}$$

$$(3.77)$$

where $f = (1 + 0.089 h_w - 0.1166 h_w \varepsilon_p)(1 + 0.07866N)$ and $C = 520(1 - 0.00051\theta^2)$ for $\theta < 70°$; if $70° \le \theta \le 90°$, use $\theta = 70°$.

Where, N is the number of glass covers, σ is Stefan-Boltzmann constant (5.67×10^{-8} Wm^{-2}K^{-4}), ε_p is emissivity of an absorber plate, ε_g is emissivity of a glass cover and θ is collector tilt angle.

The back loss coefficient U_b is given by

$$U_b = \frac{k_{ins}}{\delta_{ins}} \qquad (3.78)$$

where, k_{ins} is the thermal conductivity of insulation material (Wm^{-1}K^{-1}) and δ_{ins} is the thickness of insulation material (m).

The natural convection heat transfer coefficient of the air gap between the glass cover and absorber plate can be calculated using the equation by Hollands et al. [1976]:

$$h_c^l = \frac{k_f}{S\left\{1 + 1.44[1 - R]^* \left(1 - R(Sin(1.8\theta))^{1.6}\right) + \left[0.66416R^{-1/3} - 1\right]^*\right\}} \qquad (3.79)$$

where, $R = 1708/Ra_s Cos\theta$ and Ra_s is the air gap Rayleigh number (insert formula). The meaning of asterisk (*) is that only positive values of the terms in the square brackets are to be used i.e. use zero if the term is negative. All properties are evaluated at the air gap mean temperature, $(T_p + T_g)/2$. Here, S is the depth of the air gap between the absorber plate and the glass cover and k_f is the thermal conductivity of air.

The convection heat transfer coefficient by wind movement from the outside cover can be obtained by using the following correlation proposed by McAdams [1954]:

$$h_w = 5.7 + 3.8V_\infty \tag{3.80}$$

where, V_∞ is the wind velocity in m/s and h_w is the wind heat transfer coefficient in W m^{-2} K^{-1}.

The heat transfer coefficient between the air flow in the channel and the absorber plate can be assumed to be equal. It can be estimated using

$$h_c = \frac{Nu \times k_f}{D_h} \tag{3.81}$$

where, D_h is the equivalent diameter of the duct calculated as

$$D_h = \frac{4A}{P} = \frac{2 \times W \times H}{W + H} \tag{3.82}$$

where, W is the width and H is the height of the duct.

The Nusselt number for laminar flow $(Ra < 10^9)$ can be calculated using the following relation [Bazilian and Prasad, 2002]:

$$Nu = 0.68 + \frac{0.67Ra^{0.25}}{\left[1 + \left(\frac{0.492}{Pr}\right)^{9/16}\right]^{4/9}} \tag{3.83}$$

The Nusselt number for the turbulent flow $(Ra > 10^9)$ can be calculated using

$$Nu = 0.825 + \frac{0.39Ra^{1/6}}{\left[1 + \left(\frac{0.492}{Pr}\right)^{9/16}\right]^{8/27}} \tag{3.84}$$

All properties are evaluated at the mean air temperature, which is calculated using the following expression [Hirunlabh et al., 1999]:

$$T_f = 0.25T_i + 0.75T_o \tag{3.85}$$

The density, thermal conductivity and dynamic viscosity of air are calculated using the following relation [Bolz and Tuve, 2019]:

$$\rho = 0.3147 - 0.016082T + 2.9013 \times 10^{-5}T^2 - 1.9407 \times 10^{-8}T^3 \tag{3.86}$$

$$k = (0.0015215 + 0.097459T - 3.3322 \times 10^{-5}T^2) \times 10^{-3} \tag{3.87}$$

$$\mu = (1.6157 + 0.06523T - 3.0297 \times 10^{-5}T^2) \times 10^{-6} \tag{3.88}$$

The radiative heat transfer coefficient from the outer glass surface to the sky is obtained as

$$h_{r,g-a} = \sigma \varepsilon_g \frac{T_g^4 - T_s^4}{T_g - T_a} \tag{3.89}$$

where, the sky temperature is calculated as [Swinbank, 1964]

$$T_s = 0.0552 T_a^{1.5} \tag{3.90}$$

The radiation heat transfer coefficient between two parallel plates 1 and 2 can be expressed as

$$h_{r,1-2} = \sigma (T_1 + T_2)(T_1^2 + T_2^2) \left(\frac{1}{\varepsilon_1} + \frac{1}{\varepsilon_2} - 1 \right)^{-1} \tag{3.91}$$

where, ε_1 and ε_2 are the emissivities of plate 1 and plate 2 at temperatures T_1 and T_2, respectively.

3.8.2   AIR MASS FLOW RATE

The induced air flow rate by natural convection under the steady state can be derived by applying the energy equation from inlet to outlet

$$p_1 + \frac{\rho_1 V_1^2}{2} + \rho_1 g z_1 = p_2 + \frac{\rho_2 V_2^2}{2} + \rho_2 g z_2 + \frac{\rho f L V^2}{2D_h} \tag{3.92}$$

We can assume $p_1 = p_2$ as both vents (inlet and outlet) are open to atmosphere. The inlet air is from a reservoir, and hence we can assume $V_1 = 0$ and $V_2 = V$ from continuity. Hence, the above equation simplifies to

$$\rho_1 g z_1 - \rho_2 g z_2 = \frac{\rho_2 V^2}{2} + \frac{\rho f L V^2}{2D_h} \tag{3.93}$$

The density of air at any temperature T can be written as

$$\rho_T = \rho \beta_{Th} T \tag{3.94}$$

$$\beta_{Th} = \frac{1}{T_f} \tag{3.95}$$

$$T_f = 0.25 T_i + 0.75 T_o \tag{3.96}$$

Substituting these values in equation 3.93, we get

$$g \rho \beta_{Th} T_{in} z_1 - g \rho \beta_{Th} T_{out} z_2 = \frac{V^2}{2} \left(\rho \beta_{Th} T_{out} + \frac{\rho f L}{D_h} \right) \tag{3.97}$$

For a channel of length L aligned at angle θ with respect to horizontal, we can write

$$z_2 = z_1 sin(\theta) \tag{3.98}$$

We can write equation 3.97 as

$$g\rho\beta_{Th}Lsin\theta(T_{out} - T_{in}) = \frac{V^2}{2}\left(\rho\beta_{Th}T_{out} + \frac{\rho fL}{D_h}\right) \qquad (3.99)$$

$$\therefore \quad V = \sqrt{\frac{2g\rho\beta_{Th}LSin\theta(T_{out} - T_{in})}{\rho\beta_{Th}T_{out} + \frac{\rho fL}{D_h}}} \qquad (3.100)$$

3.8.2.1 Energy and Exergy Efficiency

The energy efficiency of a solar air collector is defined as

$$\eta_{th} = \frac{Q_{air}}{Q_c} \qquad (3.101)$$

where, Q_{air} is the heat content of the air and Q_c is the energy incident on the collector.

$$Q_{air} = \dot{m}_f c_p(T_{out} - T_{in}) \qquad (3.102)$$
$$Q_c = A_c I \qquad (3.103)$$

where, A_c is the area of the collector.

The exergetic efficiency can be defined as the ratio of absorbed exergy of air and the exergy of sun radiation on the collector

$$\eta_{ex} = \frac{Ex_{out}}{Ex_{in}} \qquad (3.104)$$

The input exergy i.e. the exergy of sun radiation is defined as [Petela, 1964]

$$Ex_{in} = IA_c\left[1 + \frac{1}{3}\left(\frac{T_a}{T_{sun}}\right)^4 - \frac{4T_a}{3T_{sun}}\right] \qquad (3.105)$$

where, T_a is the temperature of ambient and T_{sun} is the temperature of sun (5600 K).

The output exergy is defined as

$$Ex_{out} = \dot{m}\left[c_p(T_{out} - T_{in}) - T_a c_v ln\left(\frac{T_{out}}{T_{in}}\right) - Rln\left(\frac{p_{out}}{p_{in}}\right)\right] \qquad (3.106)$$

3.8.2.2 Parametric Study

The parametric study on the performance of a solar collector can be carried out using the following fixed parameters:

$$H = 0.1 \text{ m}, \quad W = 1 \text{ m}, \quad L = 2 \text{ m}, \quad \theta = 30°, \quad S = 25 \text{ mm}, \quad I = 1000 \text{ W/m}^2,$$

$$\alpha_p = 0.95, \quad \varepsilon_g = 0.88, \quad \tau_g = 0.88, \quad \delta_{ins} = 0.06 \text{ m}, \quad \alpha_g = 0.06, \quad \varepsilon_b = 0.95,$$

$$T_a = 30 \text{ °C}, \quad k_{ins} = 0.06 \text{ W/m-K}.$$

(a)

(b)

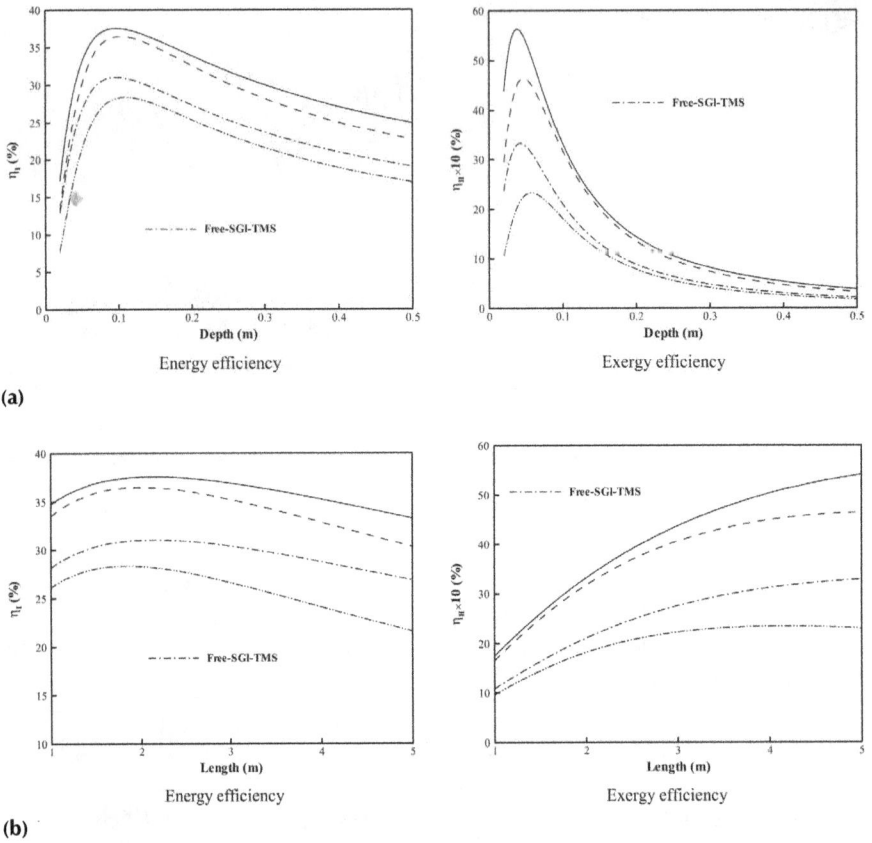

Figure 3.16 Energy and exergy efficiency variation of a solar collector as a function of depth and length. (Adapted from Bahrehmand and Ameri [2015] with permission.)

Figure 3.16 shows the energy efficiency and exergy efficiency as a function of depth and length of the collector. Both energy and exergy efficiencies increase with the increase in depth and decreases subsequently after reaching the maxima at a certain depth (see Figure 3.16a). The maxima in exergetic efficiency are observed at a lower channel depth compared to the maxima in thermal efficiency. There is an increase in air velocity and consequently heat transfer coefficient with an increase up to a certain depth and a decrease thereafter. The change in heat transfer coefficient leads to an increase in heat transfer and thus efficiency.

Both buoyancy force and friction force increase with an increase in channel length. However, the increase in buoyancy force is greater than friction losses. Therefore, there is an increase in the mass flow rate through the channel. The increase in channel length also increases the amount of solar energy received and outlet air flow temperature. The useful heat energy of air increases with the increase in mass flow rate and air temperature.

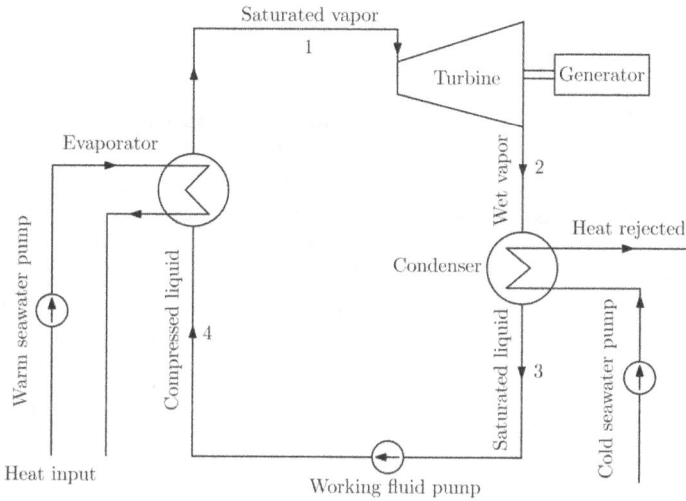

Figure 3.17 Sketch of the ORC in OTEC. (Sun et al. [2012].)

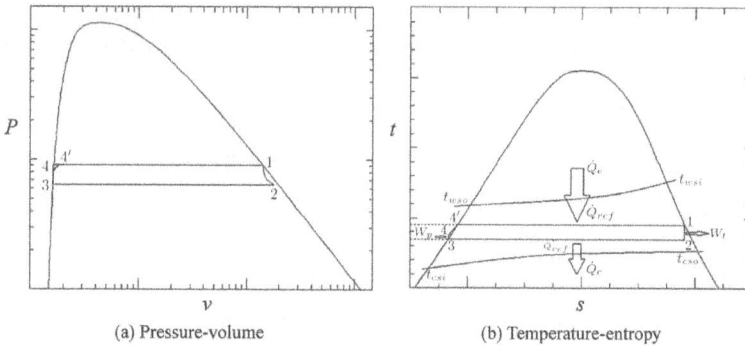

(a) Pressure-volume

(b) Temperature-entropy

Figure 3.18 ORC diagrams in OTEC. (Sun et al. [2012].)

3.9 OCEAN THERMAL ENERGY CONVERSION

Ocean thermal energy is one of the promising ways to generate clean power as a non-CO_2 emitting technology. Ocean thermal energy conversion (OTEC) generates electricity using the temperature difference between sea surface (20–30 °C) and deep sea (3–10 °C). Organic Rankine cycle (ORC) technology is an effective thermal conversion approach in low-temperature applications, which uses an organic fluid as a working medium. ORC is named due to its use of organic working fluid, which allows Rankine cycle to generate electricity from low-temperature sources. Figure 3.17 shows the schematic of an ORC in OTEC. Heat is transferred from warm seawater and converts a fraction of the heat to work and rejects the rest to cold seawater. Figure 3.18 shows the $P-v$ and $T-s$ diagram of the ORC.

The rate of heat transfer from the warm seawater can be written as

$$\dot{Q}_e = \dot{m}_{ws} c_p \Delta T_{ws} \tag{3.107}$$

where, $\dot{m}_{ws}$ is the mass flow rate of warm water, c_p is the specific heat of seawater at constant pressure and $\Delta T_{ws} = T_{ws,i} - T_{ws,o}$. Here, $T_{ws,i}$ and $T_{ws,o}$ are the warm seawater temperature at the inlet and outlet of the heat exchanger, respectively.

The rate of heat transfer rejected into the cold seawater is

$$\dot{Q}_c = \dot{m}_{cs} c_p \Delta T_{cs} \tag{3.108}$$

where, $\dot{m}_{cs}$ is the mass flow rate of cold seawater and $\Delta T_{cs} = T_{cs,i} - T_{cs,o}$. Here, $T_{cs,i}$ and $T_{cs,o}$ are the cold seawater temperature at the inlet and outlet of the heat exchanger, respectively.

The rate of heat transfer to the cycle at the evaporator can be written as

$$\dot{Q}_{ewf} = \dot{m}_{wf}(h_1 - h_{c1}) \tag{3.109}$$

where, $\dot{m}_{wf}$ is the mass flow rate of working fluid (ammonia). The rate of heat rejected from the cycle at the condenser is

$$\dot{Q}_{cwf} = \dot{m}_{wf}(h_2 - h_3) \tag{3.110}$$

The net work output is the difference between the turbine work output and the pump input is

$$\dot{W}_{net} = \dot{m}_{wf}(h_1 - h_2) - \dot{m}_{wf}(h_4 - h_3) \tag{3.111}$$

The enthalpy of saturated ammonia can be calculated by using curve fitting of the data around the operating temperature of the evaporator ($15 \leq T_e \leq 30$ °C) and condenser ($3 \leq T_c \leq 15$ °C) as given below [Sun et al., 2012]:

$$h_1 = -0.0101 T_e^2 + 1.1263 T_e + 498.7285 \text{ kJ/kg}$$

The enthalpy of the saturated ammonia liquid (point 3)

$$h_2 = 4.6821 T_c - 762.9883 \text{ kJ/kg}$$

The enthalpy at point 4 is

$$h_4 = h_3 + v_3(p_4 - p_3) \text{ kJ/kg (using } dh = Tds + vdp)$$

The enthalpy at point 2 is

$$h_2 = h_3 + (T_c + T_0)(s_2 - s_3) \text{ kJ/kg } (T_0 = 273.15) \text{ K}$$

Similarly,

$$p_1 = 0.3394 T_e^2 + 13.8111 T_e + 443.0038 \text{ (kPa)}$$
$$p_2 = 0.2733 T_c^2 + 15.7271 T_c + 428.7508 \text{ (kPa)}$$
$$v_3 = 0.0016 \text{ m}^3$$
$$s_2 = 2.9151 \times 10^{-5} T_e^2 - 0.0127 T_e + 10.3142 \text{ (kJ/kg–K)}$$
$$s_3 = 0.0165 T_c + 5.6971 \text{ (kJ/kg–K)}$$

It may be noted that the temperature T is in °C. Using the energy balance at the evaporator i.e. $\dot{Q}_e = \dot{Q}_{ewf}$, and assuming $\Delta T_{ws} = \Delta T_{cs}$, we have

$$\dot{m}_{wf} = \frac{\dot{m}_{ws} c_p \Delta T_{ws}}{h_1 - h_4} \tag{3.112}$$

The exergy efficiency of OTEC can be written as

$$\varepsilon = \frac{E_{in} - \sum I^0}{E_{in}} \tag{3.113}$$

where,

$$E_{in} = E_{in,ws} + E_{in,cs}$$
$$E_{in,ws} = \dot{m}_{ws} \left[(h_{wsi} - h_{sds}) - T_{sds} (s_{wsi} - s_{sds}) \right]$$
$$E_{in,cs} = \dot{m}_{cs} \left[(h_{csi} - h_{sds}) - T_{sds} (s_{csi} - s_{sds}) \right]$$

Here, sds corresponds to the system dead state and $T_s ds$ is in Kelvin.
The exergy at each state point can be calculated as

$$E_i = \dot{m} \dot{m}_{ws} \left[(h_i - h_{sds}) - T_{sds} (s_i - s_{sds}) \right] \tag{3.114}$$

The exergy loss in each component can be calculated as below:

$$I^0_e = E_{in,ws} - E_{out,ws} + E_{in,e} - E_{out,e} \quad \text{(Evaporator)} \tag{3.115}$$
$$I^0_t = E_{in,t} - E_{out,t} + W_t \quad \text{(Turbine)} \tag{3.116}$$
$$I^0_c = E_{in,cs} - E_{out,cs} + E_{in,c} - E_{out,c} \quad \text{(Compressor)} \tag{3.117}$$
$$I^0_{wfp} = W_{wfp} + E_{in,wfp} - E_{out,wfp} \quad \text{(Working Fluid Pump)} \tag{3.118}$$

where, W_{wfp} is the power of the working fluid pump.
The system exergy exhaust loss can be calculated as

$$I^0_s = E_{out,ws} + E_{out,cs}. \tag{3.119}$$

We can assume turbine and pump efficiency to be equal to 85%. The heat loss in piping and other auxiliary can be assumed to be negligible.
The heat transfer rate in a condenser/evaporator can be expressed as

$$\dot{Q} = UA\Delta T_m$$

where, U is the overall heat transfer coefficient, A is the cross-sectional area normal to the direction of heat transfer and ΔT_m is the logarithmic mean temperature difference expressed as

$$\Delta Tm = \frac{\Delta T_{in} - \Delta T_{out}}{\ln \frac{\Delta T_{in}}{\Delta T_{out}}}$$

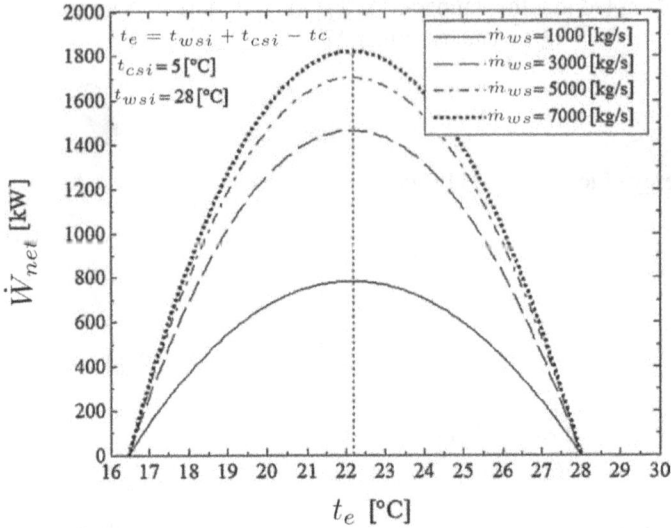

Figure 3.19 Relationship between the evaporator temperature t_e and ORC power output $\dot{W}_{net}$ at different mass flow rates of seawater. (Sun et al. [2012].)

Here, ΔT_{in} and ΔT_{out} are the temperature difference at the inlet and the outlet, respectively.

Figure 3.19 shows the relationship between the net ORC power output $(\dot{W}_{net})$ and evaporator temperature (T_e) at $\dot{m}_{ws} = 1000, 3000, 5000$ and 7000 kg/s for $T_{csi} = 5°C$ and $T_{wsi} = 28°C$. The UA value of both evaporator and condenser is taken as 10000 kW/K. It shows that there is an optimum value for the maximum power output of the OTEC. The power output also increases with the increase in $\dot{m}_{ws}$. The rate of increase is maximum at a lower value of $\dot{m}_{ws}$ compared to that at a higher value of $\dot{m}_{ws}$.

Figure 3.20 shows the variation of exergetic efficiency and exergy losses in different components i.e. evaporator, condenser, turbine, pump and exhaust as a function of $\dot{m}_{ws}$. The figure shows that maximum exergy loss takes place due to the exhaust, which increases with the increase in $\dot{m}_{ws}$. Therefore, the exergetic efficiency decreases with the increase in $\dot{m}_{ws}$ value.

3.9.1 HYDROGEN PRODUCTION USING OTEC

The OTEC plant employing a closed ORC has an energy efficiency of around 5% due to small temperature difference between surface water and deep water of the sea. It can be combined with an electrolyzer for hydrogen production. Figure 3.21 shows the schematic of an integrated hydrogen production system with OTEC [Ahmadi et al., 2013]. The working fluid can be ammonia or a freon refrigerant. Warm water from the sea is passed through the solar collector, where the working fluid is evaporated and passed through the turbine to generate electrical power. The electricity produced

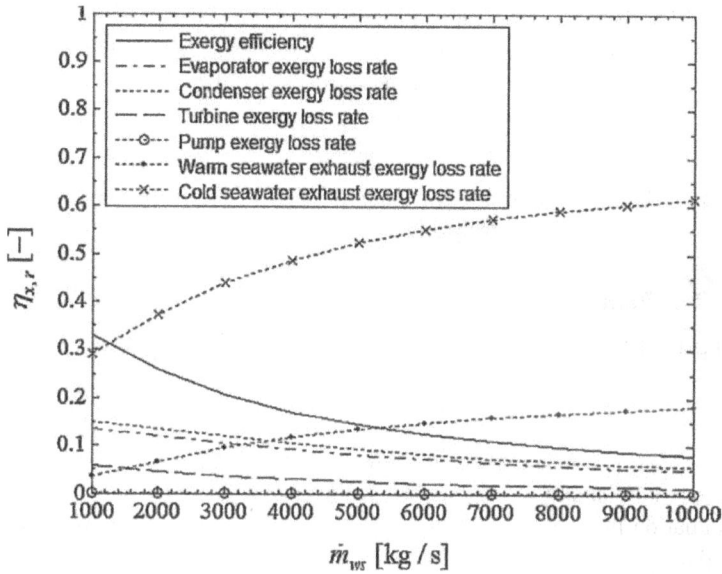

Figure 3.20   Variation in exergy efficiency $\eta_{x,s}$ with respect to flow rate and each component exergy loss rate in ORC. (Sun et al. [2012].)

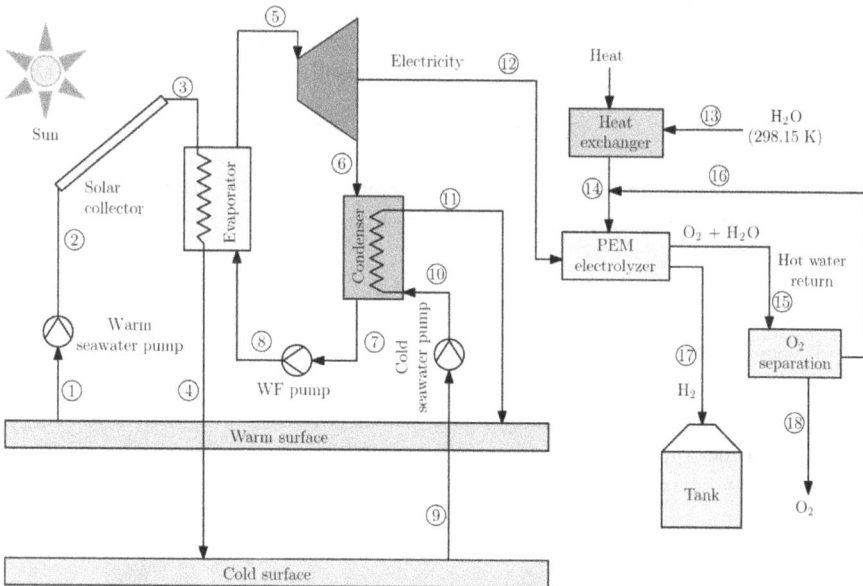

Figure 3.21   Schematic of ocean thermal energy conversion (OTEC) with a flat plate solar collector and a PEM electrolyzer for hydrogen production.

is used to drive the polymer exchange membrane (PEM) electrolyzer. The vapor is condensed in the heat exchanger driven by the cold deep seawater. The working fluid is pumped back through the warm seawater heat exchanger.

3.9.1.1 Energy Analysis

The energy balance equation for the main sections of the plant i.e. flat plate solar collector, ORC and PEM electrolyzer are described here.

3.9.1.2 Flat Plate Solar Collector

The heat gained by the seawater entering at point 2 can be written as

$$\dot{Q}_u = \dot{m}c_p(T_3 - T_2) \tag{3.120}$$

where, $\dot{m}, c_p, T_3$ and T_2 are the mass flow rate, specific heat at constant pressure, outlet temperature and inlet temperature, respectively.

The radiation flux absorbed by the absorber of a flat plate solar collector can be calculated as

$$S = I\alpha\tau$$

where, I is the solar radiation intensity and $(\alpha\tau)$ is the optical efficiency which is defined as the fraction of the incident solar radiation on the glass cover, which is transferred to the heat transfer fluid.

The Hottel–Whillier equation can be used for calculating the heat gained by the flat plate collector considering heat losses from the collector as [Farahat et al., 2009]:

$$\dot{Q}_u = A_p F_R[S - U_1(T_{in} - T_0)] \tag{3.121}$$

where, A_p is the surface area, F_R is the heat removal factor, T_0 is the ambient temperature and T_{in} is the temperature corresponding to inlet conditions. The heat removal factor, F_R is expressed as

$$F_R = \frac{\dot{m}c_p}{U_1 A_p}\left[1 - e^{-\frac{F^l U_1 A_p}{\dot{m}c_p}}\right] \tag{3.122}$$

Here, F^l is the collector efficiency factor which is assumed as 0.94. The energy efficiency of the solar flat plate collector is expressed as

$$\eta = \frac{\dot{Q}_u}{I A_p} \tag{3.123}$$

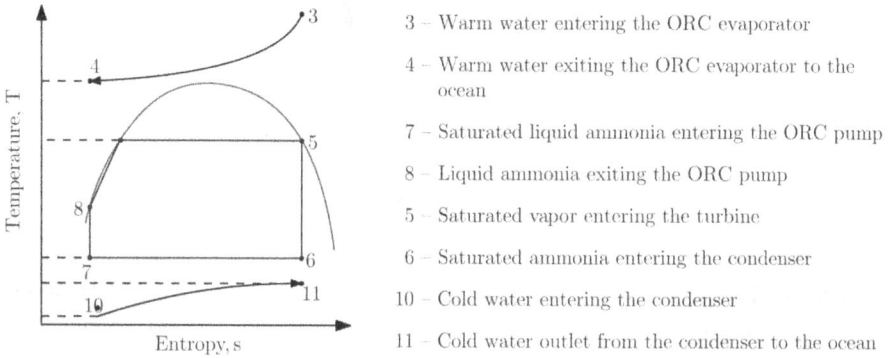

3 – Warm water entering the ORC evaporator

4 – Warm water exiting the ORC evaporator to the ocean

7 – Saturated liquid ammonia entering the ORC pump

8 – Liquid ammonia exiting the ORC pump

5 – Saturated vapor entering the turbine

6 – Saturated ammonia entering the condenser

10 – Cold water entering the condenser

11 – Cold water outlet from the condenser to the ocean

Figure 3.22 $T-s$ diagram of the OTEC system.

3.9.1.3 Organic Rankine Cycle

Figure 3.22 shows the $T-s$ diagram of the OTEC. The turbine generator power can be written using the energy balance for a control volume around the turbine as

$$\dot{W}_g = \dot{m}_f \eta_T \eta_G (h_5 - h_6) \tag{3.124}$$

where, $\dot{m}_f$ is the mass flow rate of working fluid, η_T and η_G are the turbine isentropic efficiency and generator mechanical efficiency, respectively.

The pumping power of warm seawater can be written as

$$\dot{W}_{ws} = \frac{\dot{m}_{ws} \Delta H_{ws} g}{\eta_{wsp}} \tag{3.125}$$

where,

$$\Delta H_{ws} = (\Delta H_{ws})_P + (\Delta H_{ws})_E$$

$$(\Delta H_{ws})_P = (\Delta H_{ws})_{SP} + (\Delta H_{ws})_B$$

$$(\Delta H_{ws})_{SP} = 6.82 \frac{L_{ws}}{d_{ws}^{1.17}} \left(\frac{V_{ws}}{C_{ws}}\right)^{1.85}, \quad C_{ws} = 1000$$

$$(\Delta H_{ws})_B = \sum \lambda_m \frac{V_{ws}^2}{2g}$$

where, $\dot{m}_{ws}$ is the mass flow rate of warm seawater, ΔH_{ws} is the total pump head difference of the warm seawater piping, η_{wsp} is the working fluid pump efficiency, $(\Delta H_{ws})_{SP}$ is the friction loss of straight pipe, $(\Delta H_{ws})_B$ is the bending loss on the warm seawater pipe, L_{ws} is the length of the warm seawater pipe, d_{ws} is the warm seawater inner pipe diameter and V_{ws} is the velocity of warm seawater.

The pressure difference of warm seawater in the evaporator can be expressed as

$$(\Delta H_{ws})_E = \lambda_e \frac{V_{ws}^2 L_E}{2g(D_{eq})_w} \tag{3.126}$$

where, L_E is the length of the evaporator plate and $(D_{eq})_w$ is the equivalent diameter, which is calculated as

$$D_{eq} = 2\delta$$

where, δ is the clearance.

Similarly, the cold seawater pumping power can be expressed as

$$\dot{W}_{cs} = \frac{\dot{m}_{cs}\Delta H_{cs}g}{\eta_{csp}} \tag{3.127}$$

where, $\dot{m}_{cs}$ is the mass flow rate of cold seawater, ΔH_{cs} is the total pump head of the cold seawater piping and η_{csp} is cold seawater pump efficiency. The total pump head consists of three parts i.e.

$$\Delta H_{cs} = (\Delta H_{cs})_p + (\Delta H_{cs})_c + (\Delta H_{cs})_d \tag{3.128}$$

where, $(\Delta H_{cs})_p$ is the pump head of the cold seawater pipe expressed as

$$(\Delta H_{cs})_p = (\Delta H_{cs})_{SP} + (\Delta H_{cs})_B \tag{3.129}$$

where, $(\Delta H_{cs})_{SP}$ and $(\Delta H_{cs})_B$ are the friction loss of the straight pipe and bending loss, respectively, which can be calculated similar to the warm seawater pipe.

The cold seawater pressure difference in the condenser can be written as

$$(\Delta H_{cs})_C = \lambda_c \frac{V_{cs}^2 L_{sc}}{2g(D_{eq})_C} \tag{3.130}$$

where, V_{cs} is the velocity of cold seawater, L_{sc} is the length of the condenser plate and $(D_{eq})_c$ is the equivalent diameter. The coefficient λ_c, λ_e and λ_m can be obtained from Nihous and Vega [1993].

The net power output of the system is expressed as

$$\dot{W}_{net} = \dot{W}_g - (\dot{W}_{ws} + \dot{W}_{cs} + \dot{W}_{wf}) \tag{3.131}$$

where, $\dot{W}_{wf}$ is working fluid pump power.

3.9.1.4 PEM Electrolyzer

Figure 3.23 shows the schematic of an electrolyzer explaining the physical principle (Esmaili et al. (2012)). The required energy, heat and water are provided to the electrolyzer during the electrolysis reaction. The produced hydrogen and oxygen are stored in storage tanks. Water is fed back to the system for further production cycles. The overall reaction of water splitting is given by

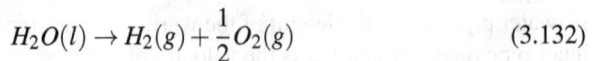

$$H_2O(l) \rightarrow H_2(g) + \frac{1}{2}O_2(g) \tag{3.132}$$

The anode is equipped with a platinum electrode, where oxidation occurs according to

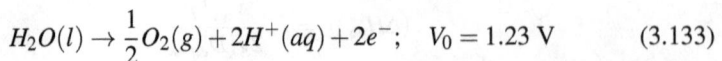

$$H_2O(l) \rightarrow \frac{1}{2}O_2(g) + 2H^+(aq) + 2e^-; \quad V_0 = 1.23 \text{ V} \tag{3.133}$$

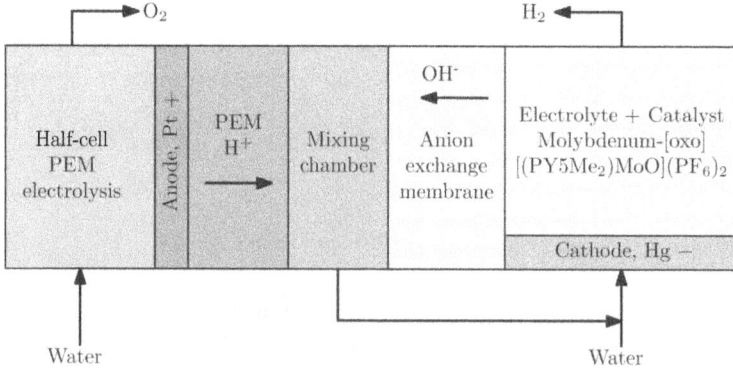

Figure 3.23 Schematic explaining the working principle of an electrolyzer.

The following reaction occurs in the cathode by a mercury pool electrode:

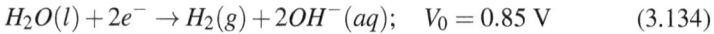

$$H_2O(l) + 2e^- \rightarrow H_2(g) + 2OH^-(aq); \quad V_0 = 0.85 \text{ V} \tag{3.134}$$

The following reaction occurs in the mixing chamber:

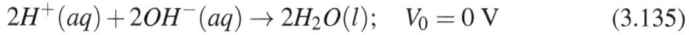

$$2H^+(aq) + 2OH^-(aq) \rightarrow 2H_2O(l); \quad V_0 = 0 \text{ V} \tag{3.135}$$

The total energy needed by the electrolyzer can be obtained based on the following equation:

$$\Delta H = \Delta G + T\Delta s \tag{3.136}$$

where, ΔG is Gibb's free energy and $T\Delta s$ is the thermal energy requirement. The value of G, s and H for hydrogen, oxygen and water can be found in thermodynamic tables. The total energy requirement is the theoretical energy required for H_2O electrolysis without any losses. The mass flow rate of hydrogen can be calculated using the following relation:

$$\dot{N}_{H_2,out} = \dot{N}_{H_2O,reacted} = \frac{J}{2F} \tag{3.137}$$

where, $\dot{N}$ is the molar mass flow rate (mol/s), J is the current density and F is the Faraday's constant (C/mol). The electrical energy input rate to the electrolyzer can be expressed as

$$E_{electric} = JV \tag{3.138}$$

where, the cell potential V is given as

$$V = V_0 + V_{act,a} + V_{act,c} + V_{ohm} \tag{3.139}$$

where, V_0 is the reversible potential, which is related to the difference in free energy between reactants and products and can be obtained by Nernst equation as follows:

$$V_0 = 1.229 - 8.5 \times 10^{-4}(T_{PEM} - 298) \tag{3.140}$$

where, T_{PEM} is the operating temperature of PEM. Here, $V_{act,a}$ and $V_{act,c}$ are the activation overpotential of anode and cathode, respectively. The ohmic overpotential V_{ohm} in the proton exchange membrane is caused by the resistance of the membrane to the hydrogen ion transporting through it.

The local ionic conductivity $\sigma(x)$ of the proton exchange membrane can be expressed as

$$\sigma_{PEM}[\lambda(x)] - [0.5139\lambda(x) - 0.326]exp\left[1268\left(\frac{1}{303} - \frac{1}{T}\right)\right] \tag{3.141}$$

where, x is the distance in the membrane measured from the cathode–membrane interface and $\lambda(x)$ is the water content at a location x in the membrane. The value of $\lambda(x)$ can be calculated using

$$\lambda(x) = \frac{\lambda_a - \lambda_c}{D}x + \lambda_c \tag{3.142}$$

where, λ_a and λ_c are the water contents at the anode–membrane and cathode–membrane interfaces, respectively, and D is the membrane thickness.

The overall ohmic resistance can be expressed as

$$R_{PEM} = \int_0^D \frac{dx}{\sigma_{PEM}[\lambda(x)]} \tag{3.143}$$

The ohmic overpotential can be written as

$$V_{ohm,PEM} = JR_{PEM} \tag{3.144}$$

The activation overpotential can be expressed as

$$V_{act,i} = \frac{RT}{F}Sinh^{-1}\left(\frac{J}{2J_{0,i}}\right); \quad i = a, c \tag{3.145}$$

where, J_0 is the exchange current density, which characterizes the electrode's capabilities in the electrochemical reaction. The exchange current density for electrolysis can be expressed as

$$J_{0,i} = J_i^{ref}exp\left(-\frac{E_{act,i}}{RT}\right); \quad i = a, c \tag{3.146}$$

where, J_i^{ref} is the pre-exponential factor and $E_{act,i}$ is the activation energy of the anode and cathode.

3.9.1.5 Energy Efficiency

The energy efficiency of the OTEC system is defined as the net power output of the system divided by the input energy at the evaporator i.e.

$$\eta = \frac{\dot{W}_{net}}{\dot{Q}_{evp}}$$

where, $\dot{Q}_{evp} = \dot{m}_{wf}(h_1 - h_4)$.

Table 3.5

Input Parameter Values Used to Model PEM Electrolysis

Parameter:	P_{O_2} (atm)	P_{O_2} (atm)	T_{PEM} (°C)	$E_{act,a}$ (kJ/mol)	$E_{act,c}$ (kJ/mol)	λ_a λ_c	D (μm)	J_a^{ref} (A/m²)	J_c^{ref} (A/m²)	F (C/mol)
Value:	1.0	1.0	80	76	18	44 10	100	1.7×10^5	4.6×10^3	96,486

3.9.1.6 Exergy Efficiency

The exergy efficiency is the ratio of the product exergy output to the exergy input given as

$$\varepsilon = \frac{\dot{W}_{net}}{Ex_{in}} = \frac{\dot{W}_{net}}{Ex_{in,ws} + Ex_{in,cs}} \qquad (3.147)$$

where,

$$Ex_{in,ws} = \dot{m}_{in,ws}\left[(h_{ws,in} - h_0) - T_0(s_{ws,in} - s_0)\right]$$
$$Ex_{in,cs} = \dot{m}_{in,cs}\left[(h_{cs,in} - h_0) - T_0(s_{cs,in} - s_0)\right]$$

3.9.2 SIMULATION RESULTS

The integrated OTEC system can be simulated using the following simplified assumption.

1. All processes operate at steady state.
2. Pure ammonia is the working fluid of an ideal saturated Rankine cycle.
3. All the components are adiabatic.
4. The pressure drop in ORC is negligible.
5. State 5 at the entry of the turbine is saturated vapor.
6. Heat losses from piping and other auxiliary components are negligible.

Table 3.5 lists the parameters used to simulate the PEM electrolyzer. Table 3.6 lists the input parameters for the OTEC system simulation. Table 3.7 lists the results of the simulation. The net power output is equal to 10 kW with the hydrogen production rate of about 1.2 kg/h. The exergy efficiency is much higher than energy efficiency. This may be due to the fact that the work is produced using a low-grade heat at the ocean surface. Figure 3.24 shows that the highest exergy destruction occurs in the condenser. Both the exergy destruction rate and exergy destruction ratio are also higher in the condenser than any other component. It suggests that it is more worthwhile to focus on the improvement of the condenser.

Table 3.6

Input Data for the System Simulation

Parameter	Value
Turbine isentropic efficiency, η_T	0.80
Generator mechanical efficiency, η_G	0.90
Working fluid pump isentropic efficiency, η_{WFP}	0.78
Seawater pump isentropic efficiency, η_P	0.8
Ambient temperature (°C)	25
Solar radiation incident on collector surface, I (W/m^2)	500
Warm seawater temperature, T_{WSI} (°C)	22
Cold seawater temperature at a depth of 1000 m, T_{CSI} (°C)	4
Warm seawater mass flow rate (kg/s)	150
Cold seawater mass flow rate (kg/s)	150
Cold seawater pipe length (m)	1000
Cold seawater pipe inner diameter (m)	0.70
Warm seawater pipe length (m)	50
Warm seawater pipe inner diameter (m)	0.70
Solar collector effective area (m^2)	5000
Electrolyzer working temperature (°C)	80

Table 3.7

Energy and Exergy Analysis Results

Parameter	Value
Net power output, W_{net} (kW)	101.96
Exergy efficiency, Ψ (%)	22.70
Energy efficiency, η (%)	3.6
Total exergy destruction rate, $\dot{E}x_{D,tot}$ (kW)	42.12
Hydrogen production rate, $\dot{m}_{H_2}$ (kg/h)	1.20
PEM electrolyzer exergy efficiency, Ψ_{PEM} (%)	56.34
Warm surface pump power, $\dot{W}_{WS}$ (kW)	1.30
Cold surface pump power, $\dot{W}_{CS}$ (kW)	3.13
Working fluid pump power, $\dot{W}_{WF}$ (kW)	0.88

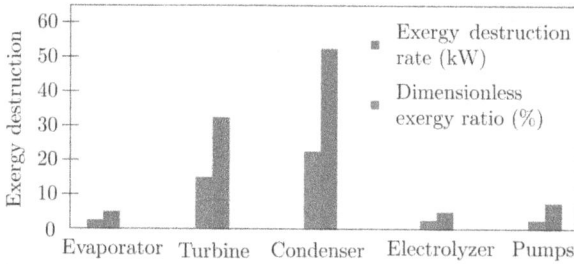

Figure 3.24 Exergy destruction rate and dimensionless exergy destruction ratio for each component of the OTEC system.

PROBLEMS

1. A lump of ice with a mass of 1.5 kg, at an initial temperature of $T_1 = 260$ K melts at a pressure of 1 bar as a result of heat transfer from the environment. After some time has elapsed, the resulting water attains the temperature of the environment i.e. $T_0 = 293$ K. Calculate the entropy production associated with this process. Take the enthalpy of fusion for water, $h_{sf} = 333.4$ kJ/kg and the isobaric specific heat capacities of ice and water as $C_{p,ice} = 2.07$ kJ/kg-K and $C_{p,w} = 4.20$ kJ/lkg-K. Ice melts at $T_m = 273.16$ K.

    (a) Calculate the specific physical exergy of CO_2 ($C_p = 0.8659$ kJ/kg-K, $R = 0.1889$ kJ/kg-K) for a state defined by $p_1 = 0.7$ bar and $T_1 = 268.15$ K.

    (b) Calculate the molar chemical exergy of CO_2. The partial pressure of CO_2 in the atmosphere is equal to 0.0003 bar. The molar ideal gas constant is 8.3144 kJ/kmol-K.

    (c) Calculate the molar chemical exergy of a mixture of gases with the following composition by volume (or by more): CO-0.15, air-0.85.

    For the environment take, $T_0 = 293.15$ K and $p_0 = 1.0$ bar.

2. A combustible mixture of CO and air containing 15% CO by volume enters a combustion chamber at 0.5 kg/s. The pressure, temperature and the mean velocity of the stream are 2.1 bar, 125 °C and 120 m/s, while the environmental pressure and temperature are 1 bar and 25 °C, respectively. Calculate the exergy rate of the stream.

3. Consider a coal gasification reactor making use of the carbon steam process in which carbon at 25 °C, 1 bar and water vapor at 316 °C, 1 bar enter the reactor and the product gas mixture exits at 927 °C and 1 bar The overall reaction is

$$C + 1.25\, H_2O \rightarrow CO + H_2 + 0.25\, H_2O$$

The energy required for this endothermic reaction is provided by an electric resistance heating unit, for operation at steady state, determine in kJ/kmol of carbon:

(a) the electricity requirement, (b) the exergy entering with the carbon, (c) the exergy entering with the steam, (d) the exergy exiting with the product gas, (e) the exergy destroyed and (f) the exergetic efficiency.

4. We wish to determine whether less exergy is destroyed when we drive a car with all windows closed and the air conditioner on than when we drive with the windows open and the air conditioner off. The car can be modeled as a blunt body with frontal area $A = 5$ m^2 traveling through air with the velocity U_∞. It is known that in the flow regime of interest, the drag coefficients are $C_D = 0.17$ when the windows are closed and $C_D = 0.51$ with the windows open. The air conditioner consumes electrical power at a constant rate, $W_{ac} = 746$ W.
   Hint: Try to show the velocity range in which one of the options is better than the other from an exergy destruction point of view.

5. A vapor compression heat pump uses R12 as the working fluid. The condenser used in the heat pump is an air-cooled counter-flow heat exchanger. The operating parameters are:

Fluid	m (kg/s)	T_1 (°C)	p_1 (bar)	T_2 (°C)	p_2 (bar)
R12	0.125	45	9.607	35	9.607
Air	1.0	18	1.045	35	1.035

Assume air to be a perfect gas with $C_p = 1.0$ kJ/kg-K and $\gamma = 1.4$. The temperature of the surrounding is 278 K. The R12 properties at inlet (1) and outlet (2) are: $S_{R1} = 0.6945$ kJ/kg-K, $S_{R2} = 0.2559$ kJ/kg-K, $h_{R1} = 207.05$ kJ/kg and $h_{R2} = 69.55$ kJ/kg. Calculate:

   a. Heat transfer rate to the environment
   b. Total irreversibility rate (exergy destruction) of a heat exchanger
   c. Irreversibility rate (exergy destruction) due to pressure losses
   d. Exergetic efficiency of the condenser

6. Cylinder (Dia, $D = 10$ cm, Length, $W = 100$ cm) arrays of a heat exchanger with an inline arrangement are exposed to a uniform velocity of $U = 2$ m/s and temperature $T_0 = 30$ °C. The Nusselt number for a cylinder in this arrangement is given by
$$Nu_D = hD/k = 0.25Re_D^{0.63}Pr^{0.36}$$
The drag coefficient for the cylinder at this Reynolds number is equal to 0.5. The cylinder surface temperature is 50 °C. The working fluid is air with the following average properties. $\rho = 1.1267$ kg/m^3; $k = 0.0271$ W/m-K; $\mu = 1.91 \times 10^{-5}$ kg/m-s and $Pr = 0.711$.

(a) Calculate the entropy generation per cylinder for this arrangement.
(b) If the diameter of the cylinder is reduced by 50%, what will be the change in entropy generation, keeping all other parameters the same?

REFERENCES

P. Ahmadi, I. Dincer, and M. A. Rosen. Energy and exergy analyses of hydrogen production via solar-boosted ocean thermal energy conversion and PEM electrolysis. *International Journal of Hydrogen Energy*, 38(4):1795–1805, Feb 2013. doi: 10.1016/j.ijhydene.2012.11.025.

D. Bahrehmand and M. Ameri. Energy and exergy analysis of different solar air collector systems with natural convection. *Renewable Energy*, 74:357–368, Feb 2015. doi: 10.1016/j.renene.2014.08.028.

M. D. Bazilian and D. Prasad. Modelling of a photovoltaic heat recovery system and its role in a design decision support tool for building professionals. *Renewable Energy*, 27(1):57–68, Sep 2002. doi: 10.1016/s0960-1481(01)00165-3.

A. Bejan, G. Tsatsaronis, and M. Moran. *Thermal Design and Optimization*. John Wiley & Sons Inc, 1995. ISBN 0-471-58467-3.

R. E. Bolz and G. L. Tuve, editors. *CRC Handbook of Tables for Applied Engineering Science*. CRC Press, 2019. ISBN 9781351829984. doi: 10.1201/9781315214092.

J. Duffie, W. Beckman, W. Beckman, and J. W. . Sons. *Solar Energy Thermal Processes*. A Wiley-Interscience publication. Wiley, 1974. ISBN 9780471223719.

P. Esmaili, I. Dincer, and G. Naterer. Energy and exergy analyses of electrolytic hydrogen production with molybdenum-oxo catalysts. *International Journal of Hydrogen Energy*, 37(9):7365–7372, May 2012. doi: 10.1016/j.ijhydene.2012.01.076.

S. Farahat, F. Sarhaddi, and H. Ajam. Exergetic optimization of flat plate solar collectors. *Renewable Energy*, 34(4):1169–1174, Apr 2009. doi: 10.1016/j.renene.2008.06.014.

J. Hirunlabh, W. Kongduang, P. Namprakai, and J. Khedari. Study of natural ventilation of houses by a metallic solar wall under tropical climate. *Renewable Energy*, 18(1):109–119, Sep 1999. doi: 10.1016/s0960-1481(98)00783-6.

K. G. T. Hollands, T. E. Unny, G. D. Raithby, and L. Konicek. Free Convective Heat Transfer Across Inclined Air Layers. *Journal of Heat Transfer*, 98(2):189–193, 1976. ISSN 0022-1481. doi: 10.1115/1.3450517.

W. McAdams. *Heat Transmission*. McGraw-Hill, 1954.

G. Nihous and L. Vega. Design of a 100 MW OTEC-hydrogen plantship. *Marine Structures*, 6(2–3):207–221, Jan 1993. doi: 10.1016/0951-8339(93)90020-4.

R. Petela. Exergy of heat radiation. *Journal of Heat Transfer*, 86(2):187–192, May 1964. doi: 10.1115/1.3687092.

F. Sun, Y. Ikegami, B. Jia, and H. Arima. Optimization design and exergy analysis of organic rankine cycle in ocean thermal energy conversion. *Applied Ocean Research*, 35:38–46, Mar 2012. doi: 10.1016/j.apor.2011.12.006.

W. C. Swinbank. Long-wave radiation from clear skies. *Quarterly Journal of the Royal Meteorological Society*, 90(386):488–493, Oct 1964. 10.1002/qj.49709038617.

4 Material Selection

4.1 MATERIAL PROPERTIES

A thermal system consists of different types of subsystems with components using a variety of materials i.e. solid, liquid and gas, having different material properties. A material can exist in different phases i.e. solid, liquid and vapor, depending on the operating conditions. The material properties change according to the respective phase condition. The materials can also exist as a mixture of two phases i.e. liquid–vapor mixture, which is expressed using mass fraction, and the properties change with respective mass fraction values. The working fluid can be a liquid–liquid mixture, which is a mixture of two or more liquid substances. The working fluid can be a mixture of gases, fuel vapor and air. Air is also a mixture of several gases i.e. nitrogen, oxygen, water vapor, argon, carbon dioxide etc. A thermal system undergoes several transport processes i.e. flow, heat and energy. Fluid structure may also be significant in a thermal system leading to transient distribution of mechanical stress on the component. Accordingly, several properties i.e. mechanical and thermal properties are required to describe transport processes. Elastic modulus, Bulk modulus, Poisson ratio, modulus of rupture, endurance/fatigue limit and hardness are typical mechanical properties used to specify the mechanical behavior. Melting temperature, specific heat, thermal conductivity, thermal diffusivity, mass diffusivity, momentum diffusivity, linear thermal expansion coefficient, emissivity and absorptivity etc. are typical thermophysical properties which are required for the design of a thermal system. In this chapter, the selection procedure for materials of a thermal system is discussed.

4.2 SOFTWARE

A large number of thermophysical properties are required for design of thermal systems. The dependence of these properties on physical state i.e. pressure, temperature and composition adds to difficulty in the comprehensive specification of material properties. Several software are available which provide thermophysical properties of many solids, liquids and gases. REFPROP is a database developed by National Institute of Standards and Technology (NIST) containing thermodynamic and transport properties of several reference fluids (www.nist.gov/srd/refprop). It is a fluid property calculator with a dynamic link (REFPROP.dll) that can be used to interface with a variety of software packages including MALAB and C++.

CoolProp (www.coolprop.org) is an open-source software package which is an alternative to REFPROP. It has an extensive database for the thermophysical properties of fluid. It also uses a dynamic link library (CoolProp.dll) that can be used to interface with various software packages.

Table 4.1 shows the list of databases and software available for material selection. Some databases are free and some are chargeable. A design engineer can also develop

DOI: 10.1201/9781003049272-4

Table 4.1

List of Material Selection Database and Software

Names	Link	Rates	Access	Contents	Attributions
MATWEB	www.matweb.com	Free	Online	Physical property data of metal, plastics, ceramic and composite. More than 115,000 kinds of materials.	USA
MATERIA	www.materia.nl	Free	Online	Information on innovative materials. More than 2600 excluding materials.	NED
Stylepark	www.stylepark.com	Free	Online	Properties of energy-efficient and eco-friendly materials used for construction and furniture manufacture.	GER
Ravara Database	www.ravara.se/Raw/Data	Free	Online	Property database of many materials.	SWE
AZOM.COM	www.zaom.com	Free	Online Software	Properties, analysis methods, detection equipment and information of materials in the industry.	AUS
IDEMAT	www.idemat.nl	Chargeable	Software	Environmental impact data of materials.	NED
MATERIALS DATA CENTER	www.techstreet.com/asm-spec	Chargeable	Software	Professional standard and data of different materials.	USA
KEY TO METALS	www.key-to-metals.com	Chargeable	Software	Property data of abundant materials.	SWZ
MPDB	www.jahm.com	Chargeable	Software	Temperature-dependent material property data.	USA
CES Selector	www.grantadesign.com	Chargeable	Software	Property data, processing technique and performance comparison of metals, plastics, composites and ceramics.	GBR
CES Edupack	www.grantadesign.com	Chargeable	Software	Education resources of materials, processes and sustainability.	GBR
GRANTA MI	www.grantadesign.com	Chargeable	Software	Management, design, simulation and analysis of materials. Data can be used by other design software.	GBR
Eco Materials Adviser	www.grantadesign.com	Chargeable	Software	Toolkit of autodesk inventor, including property and environmental impact of Eco materials.	GBR
ASM Alloy Center Database	mio.asminternational.org/ac/	Chargeable	Online	Materials property and corrosion performance data of metal and alloy.	USA

Source: Wei et al. [2018].

software and database for specific applications and link the simulation platform for design of a thermal system.

4.3 MATERIAL ATTRIBUTES

Materials are classified as families, classes, sub-classes and members. Figure 4.1 shows a typical hierarchical structure of materials available in computer-based material selection software [Ashby, 2013]. The attributes (or material record) of ALU 6061 are shown, which correspond to family of "metals", class of "Al-alloy", sub-class of "6000" and member of "6061". The attributes of a specific material i.e.

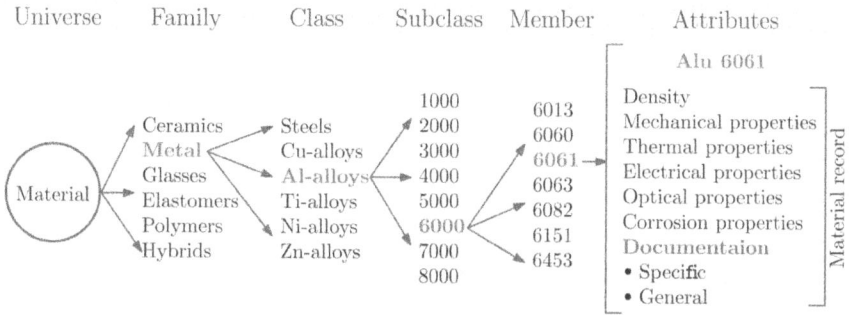

Figure 4.1   Material attributes and their hierarchical structure in a typical computer database.

mechanical, electrical, optical and chemical properties, its cost, availability and environmental consequences of its use are included in the property profile. The selection of material involves the best match of the property profiles of the material with the property profile requirement by the design.

4.4   SELECTION STRATEGIES

Figure 4.2 shows the flow chart explaining a possible material selection strategy [Ashby 2013]. It has four main steps: (1) translation, (2) screening, (3) ranking and (4) documentation. In the translation stage, a clear statement of material selection i.e. the function, constraints, objectives and free variables are specified. The function can be supporting a load, transmission of heat or containment of pressure etc. The objectives can be low weight or cheap or small size or some combination of these. The constraint can be fixed size, temperature range etc. The free variables are parameters which the designer is free to choose i.e. dimension and type of material etc. In screening, materials that cannot do the job or whose attributes lie outside the limit of constraints are eliminated. The ranking is carried out on material which passed the screening stage on the basis of how well it can perform. It can be either a single property i.e. thermal conductivity or a combination of properties i.e. specific strength, σ_f/ρ, where σ_f is the failure strength and ρ is the density etc. The property or property group that maximizes performance is called as "material index", which allows ranking of material by their ability to perform well in the given application. "Documentation" provides a detailed profile of each candidate i.e. previous case studies on its performance, failure analysis, pricing, availability etc. Based on the detailed documentation, the design engineer may not select the highest-ranked candidate material.

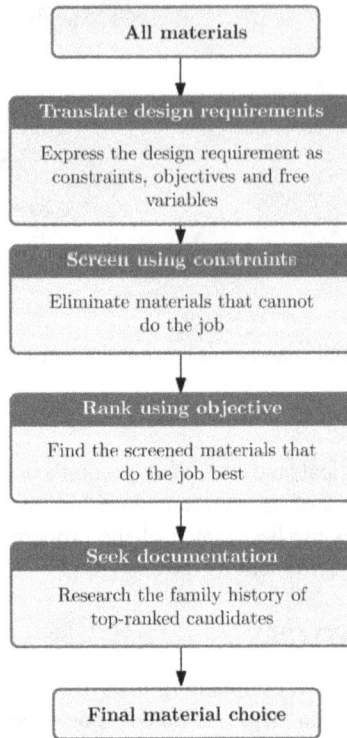

Figure 4.2   Flow chart explaining a typical material selection procedure.

4.4.1   MATERIAL INDICES

The performance P of a system can be described as

$$P = f(F,G,M) = [(\text{Functional requirements}, F), (\text{Generic parameters}, G),$$
$$\times (\text{Material properties}, M)] \tag{4.1}$$

The above functional quantities also can be separable and written as

$$P = f_1(F)f_2(G)f_3(M) \tag{4.2}$$

where, f_1, f_2 and f_3 are separate functions. The function f_3 is known as a material index, and the combination of function $f_1 \cdot f_2$ is known as a structural index. For example, let's consider the design of a tie rod which must support axial load F^*. The objective is to minimize the weight of the tie rod. Here, the free variables are area A and choice of material. We can write the objective function i.e. mass of the rod as

$$m = AL\rho \tag{4.3}$$

where, A is the cross-sectional area, l is the length of the tie rod and ρ is the density of the material. The stress of the tie rod should satisfy the failure condition i.e.

$$\frac{F^*}{A} \leq \sigma_f \qquad (4.4)$$

where, σ_f is the failure strength. We can rewrite the equation of "m" as:

$$m = \frac{F^*}{\sigma_f}L\rho = (F^*)(L)\left(\frac{\rho}{\sigma_f}\right) \qquad (4.5)$$

Here, F^* is termed as a functional index (F), L is termed as geometric index (G) and ρ/σ_f is termed as material index (m).

A few material selection case studies are discussed here.

4.5   CASE STUDIES

Several case studies on material selection for thermal design applications are discussed in this section i.e. material selection for (1) heat sink, (2) sensible thermal energy storage (TES), (3) phase change material for cold TES, (4) insulation and (5) solar power systems.

4.5.1   CASE 1: HEAT SINK MATERIAL

Heat sink transfers thermal energy from an object (electronic chip) at a relatively high temperature to a second object at a lower temperature [Reddy and Gupta, 2010]. Figure 4.3 shows a metallic sink with fins. A fan is sometimes used to transfer the thermal energy from the heat sink to the air. Use of coolants with refrigeration is also used as an interface material for cooling electronic devices. Heat sinks are important constituents of microelectromechanical systems (MEMS) devices i.e. micro heat exchanger, micro pump etc. Meticulous material selection of heat sinks can optimize the device performance. The heat sink material needs to be selected based on the material properties i.e. thermal expansion (α), Young's modulus (E), thermal conductivity (k), electrical resistivity (ρ_e) of suitable material. The heat sink must be a good electrical insulator to prevent electrical coupling between a microchip and heat sink i.e. it should have high resistivity, $\rho_e \geq 10^{19}$ μ-Ω-cm. It should also have high thermal conductivity (k) to drain away heat as fast as possible from the chip. Table 4.2 shows the details of the translation stage. Mathematically, the performance parameter of the sink material selection problem can be written as: $p = f(\rho_e, k)$. Both ρ_e and

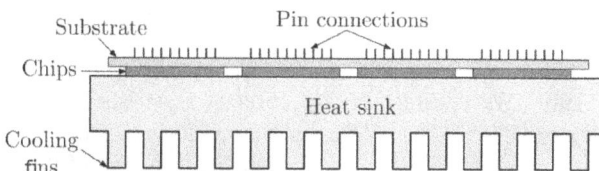

Figure 4.3   Common design of a heat sink in a metal device with many fins.

Table 4.2

Problem Specification of Material Selection for a Heat Sink

Function:	Heat sink
Constraints:	1. Materials must have $\rho_e > 10^{19}$ $\mu\Omega$-cm
	2. All dimensions are fixed
Objective:	Maximize thermal conductivity
Free variable:	Choice of material

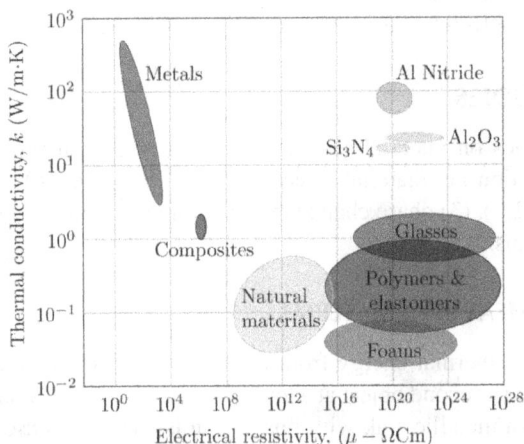

Figure 4.4 Graph showing the contour of thermal conductivity versus electrical resistivity for different classes of materials.

k should be maximized for the optimum performance of the sink. Figure 4.4 shows the thermal conductivity versus electrical resistivity of different classes of materials. Metals, composites and natural materials are screened out due to lower electrical resistivity ($\rho_e \leq 10^{19}$ μ-Ω-cm). Aluminum nitride (AlN), alumina (Al_2O_3), glass and foams satisfy the resistivity constraint. Subsequently, these materials are ranked according to thermal conductivity criteria. It is observed that AlN or Al_2O_3 satisfies the constraints, and the objective of maximum thermal conductivity is achieved with these materials.

4.5.2 CASE 2: MATERIAL FOR SENSIBLE THERMAL ENERGY STORAGE

Energy storage technologies help in the utilization of renewable energy sources and energy conservation. We consider the selection of materials for sensible TES in a temperature range of 150–200 °C for long-term storage of thermal energy. The translation stage should specify the requirements i.e. function of the design as follows. The material should meet the following constraints: (a) high energy density

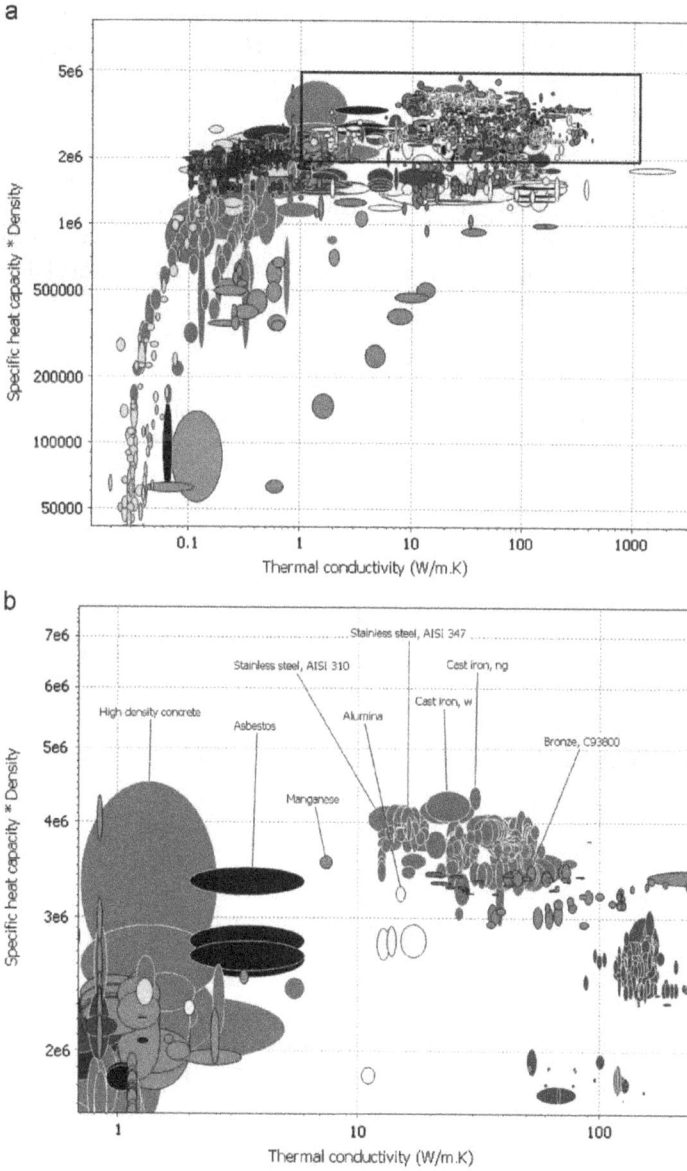

Figure 4.5 (a) Material property chart with a combination of properties i.e. energy density $(C_p\rho)$ versus thermal conductivity and (b) zoom-in the box area with constraints; materials that do not accomplish these requirements are not plotted. (Adapted from Fernandez et al. [2010] with permission.)

(heat capacity per unit volume), (b) lower limit for maximum service temperature is equal to 150 °C and (c) good thermal conductivity ($k \geq 1$ W/m-K). The primary objective of this application is to maximize the energy storage per unit of material cost, and maximizing the thermal conductivity is an additional objective. The free variables are material choice and dimension. Figure 4.5a shows summary of about 3000 engineering materials with a variation in heat capacity (ρc_p) versus thermal conductivity. The right upper part of the plot satisfies the constraint of $k \geq 1$ W/m-K. The zoomed plot of these candidate materials is shown in Figure 4.5b, which are required to satisfy the constraint of service temperature lower limit $=150$ °C and the cost.

The TES per unit volume can be expressed as

$$Q = \rho c_p \Delta T \tag{4.6}$$

The cost, C of a mass m of material with a cost per kg of Cm is

$$C = mC_m \tag{4.7}$$

Taking the ratio of equations 4.7 and 4.8, we can write:

$$Q' = \frac{c_p \Delta T}{V c_m} \tag{4.8}$$

where, Q' is the energy stored per unit volume and unit cost and V is the volume of the material. Here, ΔT is termed as a functional property, V is the geometrical property and c_p/C_m is the material property, which is defined as material index. Figure 4.6 shows the plot of c_p versus C_m, and materials with the highest material

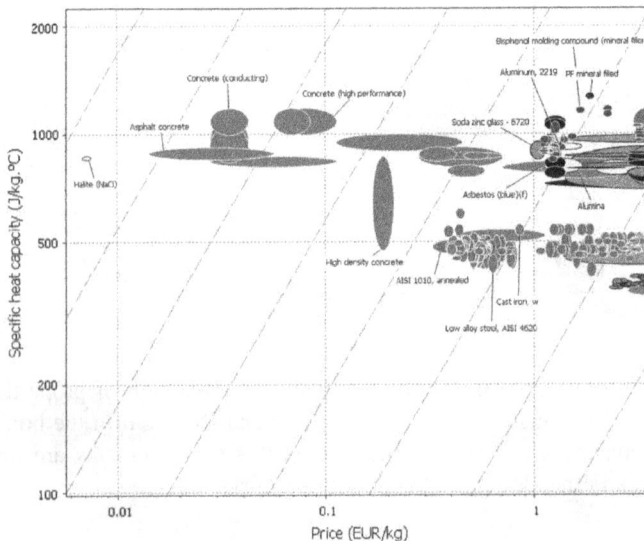

Figure 4.6 Plot of specific heat capacity versus cost per unit mass, C_m. (Fernandez et al. [2010].)

index are located at the upper left of the chart. The material with the same c_p/C_m will have a slope of unity and give the same value of Q'. The dashed lines are the lines with a slope of unity, and the materials over these lines perform better than those located below. Halite (sodium chloride) and structural ceramics such as asphalt concrete are the ideal materials with the optimal value of the objective function.

4.5.3 CASE 3: PHASE CHANGE MATERIAL FOR COLD THERMAL ENERGY STORAGE

Latent heat storage using PCMs is one of the most efficient methods to store thermal energy. It provides higher heat storage capacity and more isothermal behavior during charging and discharging compared to sensible heat storage. Cold TES systems are used in various industrial applications i.e. food storage, ice storage, transport of temperature-sensitive materials, air conditioning etc.

The energy storage density within a temperature range of $\Delta T = T_2 - T_1$ having a phase change (latent heat) can be calculated as

$$Q_{latent} = \int_{T_1}^{T_2} c_s dt + \Delta H_{ls} + \int_{T_{pc}}^{T_2} c_l dt \tag{4.9}$$

where, ΔH_{ls} is the heat of fusion at the phase change temperature T_{pc}. The subscripts s and l correspond to solid and liquid, respectively. The PCM for the TES system should have desirable thermophysical, kinetic and chemical properties listed here.

4.5.3.1 Thermophysical Properties

a) Melting temperature should be in the desired temperature range.
b) Latent heat of fusion per unit volume should be high.
c) The specific heat should be high to provide additional sensible heat storage.
d) The thermal conductivity should be high during both solid and liquid phases.
e) The volume change should be small during phase transformation.
f) The melting and freezing cycle variation should be identical for each cycle of operation.

4.5.3.2 Kinetic Properties

a) The nucleation and crystal growth rate should be high to avoid sub-cooling of the liquid phase during solidification.
b) The melting and solidification process should occur at the same temperature.

4.5.3.3 Chemical Properties

a) The freeze and melt cycle should be reversible.
b) There should be no degradation after a large number of freeze/melt cycle.
c) There should not be any corrosive effects.
d) The material should be non-toxic, non-flammable and non-explosive.

Table 4.3

Comparison of Different Thermal Conductivity Enhancement Methods

Methods	Mechanisms	Limitations
Fins and extended surfaces	Increase in heat transfer area.	Total weight increases, increasing the total cost. Properties of PCM are not changed.
PCM-embedded porous matrices	Heat transfer area increases, forming a thermal transfer network, and increasing thermal conductivity of PCM.	Porous material is expensive and it can reduce total heat storage capacity while probably increase total weight.
Dispersion of highly conductive particles within the PCM	Thermal conductivity of PCM increases by the particles with high thermal conductivity.	Sedimentation of highly conductive particles may take place.
PCM microencapsulation	Increasing heat transfer area.	Costly and will reduce the mass per unit volume of PCM.
Multiple PCM method	Increasing average temperature difference.	Only adapt to the design conditions, and maybe not useful at variable working conditions.

4.5.3.4   Economics

a) The PCM should be abundantly available.
b) The PCM should be cost-effective.
c) The PCM should be recyclable.
d) The PCM should be environmentally friendly.

Oró et al. [2012] have reported a review of different low-temperature PCMs for air conditioning applications, cold storage and transport of temperature-sensitive materials The use of nucleating and thickening agents helps in minimizing sub-cooling and phase segregation. Encapsulation of PCM can help in meeting corrosion resistance and prevent harmful interaction with the environment. When selecting PCM for practical application, it is difficult to meet all the requirements simultaneously. Therefore, after meeting the main criteria (suitable phase change temperature, large latent heat etc.), suitable techniques can be adopted to compensate for any shortcomings and deficiencies of the material. Table 4.3 presents suitable thermal conductivity enhancement methods [Wei et al., 2018]. Wei et al. [2018] also reviewed the selection principle and thermophysical properties of high-temperature PCM for TES for concentrated solar power application.

4.5.4 CASE 4: SELECTION OF INSULATION MATERIAL

Insulation is used in various energy systems for several reasons i.e. to conserve energy, to reduce heat loss or gain, to maintain the required temperature condition, to assist in maintaining a product at constant temperature, to prevent condensation, to create a comfortable environment condition and to protect personnel.

When the insulation is used to prevent heat loss from the process, the term hot insulation is used. When the insulation is used to prevent heat gain in the process, the term cold insulation is used. In addition to the heat transfer aspect, there may be special service condition requirements e.g. resistance to vibration, resistance to mechanical damage, resistance to corrosive liquid and gases, resistance to ambient conditions and so on. Some insulation may require to be protected from mechanical damage by using metal cladding or jacketing. Typical insulating materials used at temperature higher than ambient are phenolic foam, polyurethane rigid foam, polyurethane flexible foam and mineral wool (glass). In addition to the thermal conductivity value, the following additional specifications are required for the selection of an insulating material i.e. service temperature range, fire characteristics, mechanical properties i.e. compressive strength, flexural strength and tensile strength and water vapor transmission. When one has to select thermal insulation for subsea pipelines, seawater resistance, thermal conductivity and low creep under hydrostatic pressure conditions should be considered [Collins, 1989].

4.5.5 CASE 5: HEAT TRANSFER FLUIDS FOR SOLAR POWER SYSTEMS

Concentrating solar power (CSP) is pursued as one of the key technologies for alternative clean and renewable energy sources. Most of the common technologies of CSP are: (1) parabolic dish system, (2) parabolic trough collector, (3) solar power tower and (4) linear Fersnel reflector. In CSP systems, solar irradiation is concentrated onto a receiver by programmed mirrors (heliostats). The heat is collected from the receiver by a thermal energy carrier called heat transfer fluid (HTF). The HTF can be used directly to drive a turbine to produce power. Alternatively, it can be combined with a heat exchanger and a secondary cycle to generate power.

HTF is the most important component of a CSP plant as it influences the overall performance and efficiency of CSP systems. It is essential to minimize the cost of HTF while maximizing its performance. The desired characteristics of HTF are: (1) low melting point, (2) high boiling point, (3) high thermal stability, (4) low vapor pressure (< 1 atm) at high temperature, (5) low corrosion with metal alloys, (6) low viscosity, (7) high thermal conductivity, (8) high heat capacity and (9) low cost. The six main groups of HTF are: (1) air and other gases, (2) water/steam, (3) thermal oils, (4) organics, (5) molten salts and (6) liquid. Vignarooban et al. [2015] presented a comprehensive list of working temperatures of various HTFs due to high working temperature (> 500 °C), high heat capacity, low vapor pressure and low corrosive property and good thermal properties at elevated temperature. For thermal stability range, thermal conductivity, viscosity, heat capacity and corrosion rate for piping/-container alloys, different HTFs are summarized in Table 4.4.

Table 4.4
Thermal and Physical Properties of Commonly Used HTFs

Name	Composition (wt.%)	Melting Point (°C)	Stability Limit (°C)	Viscosity (Pas)	Thermal Conductivity (W m⁻¹ K⁻¹)	Heat Capacity (kJ kg⁻¹ K⁻¹)	Cost ($/kg)	Corrosion Rate (μm/year Unless Specified)	Alloy	Temperature (°C)
Air	Air	—	—	0.00003 (at 600°C)	0.06 (at 600°C)	1.12 (at 600°C)	0	7–14 g/m	Fe–Al (5.8–16.2 wt%)–Cr (1.9–9.7 wt%)	1100
Water/steam	H_2O	0	—	0.00133 (at 600°C)	0.08 (at 600°C)	2.42 (at 600°C)	0	1.7–3.5	In600	300
Thermal oils										
Mineral oil	N/A	–20	300	N/A	~0.1	N/A	0.3	N/A		
Synthetic oil	N/A	–20	350	N/A	~0.1	N/A	3	N/A		
Silicone oil	N/A	–20	400	N/A	~0.1	N/A	5	N/A		
Xceltherm 600	Paraffinic mineral oil	N/A	315	0.001085 (at 300°C)	~0.1	2.436 (at 300°C)	N/A	N/A		
Organics										
Biphenyl/Diphenyl oxide	N/A	12	393	0.00059 (at 300°C)	~0.01 (at 300°C)	1.93 (at 300°C)	100	N/A		
Molten salts										
Solar salt	$NaNO_3$ (60)–KNO_3 (40)	220	600	0.00326 (at 300°C)	0.55 (at 400°C)	1.1 (at 600°C)	0.5	5 6–15 15.9/4 60 10.4/4 47 19.8/6 88 21.7/5 94	A36 304 316 321 347 Ha230 In625	316 570 600/680 600/680 600/680 600/680
Hitec	$NaNO_3$ (7)–KNO_3 (53)–$NaNO_2$ (40)	142	535	0.00316 (at 300°C)	~0.2 (at 300°C)	1.56 (at 300°C)	0.93	2		321 570
Hitec XL	$NaNO_3$ (7)–KNO_3 (45)–$Ca(NO_3)_2$ (48)	120	500	0.00637 (at 300°C)a	0.52 (at 300°C)	1.45 (at 300°C)	1.1	6–10	304, 316	570
Na–K–Li nitrates	$NaNO_3$ (28)–KNO_3 (52)–$LiNO_3$ (20)	130	600	0.03 (at 300°C)	N/A	1.091	~1.1	N/A		

Table 4.4 (Continued)

Name	Compositions (wt.%)	Melting Point (°C)	Stability Limit (°C)	Viscosity (Pas)	Thermal Conductivity (W m^{-1} K^{-1})	Heat Capacity (kJ kg^{-1} K^{-1})	Cost ($/kg)	Corrosion Rate (μm/year Unless Specified)	Alloy	Temperature (°C)
LiNaK carbonates	Li$_2$CO$_3$ (32.1)–Na$_2$CO$_3$ (33.4)–K$_2$CO$_3$ (34.5)	~400	800–850	0.0043 (at 800 °C)	N/A	~1.4–1.5	~1.2–1.3	<1000	In600	900
K–Li–Ca nitrates	KNO$_3$ (50–80)–LiNO$_3$ (0–25)–Ca(NO$_3$)$_2$ (10–45)	<80	~500	~0.004 (at 190 °C)	0.43 (at 300 °C)	N/A	0.6–0.8	N/A		
Na–K–Li nitrates/nitrites	NaNO$_3$ (14.2)–KNO$_3$ (50.5)–LiNO$_3$ (17.5)–NaNO$_2$ (17.8)	99	430	N/A	N/A	1.66 (at 500 °C)	N/A	N/A		
Sandia mix	NaNO$_3$ (9–18)–KNO$_3$ (40–52)–LiNO$_3$ (13–21)–Ca(NO$_3$)$_2$ (20–27)	<95	500	0.005–0.007 (at 300 °C)	0.654 (at 250 °C)	1.16–1.44	N/A	N/A		
(at 247 °C)	0.62–0.81									
Halotechnics SS-500	NaNO$_3$ (6)–KNO$_3$ (23)–LiNO$_3$ (8)–CsNO$_3$ (44)–Ca(NO$_3$)$_2$ (19)	65	500	N/A	N/A	1.22 (at 150 °C)	N/A	N/A		
Li–Na–K fluorides/carbonates	N/A	~400	~900	N/A	1.17 (at 400 °C)	N/A	N/A	8–12	316L	465
Halotechnics SS-700	N/A	257	700	0.004 (at 500 °C)	0.35–0.4	0.79 (at 300 °C)	N/A	165 276 74 74 160 493	316L Ni201 IN625 IN620 Ha230 IN800H/HT	700
Na–K–Zn chlorides	NaCl (7.5)–KCl (23.9)–ZnCl2 (68.6)	204	850	0.004 (at 600–800 °C)	0.325 (at 300 °C)	0.81 (at 300–600 °C)	¡1	110–200 < 50 < 50	304 C-22 C-276	800
Liquid metals										
Na	–	98	883	0.00021 (at 600 °C)	46.0 (at 600 °C)	1.25 (at 600 °C)	2	N/A		
Na–K	Na (22.2)–K (77.8)	–12	785	0.00018 (at 600 °C)	26.2 (at 600 °C)	0.87 (at 600 °C)	2	N/A		
Pb–Bi	Pb (44.5)–Bi (55.5)	125	1533	0.00108 (at 600 °C)	12.8 (at 600 °C)	0.15 (at 600 °C)	13	25–250 >250 >250	Ferritic SS Austenitic SS Nickel alloy	800

Source: Vignarooban et al. [2015].

4.6 SUMMARY

Appropriate material selection for thermal system applications is one of the most important parts of a thermal system design. The performance of a thermal system depends on the working fluid, which may not be known at the initial stage of a design. Therefore, simulation studies may need to be carried out in an iterative manner for the proper selection of material. The flow chart presented in Figure 4.2 should be coupled with computational fluid dynamics simulation. The modeling and simulation of thermal systems will be discussed in Chapter 5.

PROBLEMS

1. Discuss the procedure of material selection for mold material for casting of aluminum. How will the material be different when it is required to cast steel instead?

2. Determine the materials currently being used for proper functioning of the following systems: (a) Solar cells, (b) Wind turbine blade, (c) Fuel cells, (d) Automobile bodies and (e) ship bodies. Mention the relevant material properties for justifying the use of these materials.

3. Discuss how material selection for thermal system can play a role in green design.

4. Discuss the ideal coolant selection for immersion cooling of data center with proper justification.

5. Review the state of the art on application of artificial intelligence for material selection related to thermal management of any electronics cooling application.

REFERENCES

M. Ashby. *Materials Selection in Mechanical Design.* Elsevier Science, 2016. ISBN 9780081006108. https://books.google.co.in/books?id=K4h4CgAAQBAJ.

M. Ashby, Y. Bréchet, D. Cebon, and L. Salvo. Selection strategies for materials and processes. *Materials & Design*, 25(1):51–67, Feb 2004. doi: 10.1016/s0261-3069(03)00159-6.

M. Collins. Thermal insulation materials for subsea flowlines. *Materials & Design*, 10(4):168–174, Jul 1989. doi: 10.1016/s0261-3069(89)80002-0.

A. Fernandez, M. Martínez, M. Segarra, I. Martorell, and L. Cabeza. Selection of materials with potential in sensible thermal energy storage. *Solar Energy Materials and Solar Cells*, 94(10):1723–1729, Oct 2010. doi: 10.1016/j.solmat.2010.05.035.

E. Oró, A. de Gracia, A. Castell, M. Farid, and L. Cabeza. Review on phase change materials (PCMs) for cold thermal energy storage applications. *Applied Energy*, 99:513–533, 2012. doi: 10.1016/j.apenergy.2012.03.058.

G. P. Reddy and N. Gupta. Material selection for microelectronic heat sinks: An application of the Ashby approach. *Materials & Design*, 31(1):113–117, Jan 2010. doi: 10.1016/j.matdes.2009.07.013.

K. Vignarooban, X. Xu, A. Arvay, K. Hsu, and A. Kannan. Heat transfer fluids for concentrating solar power systems – A review. *Applied Energy*, 146:383–396, May 2015. doi: 10.1016/j.apenergy.2015.01.125.

G. Wei, G. Wang, C. Xu, X. Ju, L. Xing, X. Du, and Y. Yang. Selection principles and thermophysical properties of high temperature phase change materials for thermal energy storage: A review. *Renewable and Sustainable Energy Reviews*, 81:1771–1786, Jan 2018. doi: 10.1016/j.rser.2017.05.271.

5 Heat Exchangers

5.1 INTRODUCTION

Process industries, power industries and several other applications i.e. heating, ventillation, air conditioning, refrigeration and cooling of electronic systems use different types of heat exchangers. This chapter introduces the theoretical background and design procedure of heat exchangers.

5.2 CLASSIFICATION OF HEAT EXCHANGER

Heat exchangers can be classified according to the following main criteria:

a) *According to the Heat Transfer Process*
   i. Direct Contact Type: In the direct contact type, the fluid streams are in direct contact with each other, which exchange heat across an interface between the two fluids. The enthalpy of phase change represents a significant portion of the total energy transfer. In these exchangers very high heat transfer rate is achievable. There is no fouling problem as there are no separating surfaces between the fluid streams.
   ii. Indirect Contact Type: In the indirect contact type, the fluid streams remain separate. The heat transfer takes place through an impervious wall between the fluid (see Figure 5.1).

Shell and tube heat exchangers are indirect contact types having applications in power and process industries. Similarly, tube fin air-cooled condensers and plate-fin cryogenic condensers are indirect contact-type extended surface heat exchangers. Pool, spray and packed columns belong to the direct contact type heat exchanger.

b) *According to Construction*
   i. Tubular: The tubular heat exchangers are built using circular, elliptical and rectangular tubes. Tubular heat exchangers are primarily used for liquid-to-liquid and liquid-to-vapor (condensing or evaporating) heat transfer applications. The exchangers can be classified as shell and tube and double pipe (see Figure 5.1).
   ii. Plate Type: Plate-type heat exchangers are built using thin plates. The plates are either smooth or have some form of corrugation. The plate heat exchangers cannot accommodate very high pressure and temperature differences. The plate heat exchangers can be either gasketed, welded or brazed depending on the leak tightness required. Figures 5.7 and 5.8 show the schematic of a gasketed plate and a frame heat exchanger.

DOI: 10.1201/9781003049272-5

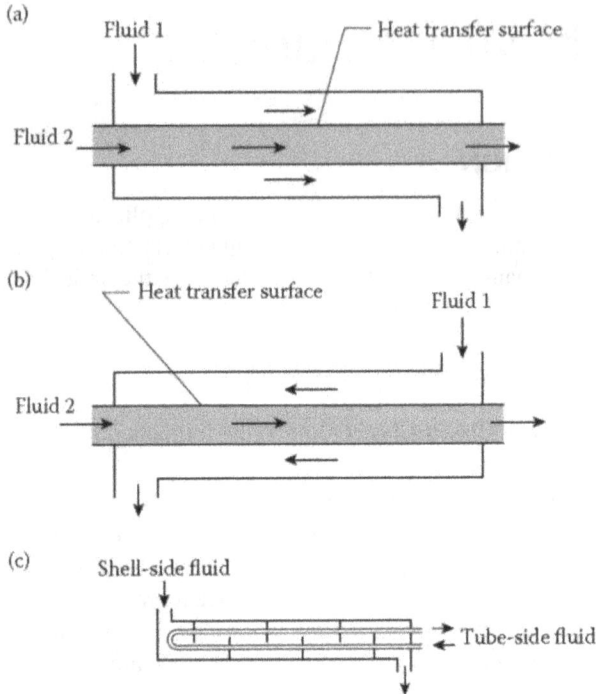

Figure 5.1   Indirect contact-type heat exchanger (a,b) double pipe and (c) shell and tube type. (Kakac et al. [2012].)

    iii. Extended surface: In an extended surface heat exchanger, fins are added to one or both fluid sides to increase the surface area for enhancing the heat transfer. The resulting heat exchanger is referred to as an extended surface heat exchanger (see Figures 5.3 and 5.4).

Double-pipe, shell and tube and spiral tube-type heat exchangers are classified as tubular heat exchangers. Gasketed plate heat exchanger, spiral plate heat exchanger and lamella heat exchanger belong to plate-type heat exchanger. Plate-fin heat exchanger (Figure 5.2) and tubular-fin heat exchanger are classified as extended heat exchangers.

c) *According to the Heat Transfer Mechanism (See Figure 5.5)*
    i. Single Phase: The single phase convection takes place in single-phase-type heat exchanger.
    ii. Two Phase: In a two-phase heat exchanger, either condensation or evaporation can take place.
d) *According to Flow Arrangements*
    i. Parallel Flow (see Figure 5.6a): The two fluid streams enter together at one end and leave together at other end in a parallel-flow heat exchanger.

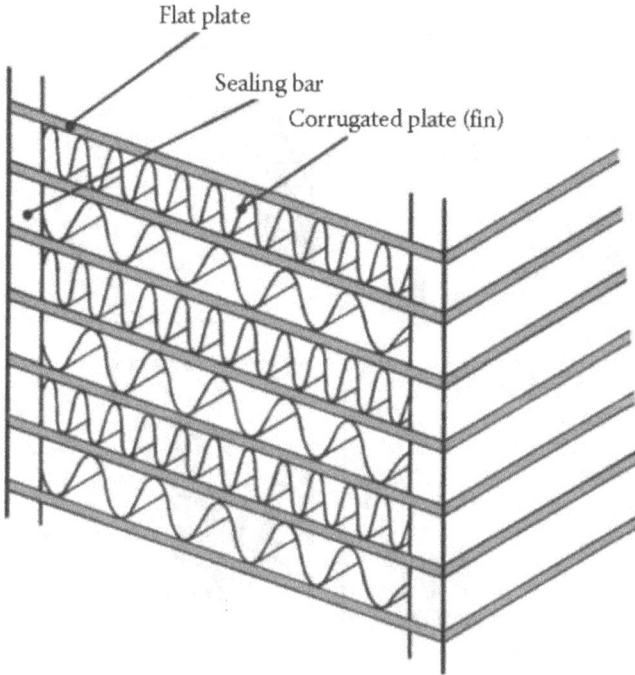

Figure 5.2 Basic construction of a plate-fin exchanger. (Kakac et al. [2012].)

ii. Counter-Flow (see Figure 5.6b): The two fluid streams flow in opposite directions in case of a counter-flow heat exchanger.

iii. Cross-Flow (see Figure 5.6c): The fluid flows at right angles to each other in case of a cross-flow heat exchanger.

e) *According to Surface Compactness*
   i. Compact
   ii. Cross-flow $(\beta < 400)$ m²/m²

Surface area density (β) is defined as the ratio of surface area to volume of a heat exchanger. For example, the surface area density of a plate heat exchanger is given as $\beta = \frac{A_h}{V_h}$ or $\frac{A_c}{V_c}$. Here, A is the heat transfer surface area with subscripts h and c for hot and cold fluid side, respectively. The volumes individually occupied by the hot and cold fluid side heat transfer surfaces are denoted, respectively, by V_h and V_c.

A gas-to-fluid heat exchanger is classified as *compact*; it has a surface area density greater than 700 m²/m³. The term micro heat exchanger is used if the surface area density is greater than 15000 m²/m³. A liquid/two-phase fluid heat exchanger is called a compact heat exchanger if the surface area density on any one fluid side is greater than 400 m²/m³. An automobile radiator has a surface area density of the order of 1870 m²/m³. Human lungs are one of the most compact heat exchangers with a surface area density of about 17500 m²/m³.

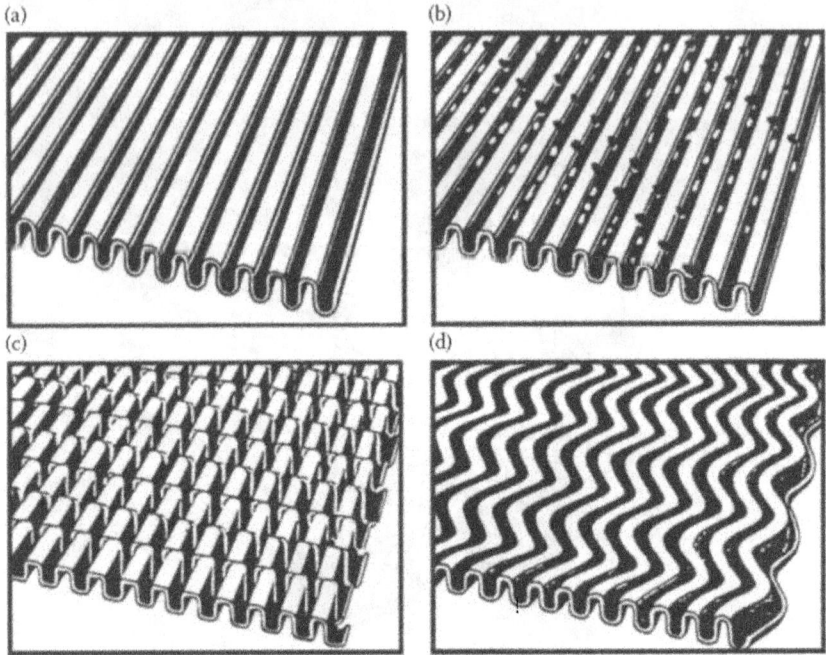

Figure 5.3 Examples of fin types in plate-fin exchangers: (a) plain, (b) perforated, (c) serrated and (d) herringbone. (Kakac et al. [2012].)

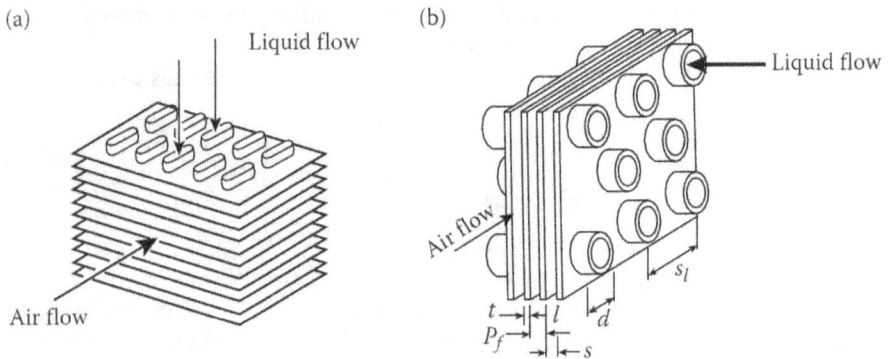

Figure 5.4 Examples of tube-fin heat exchangers: (a) flattened tube-fin and (b) round tube-fin. (Kakac et al. [2012].)

5.3 OVERALL HEAT TRANSFER COEFFICIENT

Figure 5.9 shows the schematic explaining the overall heat transfer coefficient concept using heat transfer in a finned tube heat exchanger. The overall heat transfer coefficient is used to express the total heat transfer across the heat transfer surface of

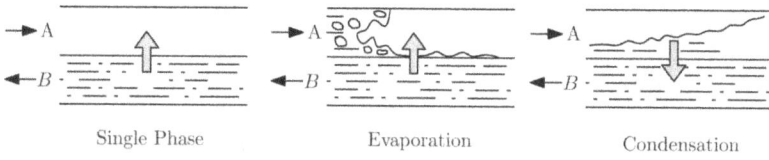

Figure 5.5 Schematic explaining single-phase and two-phase heat exchangers. (Kakac et al. [2012].)

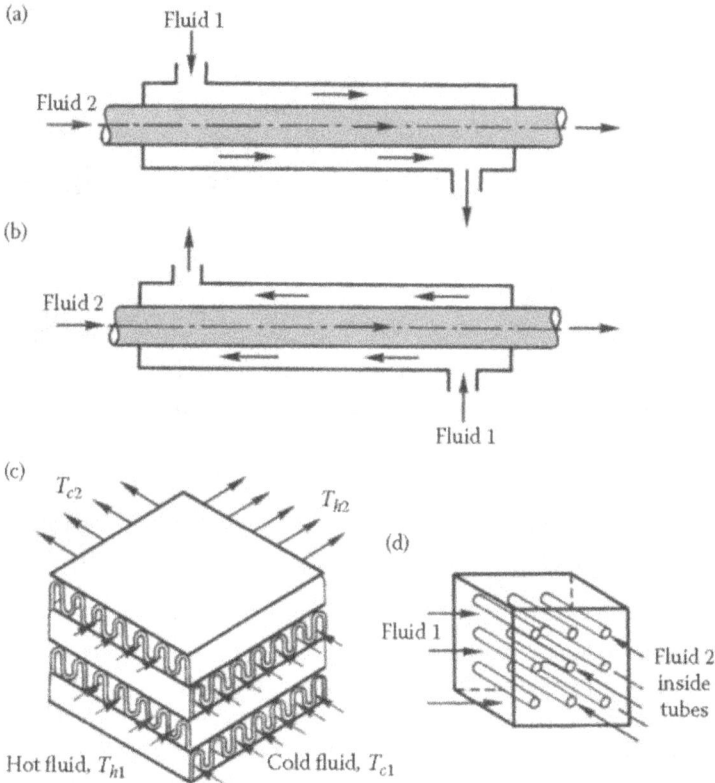

Figure 5.6 Classification of a heat exchanger according to flow arrangements: (a) parallel flow, (b) counter-flow, (c) cross-flow, both fluids unmixed, and (d) cross-flow, fluid 1 mixed, fluid 2 unmixed. (Kakac et al. [2012].)

a heat exchanger i.e.

$$Q = UA\Delta T = U_0 A_0 \Delta T_0 = U_i A_i \Delta T_i \qquad (5.1)$$

where, Q is the total heat transfer, A is the surface area and ΔT is the temperature difference between the fluid stream and base temperature. The subscripts o and i

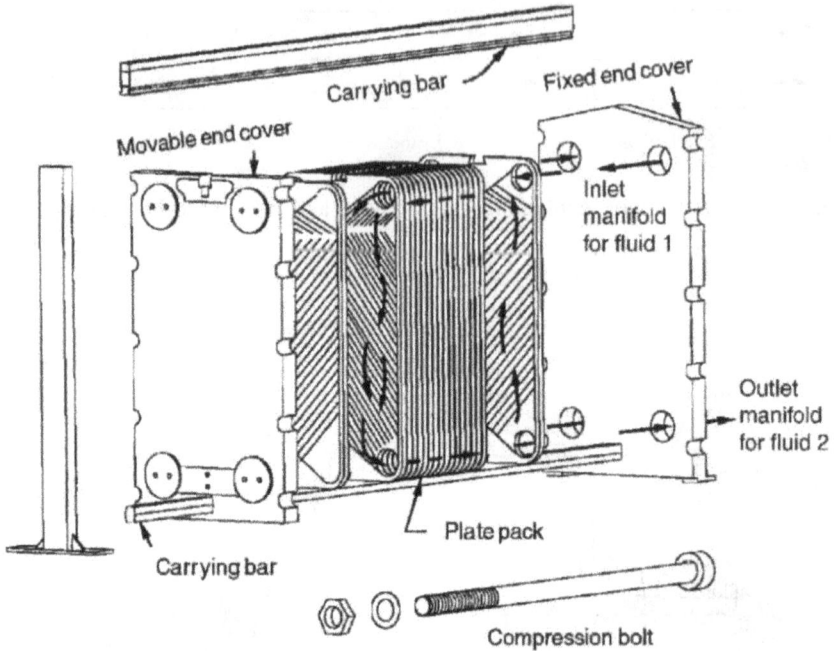

Figure 5.7 Schematic of a gasketed plate and a frame heat exchanger. (Shah and Sekulic [1985].)

correspond to the outer and inner surfaces, respectively. Here, ΔT can be either $(T_h - T_{w1})$ or $T_{w2} - T_c$ with subscripts h and c refer to hot and cold fluids, respectively, and subscript w corresponds to wall. Both or one of the inner and outer exchanger surfaces can be finned. The surface area (A) will be a summation of the finned (A_f) and unfinned part (A_u) i.e. $A = A_f + A_u$. The exchanger surface may acquire additional heat transfer resistance due to oxidation, crust deposit during operation. This additional thermal resistance is denoted as fouling resistance, R_{fi}, which depends on the type of fluid, fluid velocity, type of surface and length of the service period. The efficiency of finned surface, n_f depends on the geometry and nature of the finned material. The overall heat transfer coefficient based on the outer surface is the inverse of the total resistance offered to heat transfer i.e. summation of convective resistance at the interior surface, fouling resistance at the interior surface, conductive resistance, fouling resistance at the outer surface and convective resistance of the outer finned surface.

$$U_0 = \frac{1}{\frac{A_0}{A_i}\frac{1}{n_i h_i} + \frac{A_0 R_{fi}}{n_i A_i} + A_0 R_w + \frac{R_{fo}}{n_o} + \frac{1}{n_o h_o}} \tag{5.2}$$

The conductive resistance for a plane wall is given by

$$R_w = \frac{t}{kA} \tag{5.3}$$

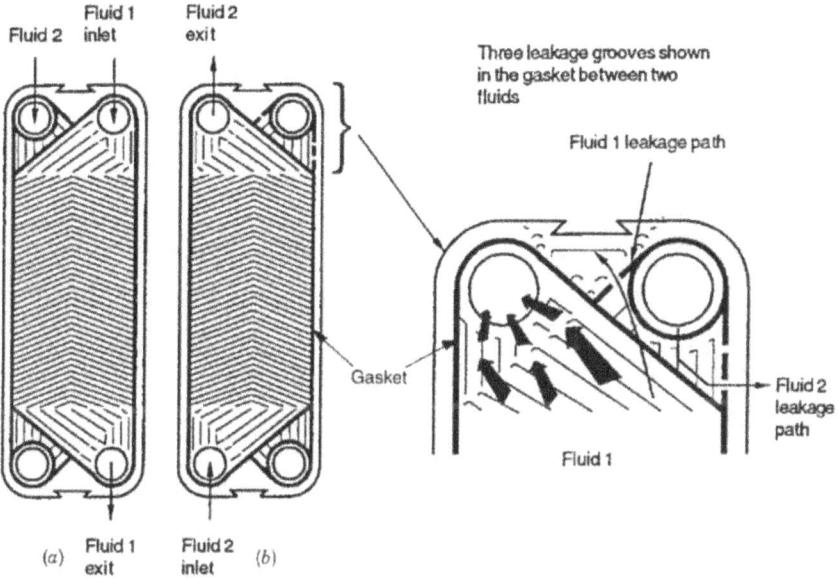

Figure 5.8 Plates showing gaskets around the ports. (Shah and Sekulic [1985].)

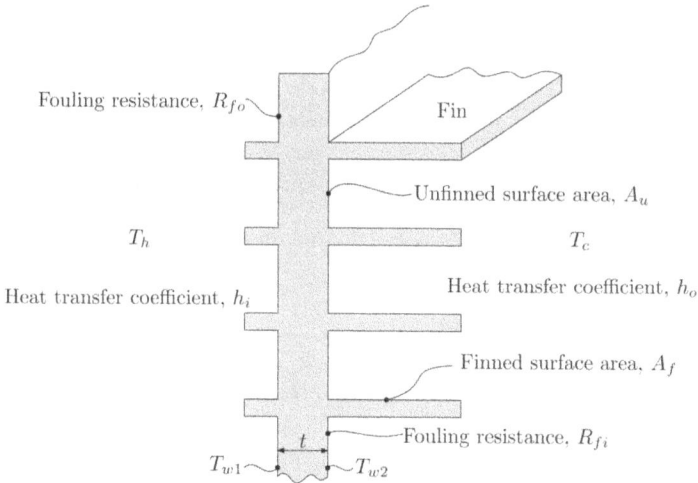

Figure 5.9 Schematic of a finned heat exchanger.

The conductive resistance for a tube wall is given by

$$Rw = \frac{ln(r_o/r_i)}{2\pi Lk} \qquad (5.4)$$

Here, k is the thermal conductivity of heat exchanger surface, t is the thickness of wall, L is the length and r is the radius of tube. The fin efficiency, n for different

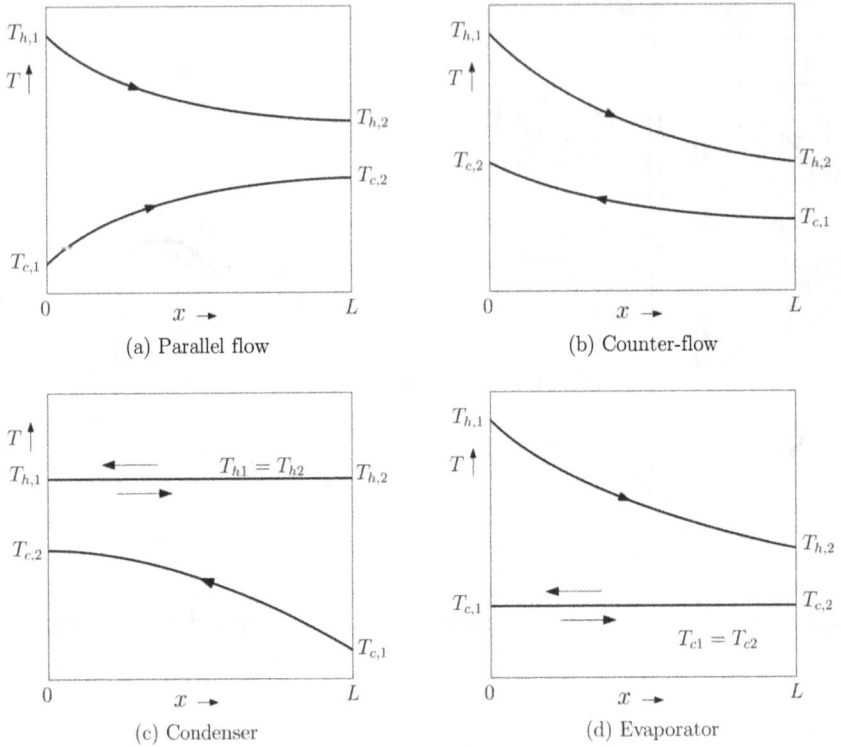

Figure 5.10   Temperature profile along the heat exchanger surface: (a) parallel flow heat exchanger, (b) counter-flow heat exchanger, (c) condenser and (d) evaporator.

types of fins is available in Table 5.1. The fouling resistance for different fluids is presented in Tables 5.2 and 5.3.

5.4   LOG MEAN TEMPERATURE DIFFERENCE (LMTD)

In the previous section, the temperature difference is assumed to be constant across an elementary length of surface for defining the overall heat transfer coefficient. However, in real situations, the temperature of working fluids changes over the length of the surface (see Figure 5.10).

The LMTD denotes the average temperature difference between the hot and cold fluids acting over the entire length of the heat exchanger. It is used to calculate the heat transfer across the heat exchanger surface i.e.

$$Q = UA\Delta T_{lm} \tag{5.5}$$

Table 5.1

Fin Efficiency for Plate–Fin and Tube–Fin Geometries of Uniform Fin Thickness

Geometry	Fin efficiency formula $m_i = \left[\dfrac{2h}{k_f\delta_i}\left(1+\dfrac{\delta_i}{\xi}\right)\right]^{1/2}$ $\quad E_i = \dfrac{\tanh{(m_i\ell_i)}}{m_i\ell_i}$ $\quad i=1,2,3,4$
Plain, wavy, or offset strip fin of rectangular cross section	$\eta_f = E_1$ $\ell_1 = \dfrac{b}{2} - \delta_1 \qquad \delta_1 = \delta$
Triangular fin heated from one side	$\eta_f = \dfrac{hA_1(T_0 - T_\infty)\dfrac{\sinh{(m_1\ell_1)}}{m_1\ell_1} + q_e}{\cosh{(m_1\ell_1)}\left[hA_1(T_0 - T_\infty) + q_e\dfrac{T_0 - T_\infty}{T_1 - T_\infty}\right]}$ $\qquad \delta_1 = \delta$
Plain, wavy, or louver fin of triangular cross section	$\eta_f = E_1$ $\ell_1 = \dfrac{\ell}{2} \qquad \delta_1 = \delta$
Double sandwich fin	$\eta_f = \dfrac{E_1\ell_1 + E_2\ell_2}{\ell_1 + \ell_2}\dfrac{1}{1 + m_1^2 E_1 E_2 \ell_1 \ell_2}$ $\delta_1 = \delta \qquad \delta_2 = \delta_3 = \delta + \delta_s$ $\ell_1 = b - \delta + \dfrac{\delta_s}{2} \qquad \ell_2 = \ell_3 = \dfrac{p_f}{2}$
Triple sandwich fin	$\eta_f = \dfrac{(E_1\ell_1 + 2\eta_{f24}\ell_{24})/(\ell_1 + 2\ell_2 + \ell_4)}{1 + 2m_1^2 E_1\ell_1\eta_{f24}\ell_{24}}$ $\eta_{f24} = \dfrac{(2E_2\ell_2 + E_4\ell_4)/(2\ell_2 + \ell_4)}{1 + m_2^2 E_2\ell_2\ell_4/2} \qquad \ell_{24} = 2\ell_2 + \ell_4$ $\delta_1 = \delta_4 = \delta \qquad \delta_2 = \delta_3 = \delta + \delta_s$ $\ell_1 = b - \delta + \dfrac{\delta_s}{2} \qquad \ell_2 = \ell_3 = \dfrac{p_f}{2} \qquad \ell_4 = \dfrac{b}{2} - \delta + \dfrac{\delta_s}{2}$
Pin fin	$\eta_f = \dfrac{\tanh{(m\ell)}}{m\ell}$ $\ell = \dfrac{b}{2} - d_o \qquad m = \left(\dfrac{4h}{k_f d_o}\right)^{1/2} \qquad \delta = \dfrac{d_o}{2}$
Circular fin	$\eta_f = \begin{cases} a(m\ell_e)^{-b} & \text{for } \Phi > 0.6 + 2.257(r^*)^{-0.445} \\ \dfrac{\tanh{\Phi}}{\Phi} & \text{for } \Phi \leq 0.6 + 2.257(r^*)^{-0.445} \end{cases}$ $a = (r^*)^{-0.246} \qquad \Phi = m\ell_e(r^*)^n \qquad n = \exp(0.13m\ell_e - 1.3863)$ $b = \begin{cases} 0.9107 + 0.0893r^* & \text{for } r^* \leq 2 \\ 0.9706 + 0.17125\ln r^* & \text{for } r^* > 2 \end{cases}$ $m = \left(\dfrac{2h}{k_f\delta}\right)^{1/2} \qquad \ell_e = \ell_f + \dfrac{\delta}{2} \qquad r^* = \dfrac{d_e}{d_o}$
Studded fin	$\eta_f = \dfrac{\tanh{(m\ell_e)}}{m\ell_e}$ $m = \left[\dfrac{2h}{k_f\delta}\left(1+\dfrac{\delta}{w}\right)\right]^{1/2} \qquad \ell_e = \ell_f + \dfrac{\delta}{2} \qquad \ell_f = \dfrac{(d_e - d_o)}{2}$

Source: Shah and Sekulic [1985].

Table 5.2

TEMA Design Fouling Resistance for Industrial Fluids

Industrial Fluids	R_f (m²·K/W)
Oils	
Fuel oil #2	0.000352
Fuel oil #6	0.000881
Transformer oil	70.000176
Engine lube oil	0.000176
Quench oil	0.000705
Gases and Vapors	
Manufactured gas	0.001761
Engine exhaust gas	0.001761
Steam (nonoil bearing)	0.000088
Exhaust steam (oil bearing)	0.000264 − 0.000352
Refrigerant vapors (oil bearing)	0.000352
Compressed air	0.000176
Ammonia vapor	0.000176
CO2 vapor	0.000176
Chlorine vapor	0.000352
Coal flue gas	0.001761
Natural gas flue gas	0.000881
Liquids	
Molten heat transfer salts	0.000088
Refrigerant liquids	0.000176
Hydraulic fluid	0.000176
Industrial organic heat transfer media	0.000352
Ammonia liquid	0.000176
Ammonia liquid (oil bearing)	0.000528
Calcium chloride solutions	0.000528
Sodium chloride solutions	0.000528
CO2 liquid	0.000176
Chlorine liquid	0.000352
Methanol solutions	0.000352
Ethanol solutions	0.000352
Ethylene glycol solutions	0.000352

Source: TEMA [2007].

Table 5.3

Fouling Resistance for Water, R_f (m²-K/W) at Different Temperatures and Velocity Condition

Temperature of Heating Medium Temperature of Water	Up to 115 °C 51.7 °C		115 to 204 °C Over 51.7 °C	
	Water Velocity (m/s)			
	0.91 and Less	Over 0.91	0.91 and Less	Over 0.91
Seawater	0.000088	0.000088	0.000176	0.000176
Brackish water	0.000352	0.000176	0.000528	0.000352
Cooling Tower and Artificial Spray Pond:				
Treated make up	0.000176	0.000176	0.000352	0.000352
Untreated	0.000528	0.000528	0.00088	0.000704
City or well water	0.000176	0.000176	0.000352	0.000352
River Water:				
Minimum	0.000352	0.000176	0.000528	0.000352
Average	0.000528	0.000352	0.000704	0.000528
Muddy or silty	0.000528	0.000352	0.000704	0.000528
Hard (over 15 Grains/Gal.)	0.000528	0.000528	0.00088	0.00088
Engine jacket	0.000176	0.000176	0.000176	0.000176
Distilled or Closed Cycle:				
Condensate	0.000088	0.000088	0.000088	0.000088
Treated boiler feedwater	0.000176	0.0005	0.000176	0.000176
Boiler slowdown	0.000352	0.000352	0.000352	0.000352

Source: TEMA [2007].

The LMTD can be derived for different flow arrangements of a heat exchanger surface as

$$\Delta T_{lm} = \frac{\Delta T_1 - \Delta T_2}{ln\frac{\Delta T_1}{\Delta T_2}} \qquad (5.6)$$

where, ΔT_1 is the temperature difference between hot stream and cold stream at $x = 0$ and ΔT_2 is the temperature difference between hot stream and cold stream at $x = L$.

This expression is not applicable for heat transfer of cross-flow and multi-pass heat exchanger. In this case, the heat transfer is obtained using integration of the energy equation leading to an integrated mean temperature difference given as

$$Q = UA\Delta T_{lm}$$

where, the mean temperature difference can be expressed in terms of P, R and $\Delta T_{lm,cf}$ defined as

$$P = \frac{T_{c2} - T_{c1}}{T_{h2} - T_{c1}} = \frac{\Delta T_c}{\Delta T_{max}} \tag{5.7}$$

$$R = \frac{C_c}{C_h} = \frac{T_{h1} - T_{h2}}{T_{c2} - T_{c1}} \tag{5.8}$$

$$\Delta T_{lm,cf} = \frac{(T_{h2} - T_{c1}) - (T_{h1} - T_{c2})}{\ln\left[(T_{h2} - T_{c1})/(T_{h1} - T_{c2})\right]} \tag{5.9}$$

where, $\Delta T_{lm,cf}$ is the LMTD for a counter-flow arrangement with the same fluid inlet and outlet temperatures, P is the temperature effectiveness of the heat exchanger on the cold–fluid side and R is the ratio of the $\dot{m}c_p$ value of the cold fluid to the hot fluid.

A correction factor can be used for calculating the total heat transfer i.e.

$$Q = UA\Delta T_{lm} = UAF\Delta T_{lm,cf} \tag{5.10}$$

The correction factors F for different shell and tube and cross-flow heat exchangers are presented in Figures 5.11 to 5.15.

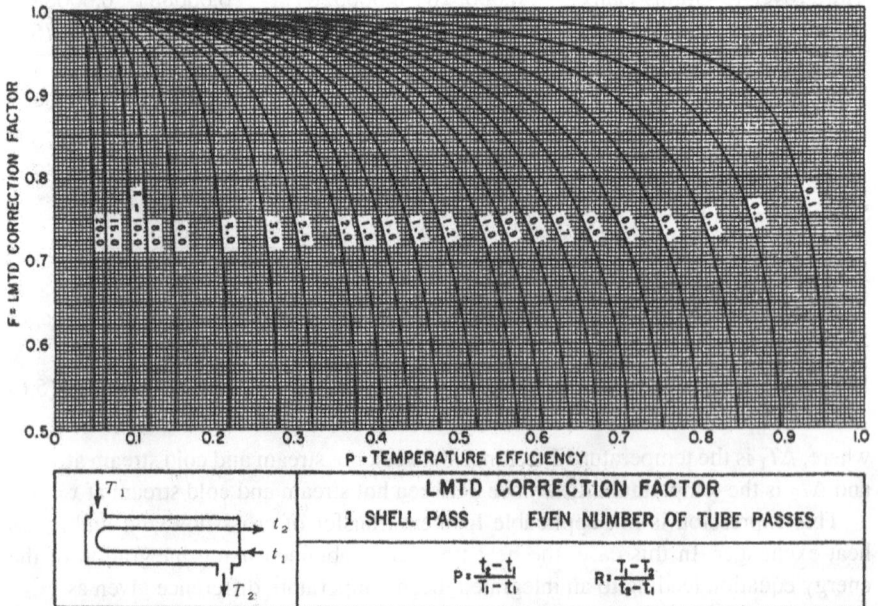

Figure 5.11   LMTD correction factor (F) for a shell and tube heat exchanger with one shell pass and two or multiple two tube passes. (TEMA [2007].)

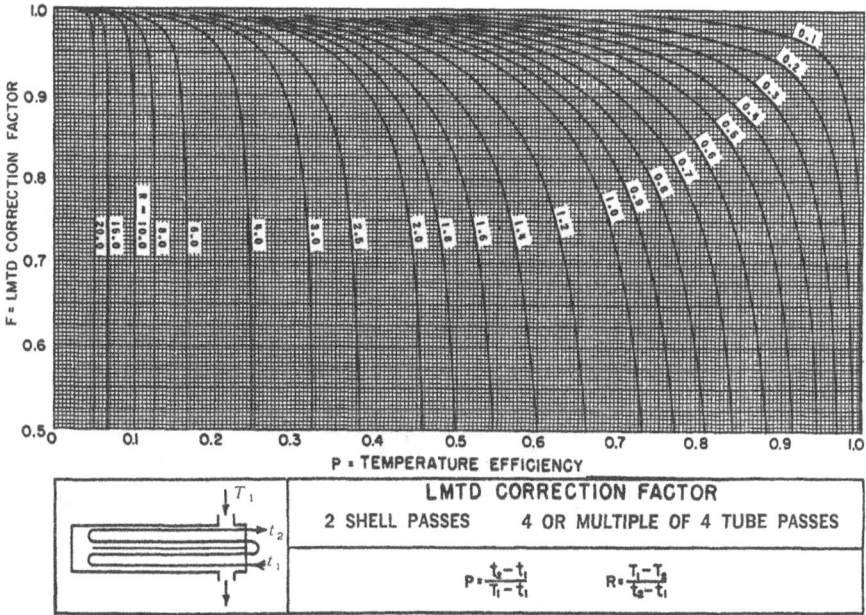

Figure 5.12 LMTD correction factor (F) for a shell and tube heat exchanger with two shell passes and four or multiple of four tube passes. (TEMA [2007].)

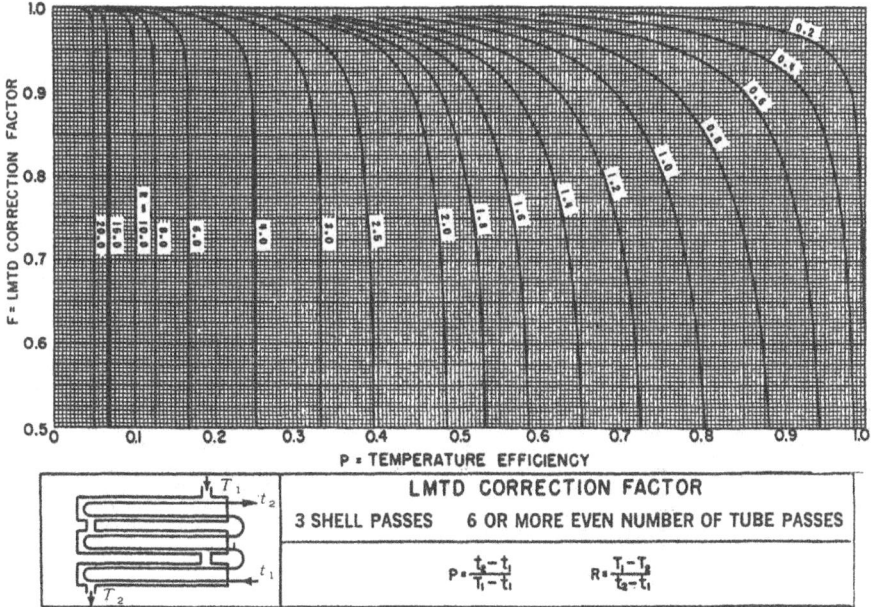

Figure 5.13 LMTD correction factor (F) for a shell and tube heat exchanger with three shell passes and six or more even-number tube passes. (TEMA [2007].)

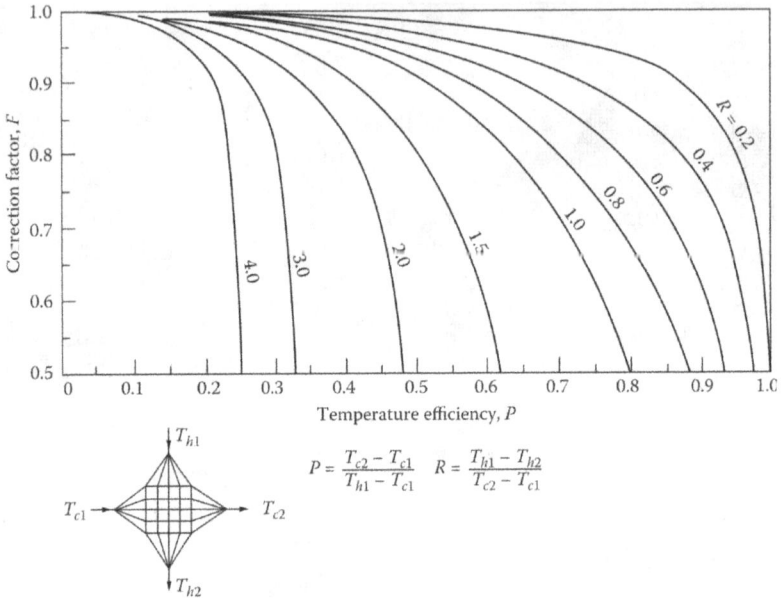

Figure 5.14 LMTD correction factor (F) for a cross-flow heat exchanger with both fluids unmixed. (Kakac et al. [2012].)

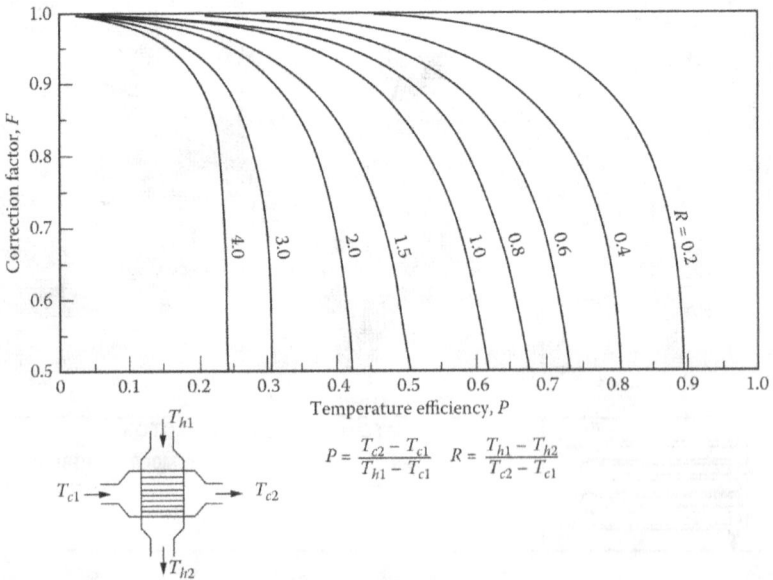

Figure 5.15 LMTD correction factor (F) for a single-pass cross-flow heat exchanger with one fluid mixed and another unmixed. (Kakac et al. [2012].)

5.5 THE ε–NTU METHOD

When the inlet or outlet temperatures of the fluid streams are unknown, the number of transfer units (NTU) based on the concept of heat exchanger effectiveness (ε) can be used. The heat transfer effectiveness of the heat exchanger is defined as the ratio of the actual heat transfer rate in a heat exchanger to the thermodynamically limited maximum possible heat transfer rate.

$$Q = \frac{Q}{Q_{max}} \tag{5.11}$$

$$= \frac{C_h(T_{h1} - T_{h2})}{C_{min}(T_{h1} - T_{c1})} \quad (\text{when } C_h = C_{min}) \tag{5.12}$$

$$= \frac{C_c(T_{c2} - T_{c1})}{C_{min}(T_{h1} - T_{c1})} \quad (\text{when } C_c = C_{min}) \tag{5.13}$$

NTU designates the non-dimensional heat transfer of a heat exchanger defined as

$$\text{NTU} = \frac{UA}{C_{min}} = \frac{1}{C_{min}} \int_A U \, dA \tag{5.14}$$

The ε–NTU relationship has been derived for different types of heat exchangers in heat transfer books, which can be broadly written as

$$\varepsilon = f(\text{NTU}, C^*, \text{flow arrangement}) \tag{5.15}$$

where, C^* is the capacity rate ratio defined as

$$C^* = \frac{C_{max}}{C_{min}} \tag{5.16}$$

The ε–NTU relationships for various types of heat exchangers are summarized in Table 5.4.

5.6 VARIABLE OVERALL HEAT TRANSFER COEFFICIENT

In the previous section, we have assumed a constant overall heat transfer coefficient along the length of a heat exchanger. However, the overall heat transfer coefficient strongly depends on Reynolds number, heat transfer surface geometry and physical properties of a fluid. Figure 5.16 shows the typical cases of a condenser and an evaporator with a variable overall heat transfer coefficient.

In the condenser, the fluid vapor enters at a temperature greater than the saturation temperature and the condensed liquid becomes subcooled before leaving the condenser. In the evaporator, the cold fluid enters as a subcooled liquid, is heated,

Table 5.4
The ε–NTU Expression

Heat Exchanger	$\varepsilon(\text{NTU}, C^*)$	$\varepsilon(\text{NTU}, C^* = 0)$	$\varepsilon(\text{NTU}, C^* \to \infty)$	NTU (ε, C^*)
Counter flow	$\varepsilon = \dfrac{1 - e^{-\text{NTU}(1-C^*)}}{1 - C^* e^{-\text{NTU}(1-C^*)}}$	$\varepsilon = 1 - e^{-\text{NTU}}$	$\varepsilon = 1$ for all C^*	$\text{NTU} = \dfrac{1}{1-C^*} \ln\left(\dfrac{1-\varepsilon C^*}{1-\varepsilon}\right)$
Parallel flow	$\varepsilon = \dfrac{1}{1+C^*}\left[1 - e^{-\text{NTU}(1+C^*)}\right]$	$\varepsilon = 1 - e^{-\text{NTU}}$	$\varepsilon = \dfrac{1}{1+C^*}$	$\text{NTU} = -\dfrac{1}{1+C^*}\ln(1 + \varepsilon(1+C^*))$
Cross-flow C_{min} mixed & C_{max} unmixed	$\varepsilon = 1 - e^{-\frac{1-e^{-C^*\text{NTU}}}{C^*}}$	$\varepsilon = 1 - e^{-\text{NTU}}$	$\varepsilon = 1 - e^{-1/C^*}$	$\text{NTU} = -\dfrac{1}{C^*}\ln(1 + C^*\ln(1-\varepsilon))$
Cross-flow C_{max} mixed & C_{min} unmixed	$\varepsilon = \dfrac{1}{C^*}\left[1 - e^{-C^*\left(1-e^{-\text{NTU}}\right)}\right]$	$\varepsilon = 1 - e^{-\text{NTU}}$	$\varepsilon = \dfrac{1}{C^*}\left(1 - e^{-C^*}\right)$	$\text{NTU} = -\ln\left(1 + \dfrac{1}{C^*}\text{lr}(1 - \varepsilon C^*)\right)$
1–2 shell & tube heat exchanger TEMA-E	$\varepsilon = \dfrac{2}{1+C^*+\sqrt{1+C^{*2}}\dfrac{1+e^{-\text{NTU}\sqrt{1+C^{*2}}}}{1-e^{-\text{NTU}\sqrt{1+C^{*2}}}}}$	$\varepsilon = 1 - e^{-\text{NTU}}$	$\varepsilon = \dfrac{2}{1+C^*+\sqrt{1+C^{*2}}}$	$\text{NTU} = \dfrac{1}{\sqrt{1+C^{*2}}}\ln\left[\dfrac{2-\varepsilon\left(1+C^*-\sqrt{1+C^{*2}}\right)}{2-\varepsilon\left(1+C^*+\sqrt{1+C^{*2}}\right)}\right]$

Source: Kakac et al. [2012].

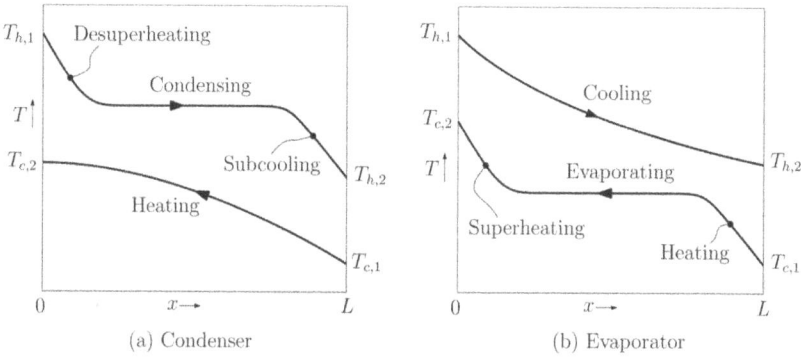

(a) Condenser   (b) Evaporator

Figure 5.16 Typical case of a heat exchanger with variable U: (a) condenser and (b) evaporator.

evaporated and then superheated. In the above two cases, the heat exchnager can be treated as three different heat exchangers in series. The heat exchanger is divided into many segments, and different values of U are assigned to each segment for arbitrary variation in U through the heat exchanger. This analysis can be best performed by numerical or computational fluid dynamics (CFD) technique. The numerical analysis can be carried out using the following steps:

(a) Divide the heat exchanger domain to a small value i.e. ΔA_i.
(b) Calculate the inner and outer heat transfer coefficients and U for inlet conditions.
(c) Calculate ΔQ_i for the increment region.
(d) Calculate the value of T_h, T_c for the next increment using

$$\Delta Q_i = (\dot{m} c_p)_{hi}(T_{h,i+1} - T_{h,i}) = (\dot{m} c_p)_{ci}(T_{c,i+1} - T_{c,i}) \qquad (5.17)$$

(e) The total heat transfer rate is then calculated using

$$Q_i = \sum_{i=1}^{n} \Delta Q_i \qquad (5.18)$$

5.7   HEAT EXCHANGER THERMAL DESIGN

The most important heat exchanger thermal design problems can be classified as (1) rating problem and (2) sizing problem.

5.7.1   RATING PROBLEM

Rating problem determines the heat exchanger and pressure drop performance of an existing heat exchanger (see Figure 5.17). The heat exchanger construction, flow arrangement, overall dimensions, material details, surface geometries on both sides, fluid flow rates, inlet temperature and fouling factors are provided as input. The outlet temperature of fluid, total heat transfer rate and pressure drop on each side of the heat exchanger are determined in the rating problem.

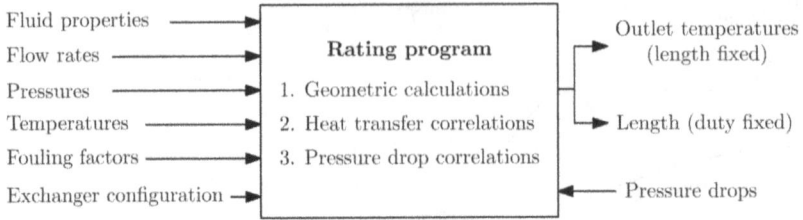

Figure 5.17   Flow chart explaining a typical rating program. (Kakac et al. [2012].)

5.7.2 SIZING PROBLEM

Inputs to the sizing problem are surface geometries, fluid flow rates, inlet and outlet fluid temperatures, fouling factor and pressure drop in each fluid side. It determines the heat exchanger construction type, flow arrangement, tube/plate and fin material and physical size of the heat exchanger to meet the specified heat transfer and pressure drop. The sizing problem for a shell and tube heat exchanger refers to the determination of shell type, diameter and length of the shell, tube diameter and number of tubes, tube layout, pass arrangement and so on. The sizing problem for a plate-type heat exchanger refers to the selection of plate type and size, number of plates, pass arrangements, gasket type and so on.

5.8 FORCED CONVECTION CORRELATION FOR SINGLE–PHASE SIDE OF A HEAT EXCHANGER

The single-phase fluid flow inside a heat exchanger can be classified as either laminar or turbulent flow depending on the values of the Reynolds number. The heat transfer problem can have four different regimes depending on the hydrodynamic and thermal boundary layer thickness with respect to the size of the duct i.e. (1) hydro-dynamically and thermally developed, (2) hydrodynamically fully developed and thermally developing, (3) thermally developed and hydrodynamically developing and (4) simultaneously developing. Table 5.5 presents the laminar forced convection correlations for four different flow regimes with isothermal wall and constant heat flux boundary conditions. Here, the mean value of the Nusselt number is defined as

$$Nu = \frac{hL}{k} = \frac{1}{L}\int_0^L Nu_x dx \qquad (5.19)$$

where, h is the mean value of the heat transfer coefficient, L is the length of duct and k is the thermal conductivity of the fluid.

Re is the Reynolds number defined as $\frac{u_m d}{\nu}$, Pr is the Prandtl number $(\mu c_p / k)$ and $Pe = Re \times Pr$. All physical properties are evaluated at bulk mean fluid temperature T_b given as

$$T_b = \frac{T_i + T_o}{2} \qquad (5.20)$$

Table 5.5

Laminar Forced Convection Correlations in Smooth Circular Ducts for Different Boundary and Flow Conditions

Sr. No.	Correlations	Limitations and Remarks
1	$Nu_T = 1.61(Pe_b d/L)^{1/3}$	$Pe_b d/L > 10^3$, constant wall temperature
	$Nu_T = 3.66$	$Pe_b d/L < 10^2$, fully developed flow in a circular duct, constant wall temperature
2	$Nu_T = \left[3.66^3 + 1.61^3 Pe_b(d/L)\right]^{1/3}$	Superposition of two asymptotics given in case 1 for the mean Nusselt number, $0.1 < Pe_b d/L < 10^4$
3	$Nu_T = 3.66 + \frac{0.19(Pe_b d/L)^{0.8}}{1+0.177(Pe_b d/L)^{0.467}}$	Thermal entrance region, constant wall temperature, $0.1 < Pe_b d/L < 10^4$
4	$Nu_H = 1.953(Pe_b d/L)^{1/3}$	$Pe_b d/L > 10^2$, constant heat flux
	$Nu_H = 4.36$	$Pe_b d/L > 10$, fully developed flow in a circular duct, constant heat flux
5	$Nu_T = 0.664 \frac{(Pe_b d/L)^{1/2}}{Pr^{1/6}}$	$Pe_b d/L > 10^4$, $0.5 < Pr < 500$, simultaneously developing flow
6	$Nu_T = Nu + \phi\left(\frac{d_o}{D_i}\right)\frac{0.19(PeD_h/L)^{0.8}}{1+0.117(PeD_h/L)^{0.467}}$	Circular annular duct, constant wall temperature, thermal entrance region
	$\phi(d_o/D_i) = 1 + 0.14(d_o/D_i)^{-1/2}$	Outer wall is insulated, heat transfer through inner wall
	$\phi(d_o/D_i) = 1 + 0.14(d_o/D_i)^{0.1}$	Heat transfer through outer and inner walls
7	$Nu_T = 1.86(Re_b Pr_b d/L)^{1/3}(\mu_b/\mu_w)^{0.14}$	Thermal entrance region, constant wall temperature, $0.48 < Pr_b < 16700$, $4.4 \times 10^{-3} < \mu_b/\mu_w < 9.75$, $(Re_b Pr_b d/L)^{1/3}(\mu_b/\mu_w)^{0.14} > 2$
8	$Nu_H = 1.86(Re_b Pr_b d/L)^{1/3}(\mu_b/\mu_w)^{0.152}$	Thermal entrance region, constant wall heat flux, for oils $0.8 \times 10^3 < Re_b < 1.8 \times 10^3$, $1 < (T_w/T_b) < 3$
9	$Nu_H = 1.23(Re_b Pr_b d/L)^{0.4}(\mu_b/\mu_w)^{1/6}$	Thermal entrance region, constant heat flux, $400 < Re_b < 1900$, $170 < Pr_b < 640$, for oils
10	$Nu_b = 1.4(Re_b Pr_b d/L)^{1/3}(\mu_b/\mu_w)^n$	Thermal entrance region, $n = 0.05$ for heating liquids, $n = 1/3$ for cooling liquids

Source: Kakac et al. [2012].

Note: Here, the fluid properties are evaluated at the bulk fluid temperature, $Tb = (Ti + To)/2$.

Table 5.6

Correlations for Fully Developed Turbulent Forced Convection through a Circular Duct with Constant Properties

Sr. No.	Correlations	Remarks and Limitations
1	$Nu_b = \dfrac{(f/2)Re_bPr_b}{1+8.7(f/2)^{1/2}(Pr_b-1)}$	Based on a three-layer turbulent boundary layer model, $Pr > 0.5$
2	$Nu_b = 0.021Re_b^{0.8}Pr_b^{0.4}$	Based on data for common gases; recommended for $Pr \approx 0.7$
3	$Nu_b = \dfrac{(f/2)Re_bPr_b}{1.07+12.7(f/2)^{1/2}(Pr_b^{2/3}-1)}$	Based on a three-layer model with constants adjusted to match experimental data $0.5 < Pr_b < 2000$, $10^4 < Re_b < 5 \times 10^6$
4	$Nu_b = \dfrac{(f/2)Re_bPr_b}{1.07+9(f/2)^{1/2}(Pr_b-1)Pr_b^{2/3}Pr_b^{-1/4}}$	Theoretically based; Webb found case 3 better at high Pr and this one the same at other Pr
5	$Nu_b = 5+0.015Re_b^m Pr_b^n$ $m = 0.88-0.24/(4+Pr_b)$ $n = 1/3+0.5exp(-0.6Pr_b)$ $Nu_b = 5+0.012Re_b0.87(Pr_b+0.29)$	Based on numerical results obtained for $0.1 < Pr_b < 10^4$, $10^4 < Re_b < 10^6$ Within 10% of case 6 for $Re_b > 10^4$ Simplified correlation for gases, $0.6 < Pr_b < 0.9$
6	$Nu_b = \dfrac{(f/2)(Re_b-1000)Pr_b}{1+12.7(f/2)^{1/2}(Pr_b^{2/3}-1)}$ $f = (1.58 \ln Re_b-3.28)^{-2}$ $Nu_b = 0.012(Re_b^{0.8}-100)Pr_b^{0.4}$ $Nu_b = 0.012(Re_b^{0.87}-280)Pr_b^{0.4}$	Modification of case 3 to fit experimental data at low Re ($2300 < Re_b < 10^4$) Valid for $2300 < Re_b < 5 \times 10^6$ and $0.5 < Pr_b < 2000$ Simplified correlation for $0.5 < Pr < 1.5$; agrees with case 4 within –6% and +4% Simplified correlation for $1.5 < Pr < 500$; agrees with case 4 within –10% and +0% for $3 \times 10^3 < Re_b < 10^6$
7	$Nu_b = 0.022Re_b^{0.8}Pr_b^{0.5}$	Modified Dittus–Boelter correlation for gases ($Pr \approx 0.5–1.0$); agrees with case 6 within 0 to 4% for $Re_b \geq 5000$

Source: Kakac et al. [2012].

where, T_i and T_o are bulk mean temperature at the inlet and outlet of duct, respectively. The bulk mean temperature for incompressible flow is defined as

$$T_b = \frac{1}{A_c u_m} \int_{A_c} uT \, dA_c \qquad (5.21)$$

where, u_m is the mean velocity and A_c is the flow cross section. The correlations for fully developed turbulent forced convection through a circular duct with constant properties are compiled in Tables 5.6 and 5.7. The effect of thermal boundary

Table 5.7

Turbulent Flow Isothermal Fanning Friction Factor Correlations for Smooth Circular Ducts

Sr. No.	Correlations	Remarks and Limitations
1	$f = \frac{\tau_w}{\frac{1}{2}\rho u_m^2} = 0.0791 Re^{-1/4}$	This approximate explicit equation agrees with case 3 within $\pm2.5\%$, $4 \times 10^3 < Re < 10^5$
2	$f = 0.00140 + 0.125 Re^{-0.32}$	This correlation agrees with case 3 within -0.5% and $+3\%$, $4 \times 10^3 < Re < 5 \times 10^6$
3	$1/\sqrt{f} = 1.737 \ln\left(Re\sqrt{f}\right) - 0.4$	Von Karman's theoretical equation with the constants adjusted to best fit Nikuradse's experimental data, also referred to as the Prandtl correlation, should be valid for very high values of Re. $4 \times 10^3 < Re < 3 \times 10^6$
	or $1/\sqrt{f} = 4\log\left(Re\sqrt{f}\right) - 0.4$ approximated as $f = \left(3.64\log Re^{3.28}\right)^{-2}$ $f = 0.046 Re^{-0.25}$	This approximate explicit equation agrees with the preceding within -0.4 and $+2.2\%$ for $3 \times 10^4 < Re < 10^6$
4	$f = 1/(1.58 \ln Re - 3.28)^2$	Agrees with case 3 within $\pm0.5\%$ for $3 \times 10^4 < Re < 10^7$ and within $\pm1.8\%$ at $Re = 10^4$. $10^4 < Re < 5 \times 10^5$
5	$1/f = \left(1.7372 \ln \frac{Re}{1.964 \ln Re - 3.8215}\right)$	An explicit form of case 3; agrees with it within $\pm0.1\%$, $10^4 < Re < 2.5 \times 10^8$

Source: Kakac et al. [2012].

conditions is negligible for turbulent flow. Therefore, these conditions can be used for both isothermal and constant wall heat flux boundary conditions.

5.9 EFFECT OF VARIABLE PROPERTIES

The transport properties of most fluids vary with temperature, which influences the velocity and temperature distributions inside the boundary layer. Therefore, the correlations based on bulk fluid mean temperature can cause significant errors when there is a large temperature difference between the wall and the bulk mean fluid temperature. Corrections based on the property ratio can be used with correlations based on the constant property for variable physical property situation using the following relations:

Table 5.8

Exponents *n* and *m* Associated with Equations 5.22 and 5.23 for Laminar Forced Convection through Circular Ducts for *Pr* > 0.5

Fluid	Condition (Laminar)	n	m[a]	Limitations
Liquid	Heating	0.14	−0.58	Full developed flow
	Cooling	0.14	−0.50	$q_w'' = $ constant, $Pr > 0.6, \mu/\mu_w = (T/T_w)^{-1.6}$
	Heating	0.11	—	Developing and fully developed regions of a circular duct, $T_w = $ constant $q_w'' = $ constant
Gas	Heating	0	1.0	Developing and fully developed regions, $q_w'' = $ constant, $T_w = $ constant, $1 < (T_w/T_b) < 3$
	Cooling	0	0.81	$T_w = $ constant, $0.5 < (T_w/T_b) < 1$

Source: Kakac et al. [2012].
[a] Fanning friction factor f is defined as $f = 2\tau_w/(\rho u_m^2)$ and for hydrodynamically developed isothermal laminar flow as $f = 16/Re$.

5.9.1   FOR LIQUIDS

$$\frac{Nu}{Nu_{cp}} = \left(\frac{\mu_b}{\mu_w}\right)^n \tag{5.22}$$

$$\frac{f}{f_{cp}} = \left(\frac{\mu_b}{\mu_w}\right)^m \tag{5.23}$$

5.9.2   FOR GASES

$$\frac{Nu}{Nu_{cp}} = \left(\frac{T_w}{T_b}\right)^n \tag{5.24}$$

$$\frac{f}{f_{cp}} = \left(\frac{T_w}{T_b}\right)^m \tag{5.25}$$

where, subscript *cp* corresponds to constant property assumption, *b* corresponds to bulk mean temperature and *w* corresponds to wall temperature. Here, *n* and *m* are exponents.

It may be noted that variation of viscosity with temperature is responsible for most of the property effects in liquids and the correction uses the temperature effects on viscosity. However, for gases, the viscosity, thermal conductivity and density vary with absolute temperature, and therefore, corrections are specified as a function of

Table 5.9

Exponents *n* and *m* Associated with Equations 5.24 and 5.25 for Turbulent Forced Convection through Circular Ducts

Fluid	Condition (Turbulent)	*n*	*m*	Limitations
Liquid	Heating	0.11	—	$10^4 < Re_b < 1.25 \times 10^5$, $2 < Pr_b < 140$, $0.08 < \mu_w/\mu_b < 1$
	Cooling	0.25	—	$1 < \mu_w/\mu_b < 40$
	Heating	—	−0.25	$10^4 < Re_b < 23 \times 10^4$, $1.3 < Pr_b < 10^4$, $0.35 < \mu_w/\mu_b < 1$
	Cooling	—	−0.24	$1 < \mu_w/\mu_b < 2$
Gas	Heating	−0.47	—	$10^4 < Re_b < 4.3 \times 10^6$, $1 < T_w/T_b < 3.1$
	Cooling	−0.36	—	$0.37 < T_w/T_b < 1$
	Heating	—	−0.52	$14 \times 10^4 < Re_w \leq 10^6$, $1 < T_w/T_b < 3.7$
	Cooling	—	−0.38	$0.37 < T_w/T_b < 1$
	Heating	—	−0.264	$1 \leq T_w/T_b \leq 4$
	Heating	—	−0.1	$1 < T_w/T_b < 2.4$

Source: Kakac et al. [2012].

temperature. The exponents *n* and *m* for laminar forced convection through circular ducts are summarized in Table 5.8. The exponents for turbulent forced convection through circular ducts are summarized in Table 5.9.

5.10 FLOW IN SMOOTH STRAIGHT NON-CIRCULAR DUCTS

The common practice for estimating N_u and f in non-circular duct is to use hydraulic diameter, D_h in the circular duct correlations.

$$D_h = 4 \times \frac{\text{Free flow area}}{\text{Wetted (heat transfer) perimeter}} \qquad (5.26)$$

The wetted perimeter for the pressure drop calculation in the annulus (see Figure 5.18) is calculated as

$$P_w = \pi(D_i + d_o) \qquad (5.27)$$

Fluid friction acts both on the inner surface of the shell and the outer wall of the tube. The heat transfer perimeter for the annulus is calculated as

$$P_h = \pi d_o \qquad (5.28)$$

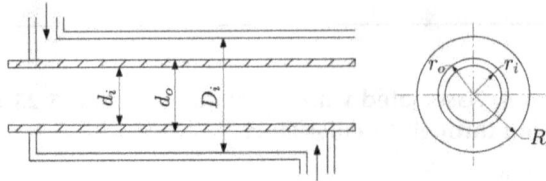

Figure 5.18   Schematic of an annular tube.

The net free flow area of the annulus is

$$A_c = \frac{\pi}{4}(D_i^2 - d_o^2) \tag{5.29}$$

The hydraulic diameter for pressure drop calculation is calculated as

$$D_w = \frac{4A_c}{p_w} \tag{5.30}$$

The hydraulic diameter for the heat transfer perimeter is calculated as

$$D_h = \frac{4A_c}{p_h} \tag{5.31}$$

It may be noted that D_h is used for calculation of heat transfer coefficient from the Nusselt number. The hydraulic diameter, D_w is used for the calculation of the Reynolds number.

5.11   HEAT TRANSFER FROM SMOOTH TUBE BUNDLES

Circular tubular arrays are commonly used as heat transfer surfaces of a heat exchanger. The most common tube arrays are inline and staggered as shown in Figure 5.19. The bundle is characterized by cylinder diameter (d_o), longitudinal spacing

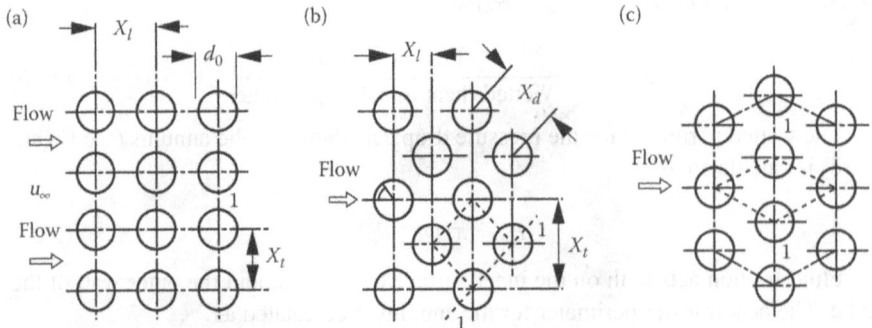

Figure 5.19   Tube bundle arrangement: (a) inline array, (b) and (c) staggered array. Minimum tube spacing at section I–I between two tubes. (Kakac et al. [2012].)

between consecutive rows (X_l) and transverse spacing between consecutive rows (X_t) of cylinders. The array average Nusselt number $\left(Nu = \frac{\bar{h}d_o}{k}\right)$ is a function of Reynolds number $\left(Re_b = \frac{\rho u_{max} d_o}{\mu}\right)$ and Prandtl number, Pr_w evaluated at the surface temperature of the tube and Prandtl number, Pr_b evaluated at bulk mean temperature of the fluid that flows around the cylinder in the bundle. The maximum velocity, u_{max} is calculated from the narrowed cross section of the bundle using mass conservation principle. The narrowed cross section for an inline array is $(x_l - d_o)$. The narrowest cross section for a staggered array is either $(x_l - d_o)$ or $2(x_d - d_o)$.

The correlation for the array averaged Nusselt number of inline tube bundles in cross-flow is given by the following relations:

$$Nu_b = 0.9 C_n Re_b^{0.4} Pr_b^{0.36} \left(\frac{Pr_b}{Pr_w}\right)^{0.25} \qquad \text{for } Re_b = 1 - 100 \qquad (5.32)$$

$$Nu_b = 0.52 C_n Re_b^{0.5} Pr_b^{0.36} \left(\frac{Pr_b}{Pr_w}\right)^{0.25} \qquad \text{for } Re_b = 100 - 1000 \qquad (5.33)$$

$$Nu_b = 0.27 C_n Re_b^{0.63} Pr_b^{0.36} \left(\frac{Pr_b}{Pr_w}\right)^{0.25} \qquad \text{for } Re_b = 1000 - 2 \times 10^5 \qquad (5.34)$$

$$Nu_b = 0.033 C_n Re_b^{0.8} Pr_b^{0.4} \left(\frac{Pr_b}{Pr_w}\right)^{0.25} \qquad \text{for } Re_b = 2 \times 10^5 - 2 \times 10^6 \qquad (5.35)$$

The correlation for staggered tube bundles in cross-flow is

$$Nu_b = 1.04 C_n Re_b^{0.4} Pr_b^{0.36} \left(\frac{Pr_b}{Pr_w}\right)^{0.25} \qquad \text{for } Re_b = 1 - 500 \qquad (5.36)$$

$$Nu_b = 0.71 C_n Re_b^{0.5} Pr_b^{0.36} \left(\frac{Pr_b}{Pr_w}\right)^{0.25} \qquad \text{for } Re_b = 500 - 10^3 \qquad (5.37)$$

$$Nu_b = 0.35 C_n Re_b^{0.6} Pr_b^{0.36} \left(\frac{Pr_b}{Pr_w}\right)^{0.25} \qquad \text{for } Re_b = 10^3 - 2 \times 10^5 \qquad (5.38)$$

$$Nu_b = 0.031 C_n Re_b^{0.8} Pr_b^{0.4} \left(\frac{Pr_b}{Pr_w}\right)^{0.25} \qquad \text{for } Re_b = 2 \times 10^5 - 2 \times 10^6 \qquad (5.39)$$

All physical properties except Pr_w are evaluated at the bulk mean temperature of the fluid that flows around the cylinders in the bundle; Pr_w is the Prandtl number evaluated at the tube surface temperature. Here, C_n is the correction factor for the number of tube rows because of shorter bundles. Figure 5.20 shows the correlation factor for shorter bundles. The correction factor becomes negligible when the number of bundles $n > 16$.

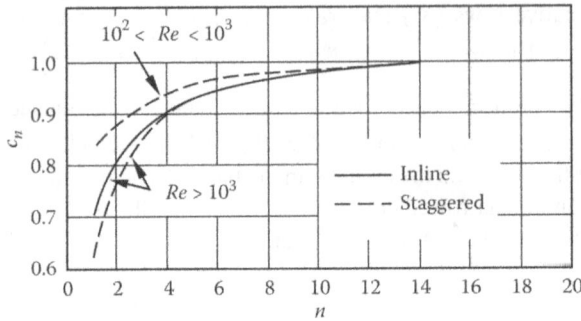

Figure 5.20 Correlation factor for the number of rows for the average heat transfer from tube bundles. (Kakac et al. [2012].)

5.12 PRESSURE DROP IN TUBE BUNDLES IN CROSS-FLOW

Similar to the average heat transfer for tube bundles discussed in the previous section, the pressure drop of a multirow bundle can be calculated by using the following expression:

$$\Delta p = \left(\frac{Eu}{x}\right)\frac{\chi}{2}u_{max}^2 n \qquad (5.40)$$

where, the Euler number, $Eu = 2\Delta p/\left(\rho u_{max}^2 n\right)$ is defined per tube row and n is number of tube rows counted in the flow direction. Figures 5.21 and 5.22 show the Euler number ($\frac{Eu}{x}$) curve as a function of Reynolds number for inline and staggered bundles, respectively. The correction factor x is provided in the upper right inset of the figure. The Reynolds number is based on the maximum average velocity.

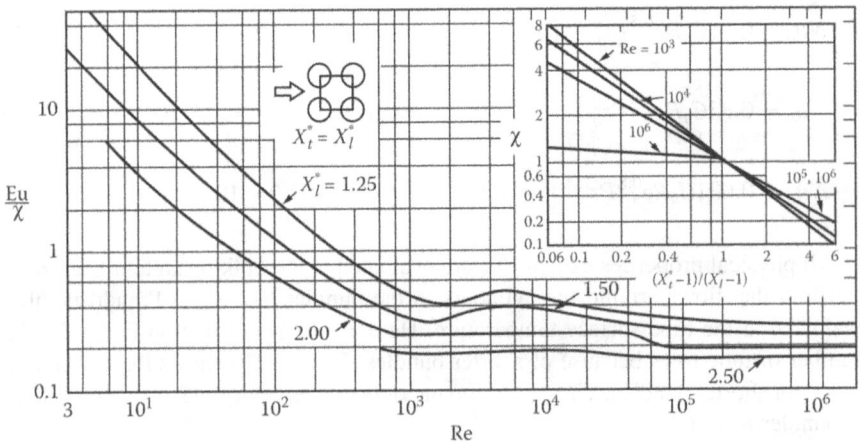

Figure 5.21 The hydraulic drag coefficient for inline bundles for the number of tube rows, $n > 9$. (Kakac et al. [2012].)

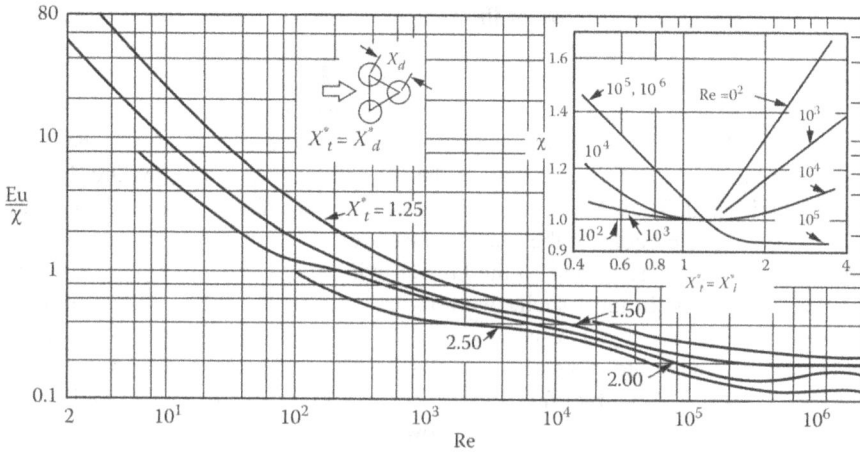

Figure 5.22   The hydraulic drag coefficient for staggered bundles for $n > 9$. (Kakac et al. [2012].)

5.13   SHELL AND TUBE HEAT EXCHANGERS

Shell and tube heat exchanger offers great flexibility to meet different service requirement. It has a relatively large heat transfer area-to-volume ratio. It can be designed for high-pressure situation relative to the environment. Shell and tube heat exchangers are used in conventional and nuclear power stations as condensers, steam generators in pressurized water reactor power plant, feed water heaters and in some air conditioning and refrigeration systems. The major components of this heat exchanger are shell, tube bundle baffles, front-end stationary head and rear-end head types. The different front and rear head types and shell types have been standardized by the Tubular Exchanger Manufacturers Association (TEMA) and identified by an alphabetic character. Figure 5.23 shows the most common TEMA shell types. The

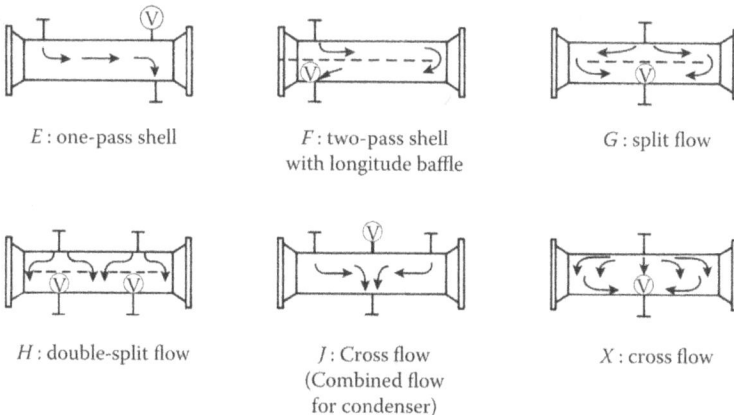

E : one-pass shell F : two-pass shell with longitude baffle G : split flow

H : double-split flow J : Cross flow (Combined flow for condenser) X : cross flow

Figure 5.23   Schematics of most common TEMA shell types. (Kakac et al. [2012].)

V-symbol indicates the location of the vent. In the E-shell, fluid enters at one end of the shell and leaves at the other end i.e. there is one pass. The tubes may have single or multi passes and are supported by transverse baffles. The F-shell provides a counter-flow arrangement with superior heat exchanger effectiveness using a longitudinal baffle and resulting in two shell passes. The pressure drop is higher than a comparable E-shell. The split flow in the G-shell has the same pressure drop as that of the E-shell. The LMTD factor, F and exchanger effectiveness of the G-shell is higher than the E-shell for the same surface area and flow rates due to increased mixing. The H-shell is similar to the G-shell with two horizontal baffles and two inlet/outlet nozzles. In the J-shell, two nozzles are near the end and one nozzle is centrally located. When it is used as a condenser, vapor phase enters at two inlets and leaves as condensate at the central outlet. The X-shell has centrally located fluid entry, and outlet and the two fluids are in cross-flow arrangements. No baffles are used, and the pressure drop is extremely low. It is used for vacuum condensers and low-pressure gases.

5.14   TUBE PASSES

More number of tube passes are generally used to increase tube-side fluid velocity and heat transfer coefficient and minimize fouling. Small tube diameters are preferred due to higher area-to-volume density. However, it limits the cleaning process. The tube may be finned or unfinned depending on the heat transfer coefficient requirement.

5.15   TUBE LAYOUT

Figure 5.24 shows different tube layouts used in a heat exchanger. The highest tube density is obtained in a 30° layout. Cleaning of the surface is easier in 45° and 90° layout. The number of tubes i.e. tube counts that can be placed within a shell depends on shell layout, tube outside diameter, pitch, number of passes and the shell diameter. Table 5.10 gives the maximum number of tubes that can be accommodated under specified conditions. For example, 97 tube counts can be accommodated inside a 13.25″ shell ID with one pass (1–P) having a tube outer diameter (OD) of 3/4″ in

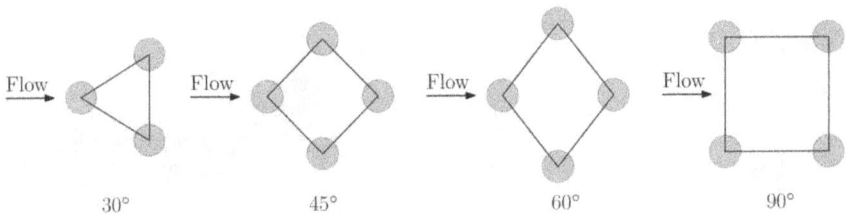

Figure 5.24   Schematic of different tube layouts used in a heat exchanger.

$1''$ square pitch. Similarly, $3/4''$ OD tube with $1''$ square pitch in a shell diameter of $13\frac{1}{4}''$ with two pass (2P) gives a maximum tube count of 90.

5.16 BAFFLE TYPE

Baffle serves two important functions i.e. (1) to support the tubes for structural rigidity for preventing tube vibration and sagging and (2) to divert the flow across the bundle for obtaining a higher heat transfer coefficient. Baffles can be either transverse or longitudinal types (F-shell). Figure 5.25 shows the schematic of the most commonly used baffle type. The optimum baffle spacing is between 0.4 and 0.6 of the shell diameter. A baffle cut of 25% to 35% is usually recommended.

5.17 TUBE-SIDE PRESSURE DROP

The pressure drop of a tube side can be calculated using the following expression:

$$\Delta p_t = \frac{4fLN_p}{d_i}\frac{\rho u_m^2}{2} \tag{5.41}$$

where, N_p is the number of tube passes and L is the length of the heat exchange. The change in direction in the passes introduces additional pressure drop of the tube flow due to secondary motion, which can be calculated using

$$\Delta p_r = 4N_p\frac{\rho u_m^2}{2} \tag{5.42}$$

Hence, the total pressure drop of the tube side is

$$\Delta p_{total} = \Delta p_t + \Delta p_r = \left(\frac{4fLN_p}{d_i} + 4N_p\right)\frac{\rho u_m^2}{2} \tag{5.43}$$

5.18 BELL–DELAWARE METHOD

The Bell–Delaware method can be used for shell-side heat transfer and pressure drop analysis. Shell-side flow can be assumed to be a combination of five different streams. Figure 5.26 indicates different flow paths of streams in the shell region. A-stream is the leaking flow through the clearance between the tubes and baffle. B-stream is the main stream across the bundle. The C-stream is the stream flow between the outermost tubes in the bundle and inside the shell. The E-stream is the baffle-to-shell leakage stream flowing through the clearance between the baffles and the shell's inside diameter. The F-stream is the flow through the channel within the tube bundle caused by a pass divider for multiple tube passes. Among all those streams, B-stream is the primary stream. The other streams reduce the B-stream and alter the temperature profile of the shell side resulting in decrease in heat transfer coefficient.

Table 5.10
The Number of Tubes (Tube Counts) for Different Layout

Shell ID (in.)	1-P	2-P	4-P	6-P	8-P
1 1/2-in. OD tubes on 1 7/8-in. square pitch					
12	16	16	12	12	
13 ¼	22	22	16	16	
15 ¼	29	29	24	24	22
17 ¼	29	39	34	32	29
19 ¼	50	48	45	43	39
21 ¼	62	60	57	54	50
23 ¼	78	74	70	66	62
25	94	90	86	84	78
27	112	108	102	98	94
29	131	127	120	116	112
31	151	146	141	138	131
33	176	170	164	160	151
35	202	196	188	182	176
37	224	220	217	210	202
39	252	246	237	230	224
1 1/2-in. OD tubes on 1 7/8-in. triangular pitch					
12	18	14	14	12	12
13 ¼	27	22	18	16	14
15 ¼	26	34	32	30	27
17 ¼	48	44	42	38	36
19 ¼	61	58	55	51	48
21 ¼	76	78	70	66	61
23 ¼	95	91	86	80	76
25	115	110	105	98	95
27	136	131	125	118	115
29	160	154	147	141	136
31	184	177	172	165	160
33	215	206	200	190	184
35	246	238	230	220	215
37	275	268	260	252	246
39	307	299	290	284	275
1 1/4-in. OD tubes on 9/16-in. triangular pitch					
10					
10	20	18	14		
12 ¼	32	30	26	22	20
13 ¼	38	36	32	28	26

Source: Kakac et al. [2012].

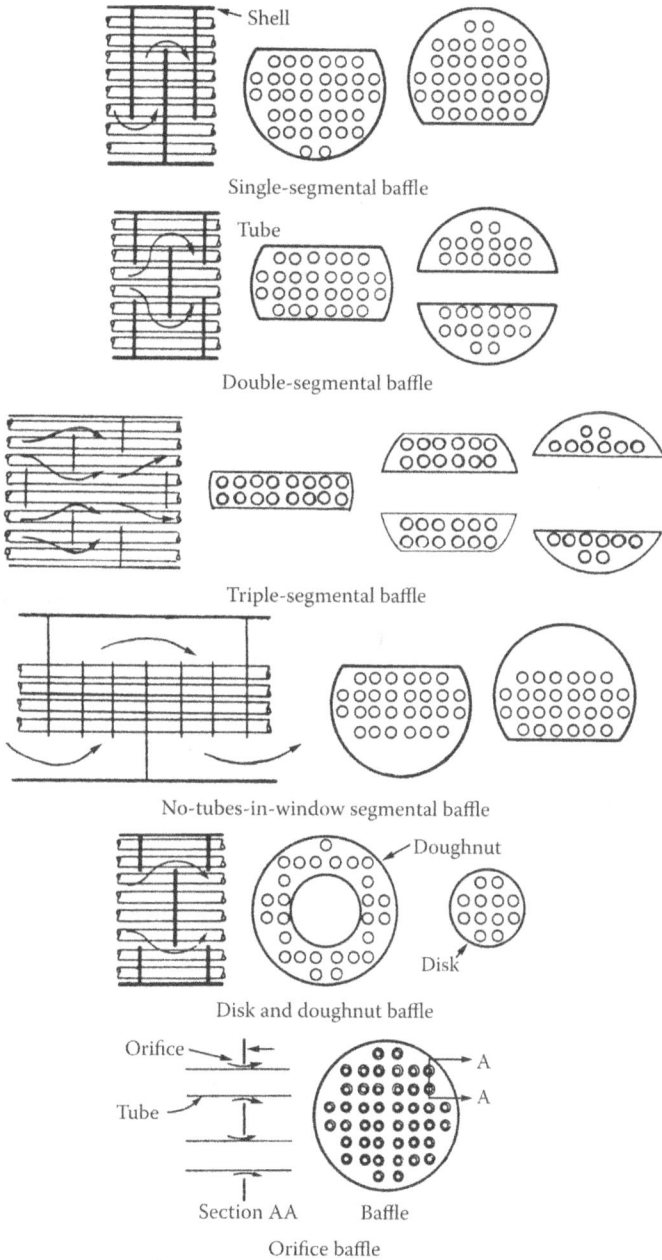

Single-segmental baffle

Double-segmental baffle

Triple-segmental baffle

No-tubes-in-window segmental baffle

Disk and doughnut baffle

Orifice baffle

Figure 5.25   Flat baffle types. (Kakac et al. [2012].)

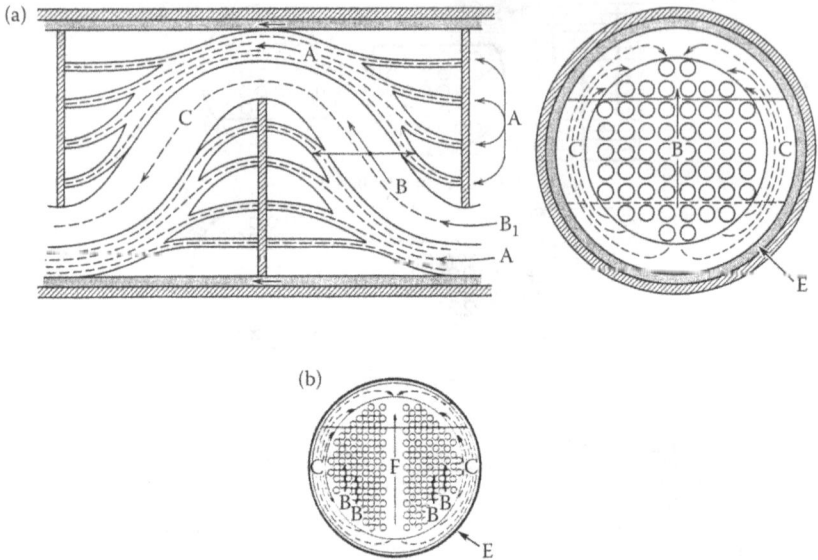

Figure 5.26 (a) Diagram indicating leaking paths for flow bypassing the tube matrix, through both the baffle clearance between the tube matrix and shell, (b) F-stream for a two-tube pass exchanger. (Kakac et al. [2012].)

5.18.1   SHELL-SIDE HEAT TRANSFER COEFFICIENT

The average shell-side heat transfer coefficient is given by

$$h_o = h_{id} J_c J_l J_b J_s J_r \tag{5.44}$$

where, h_{id} is the ideal heat transfer coefficient for pure cross-flow in ideal tube bank, J_c is the correction factor for baffle-cut and spacing, J_l is the correction factor for baffle leakage effects including tube-to-baffle and shell-to-baffle leakage, J_b is the correction factor due to the clearance between the outermost tubes and shell and pass dividers (C- and F-streams), J_s is the correction factor for variable baffle spacing at the inlet and outlet and the change in local velocity and J_r is the correction factor due to Reynolds number effect.

The ideal heat transfer coefficient, h_{id} is calculated from

$$h_{id} = j_i c_{ps} \left(\frac{\dot{m}_s}{A_s} \right) \left(\frac{k_s}{c_{ps} \mu_s} \right)^{2/3} \left(\frac{\mu_s}{\mu_{s,w}} \right)^{0.14} \tag{5.45}$$

where, the Colburn j-factor for the tube bank is given by the curve fit correlation:

$$j_i = a_1 \left(\frac{1.33}{P_t/d_o} \right)^a (Re_s)^{a_2} \tag{5.46}$$

where,

$$a = \frac{a_3}{1+0.014(Re_s)^{a_4}} \tag{5.47}$$

The f-factor is given as

$$f_i = b_1 \left(\frac{1.33}{P_t/d_o}\right)^b (Re_s)^{b_2} \tag{5.48}$$

where,

$$b = \frac{b_3}{1+0.014(Re_s)^{b_4}} \tag{5.49}$$

Table 5.11 presents the coefficients $a_1, a_2, a_3, a_4, b_1, b_2, b_3$ and b_4 for different layout angles as a function of the Reynolds number.

Here, subscript s stands for shell and A_s is the cross-flow area at the centerline of the shell for one cross-flow between two baffles. The bundle cross-flow area A_s at the center of the shell is

$$A_s = \frac{D_s CB}{P_t} \tag{5.50}$$

where, D_s is the inside diameter of the shell, C is the clearance between adjacent tubes, P_t is the pitch and B is the baffle spacing (see Figure 5.27). The Reynolds

Table 5.11
Correlation Coefficients for j_i and f_i (Equations 5.46–5.49)

Layout Angle	Re	a_1	a_2	a_3	a_4	b_1	b_2	b_3	b_4
30	$10^4 - 10^5$	0.321	−0.388			0.372	−0.123		
	$10^3 - 10^4$	0.321	−0.388			0.486	−0.152		
	$10^2 - 10^3$	0.593	−0.477	1.450	0.519	4.570	−0.476	7.000	0.500
	$10 - 10^2$	1.360	−0.657			45.100	−0.973		
	<10	1.400	−0.667			48.000	−1.000		
45	$10^4 - 10^5$	0.370	−0.396			0.303	−0.126		
	$10^3 - 10^4$	0.370	−0.396			0.330	−0.136		
	$10^2 - 10^3$	0.730	−0.500	1.930	0.500	3.500	−0.476	6.590	0.520
	$10 - 10^2$	0.498	−0.656			26.200	−0.913		
	<10	1.550	−0.667			32.000	−1.000		
90	$10^4 - 10^5$	0.370	−0.395			0.391	−0.148		
	$10^3 - 10^4$	0.107	−0.266			0.082	0.022		
	$10^2 - 10^3$	0.408	−0.460	1.187	0.370	6.090	−0.602	6.300	0.378
	$10 - 10^2$	0.900	−0.631			32.100	−0.963		
	<10	0.970	−0.667			35.000	−1.000		

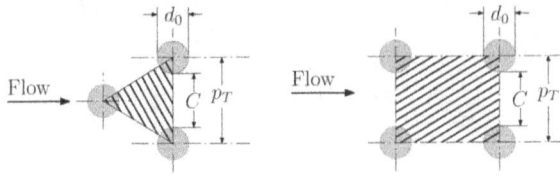

Figure 5.27 Types of pitch–tube layouts, triangular (left) and square (right).

number based on the outside tube diameter and minimum cross-sectional flow area at the shell diameter is given by

$$Re_s = \frac{d_o \dot{m}_s}{\mu_s A_s} \tag{5.51}$$

The correction factor J_c depends on the shell diameter and the baffle cut distance from the baffle tip to the shell's inside diameter. It is equal to 1.0 for a heat exchanger with no tubes in the window and for a large baffle cut; this value may decrease to a value of 0.53. It may increase to 1.15 for small windows with high velocity. The correction factor, J_l is function of the ratio of the total leakage area per baffle to the cross-flow area between adjacent baffles and also the ratio of the shell-to-baffle leakage area to the tube-to-baffle leakage area. The typical value of J_l is in the range of 0.7 to 0.8. The correction factor, J_b is about 0.90 for a relatively small clearance between the outermost tubes and the shell. For larger clearance, the value decreases to about 0.7. The correction factor J_s, which is a function of inlet and outlet spacings, is usually between 0.85 and 1.0. The correction factor J_r is significant for $Re_s < 100$ and is equal to 1.0 for $Re_s > 100$. The combined effect of all these corrections is of the order of 0.60.

5.18.2 SHELL-SIDE PRESSURE DROP

The shell-side pressure drop is calculated as the sum of pressure drop in three components i.e. entrance, internal and window (see Figure 5.28) written as

$$\Delta p_s = \Delta p_e + \Delta p_i + \Delta p_w \tag{5.52}$$

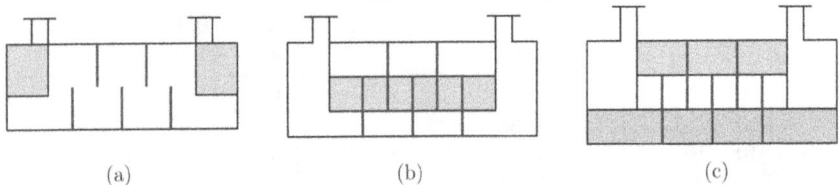

Figure 5.28 (a) Entrance, (b) internal and (c) window section of a shell.

The combined pressure drop for the entrance and exit section is given by

$$\Delta p_e = 2\Delta p_{bi} \frac{N_c + N_w}{N_c} R_b R_s \tag{5.53}$$

where, N_c is the number of tube rows crossed during flow through one cross-flow in the exchanger and N_w is the number of tubes crossed in each baffle window. The correction factors R_s for unequal baffle spacing of inlet and exit baffle section and R_b for baffle to shell bypass streams are available in graphical forms in Taborek [1991] and Bell [1988a].

The pressure drop Δp_{bi}, which is the pressure drop in tube bank in one baffle compartment, is calculated from

$$\Delta p_{bi} = 4 f_i \frac{G_s^2}{2\rho_s} \left(\frac{\mu_{s,w}}{\mu_s} \right)^{0.14} \tag{5.54}$$

where,

$$G_s = \frac{\dot{m}}{A_s} \qquad D_e = \frac{4 \times \text{free flow area}}{\text{wetted perimeter}}$$

$$f_i = exp(0.576 - 0.19 ln(Re_s)) \qquad \left(400 < Re_s = \frac{G_s D_e}{\mu} \leq 1 \times 10^6 \right)$$

With reference to Figure 5.27 for square pitch, we have

$$D_e = \frac{4(P_t^2 - \pi d_o^2/4)}{\pi d_o} \tag{5.55}$$

For triangular pitch,

$$D_e = \frac{4 \left(\frac{P_t^2 \sqrt{3}}{4} - \frac{\pi d_o^2}{8} \right)}{\pi d_o / 2} \tag{5.56}$$

The pressured drop in the interior cross-flow section is

$$\Delta p_i = \Delta p_{bi}(N_b - 1) R_l R_b \tag{5.57}$$

where, N_b is the number of baffles and R_l is the correction factor for baffle leakage effects, which typically ranges between 0.4 and 0.5. The correction factor R_b is for bypass flow (C and F streams) and is typically equal to 0.5 to 0.8.

The pressure drop in the window is affected by leakage and is calculated from

$$\Delta p_w = \Delta p_{wi} N_b R_l \tag{5.58}$$

The pressure drop in one window section is calculated using

$$\Delta p_{wi} = \frac{\dot{m}_s^2 (2 + 0.6 N_{cw})}{2 \rho_s A_s A_w} \qquad (\text{if, } Re_s \leq 100) \tag{5.59}$$

$$\Delta p_{wi} = \frac{26 \mu_s \dot{m}_s}{\sqrt{\rho_s A_s A_w}} + \left(\frac{N_{cw}}{P_t - d_o} + \frac{B}{D_w^2} \right) \frac{\dot{m}_s}{\rho_s A_s A_w} \qquad (\text{if, } Re_s < 100) \tag{5.60}$$

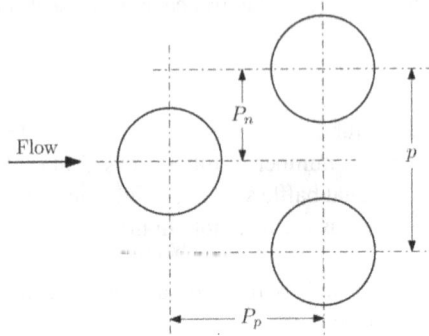

Figure 5.29 Tube pitches parallel and normal to flow (equilateral triangular arrangement shown). (Kakac et al. [2012].)

Here, B is the baffle spacing, A_s is the cross-flow area near the shell centerline, A_w is the flow area through the window zone and N_{cw} is the number of effective rows crossed in one window zone. The number of tube rows in one cross-flow can be estimated using:

$$N_c = \frac{d_i \left(1 - 2\frac{L_c}{D_s}\right)}{P_p}$$

(5.61)

where, L_c is the baffle cut distance from the baffle tip to the inside of the shell and P_b is defined in Figure 5.29.

The number of effective cross-flow rows in each window, N_{cw} can be estimated from

$$N_{cw} = \frac{0.8L_c}{P_p}$$

(5.62)

The number of baffles, N_b can be estimated from

$$N_b = \frac{L}{B} - 1$$

(5.63)

where, L is the length of the heat exchanger and B is the baffle spacing.

5.19   KERN METHOD

A simplified method known as the Kern method can also be used for shell-side heat transfer and pressure drop estimation.

5.19.1   SHELL-SIDE HEAT TRANSFER COEFFICIENT

The shell-side heat transfer coefficient is the heat transfer coefficient outside the tube bundles. The heat transfer coefficient is higher with baffles than that without baffles due to increased turbulence. The correlations obtained for flow in tubes are not applicable. The following correlation can be used for the calculation of shell-side heat transfer coefficient (h_o):

$$\frac{h_o D_e}{k} = 0.36 \left(\frac{D_e G_s}{\mu}\right)^{0.55} \left(\frac{\mu c_p}{k}\right)^{1/3} \left(\frac{\mu_b}{\mu_w}\right)^{1/4}$$

$$\left(\text{for, } 2 \times 10^3 < Re_s = \frac{G_s D_e}{\mu} < 1 \times 10^6\right)$$

(5.64)

Here, D_e is the equivalent diameter on the shell side and G_s is the shell-side mass velocity. It may be noted that mass velocity (G_s) is a function of bundle cross-flow area (A_s), which is a function of baffle spacing. The calculations of these quantities have been described in previous sections.

5.19.2 SHELL-SIDE PRESSURE DROP

The pressure drop on the shell side is calculated using the following expression:

$$\Delta p_s = \frac{f G_s^2 (N_b + 1) D_s}{2\rho D_e \phi_s}$$

(5.65)

where, $\phi = (\mu_b/\mu_w)$.

5.20   BASIC DESIGN PROCESS

Figure 5.30 shows the flow chart explaining the basic design procedure of a heat exchanger. In the first step, the exact requirement i.e. specification of heat exchanger i.e. inlet and outlet temperature, pressure, flow rate and size should be specified. Subsequently, the tentative configuration of the heat exchanger is selected. The next step is to select the tentative set of exchanger design parameters. Accordingly, a tentative exchanger size can be estimated. The initial design will be used to calculate the thermal performance and pressure drop for both streams. If the output is not acceptable, a new geometrical modification will be made. Suppose the initial design cannot meet the amount of heat transfer requirement. Then, either the area of the exchanger will be increased or ways to increase the heat transfer coefficient will be explored. To increase the tube-side heat transfer coefficient, the tube-side velocity can be increased. One can also decrease the baffle spacing or decrease the baffle cut to increase the shell-side heat transfer coefficient. One can either increase the length of the heat exchanger or increase the shell diameter for increasing the area. If the pressure drop is greater than requirement, either the number of tube passes can be decreased or the tube diameter can be increased. If the shell-side pressure drop is higher than the requirement, the baffle spacing, tube pitch or baffle cut can be increased.

5.21   PRELIMINARY DESIGN ESTIMATION

If a heat load is provided, the temperature difference can be estimated using

$$Q = \dot{m} c_p (T_{c2} - T_{c1}) = \dot{m} c_p (T_{h1} - T_{h2})$$

(5.66)

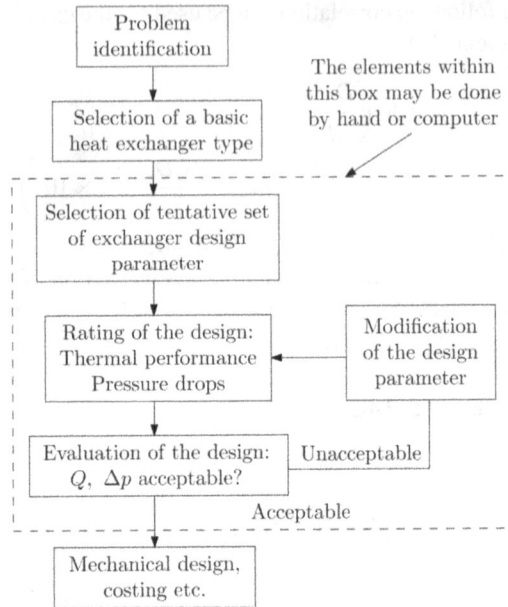

Figure 5.30 Basic logic structure for process heat exchanger design.

Subsequently, the LMTD can be estimated using the expression and F-factor data discussed in Section 5.4. In case the temperature data are provided, the heat transfer rate can be estimated using equation 5.66. The overall heat transfer coefficient based on the outside diameter of the tubes can be estimated using

$$\frac{1}{U_0} = \frac{A_o}{A_i}\left(\frac{1}{n_i h_i} + \frac{R_{fi}}{n_i}\right) + A_o R_w + \frac{R_{fo}}{n_o} + \frac{1}{n_o h_o} \tag{5.67}$$

The outside heat transfer surface area can be estimated using

$$A_o = \frac{Q}{U_0 F \Delta T_{lm}} \tag{5.68}$$

The typical heat transfer coefficients h_i and h_o can be assumed based on Tables 5.5, 5.6, 5.8 and 5.9. If N_t is the number of tubes with a given tube length L, we have

$$A_o = N_t \pi d_o L \tag{5.69}$$

If A_l is the area of a single tube and D_s is the shell diameter, we can also write

$$N_t = CTP\frac{\pi D_s^2}{4A_l} \tag{5.70}$$

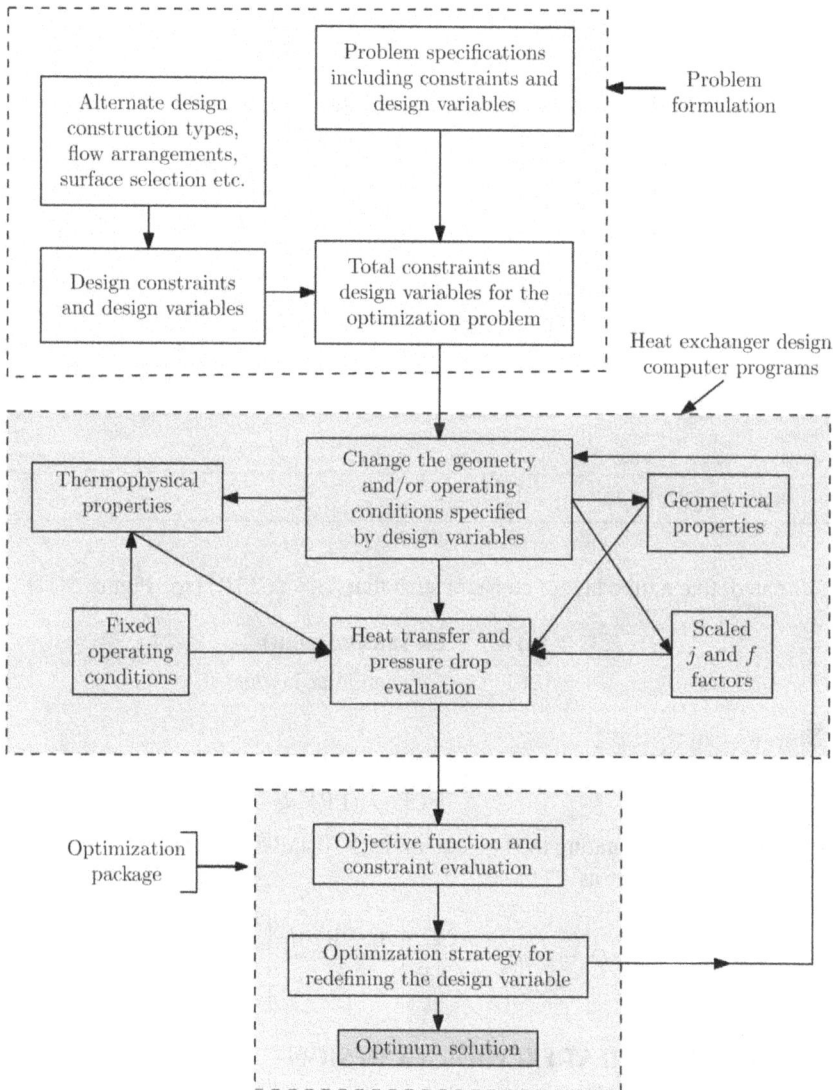

Figure 5.31   A sample flow chart for heat exchanger optimization.

where, CTP is the tube count calculation constant which incorporates the effect of incomplete coverage of the shell diameter by the tubes due to necessary clearance between the shell and outer tube circle.

$$\begin{aligned} \text{CTP} &= 0.43 \quad \text{(One-tube pass)} \\ &= 0.90 \quad \text{(Two-tube pass)} \\ &= 0.85 \quad \text{(Three-tube pass)} \end{aligned}$$

Table 5.12

Heat Transfer Matrix Geometries for Plate Plain-Fin and Fin Flat Tubes

Surface Designation	Fins (per cm)	Hydraulic Diameter (D_h, cm)	Plate Spacing (b, cm)	Tube or Fin Thickness (cm)	Extended Area — Total Area	Area — Volume between Plates (β, m²/m³)	Area — Core Volume (β, m²/m³)	Free Flow Area — Frontal Area (σ)
Plate plain-fin type								
5.30	13.46	0.031	1.191	0.0152	0.719	511.8		
11.10	28.19	0.257	0.635	0.0152	0.730	1095.8		
14.77	37.52	0.215	0.838	0.0152	0.831	1210.6		
19.86	50.44	0.152	0.635	0.0152	0.833	1493.0		
Fin flat-tube type								
9.68-0.870	24.587	0.2997		0.0102	0.795		751.3	0.697
9.68-0.870-R	24.587	0.2997		0.0102	0.795		751.3	0.697
9.1-0.737-S	23.114	0.3565		0.0102	0.813		734.9	0.788
9.29-0.737-S-R	28.753	0.3510		0.0102	0.845		885.8	0.788
11.32-0.737-S-R	23.596	0.3434		0.0102	0.814		748.0	0.780

Source: Kakac et al. [2012].

We can define a tube layout constant such that $A_l = (CL)P_t^2$ (see Figure 5.27)

$$CL = 1.0 \qquad \text{(for square layout)}$$
$$= 0.87 \qquad \text{(for triangular layout)}$$

Now, we can write

$$N_t = 0.785 \left(\frac{\text{CTP}}{\text{CL}} \right) \frac{D_s^2}{(\text{PR})^2 d_o^2} \tag{5.71}$$

where, $\text{PR} = P_t/d_o$. Equating the two equations 5.70 and 5.71 for N_t, we can approximate the shell diameter as

$$D_s = 0.637 \sqrt{\frac{\text{CL}}{\text{CTP}}} \left[\frac{A_o(\text{PR})^2 d_o}{L} \right]^{1/2} \tag{5.72}$$

5.22   COMPACT HEAT EXCHANGER DESIGN

Compact heat exchangers are widely used in industry i.e. condensers and evaporators in air-conditioning and refrigeration industry, automotive radiators, intercooler of compressor, aircraft and space applications. We will review the design of tube and plate fin heat exchangers in the following sections (Figure 5.31).

5.22.1   HEAT TRANSFER AND PRESSURE DROP

Compact heat exchangers are available in a wide variety of configurations. The heat transfer and pressure drop characteristics of these heat exchangers were studied by Kays and London [1984]. The heat transfer and pressure drop characteristics of various configurations of heat exchangers are determined experimentally. Table 5.12

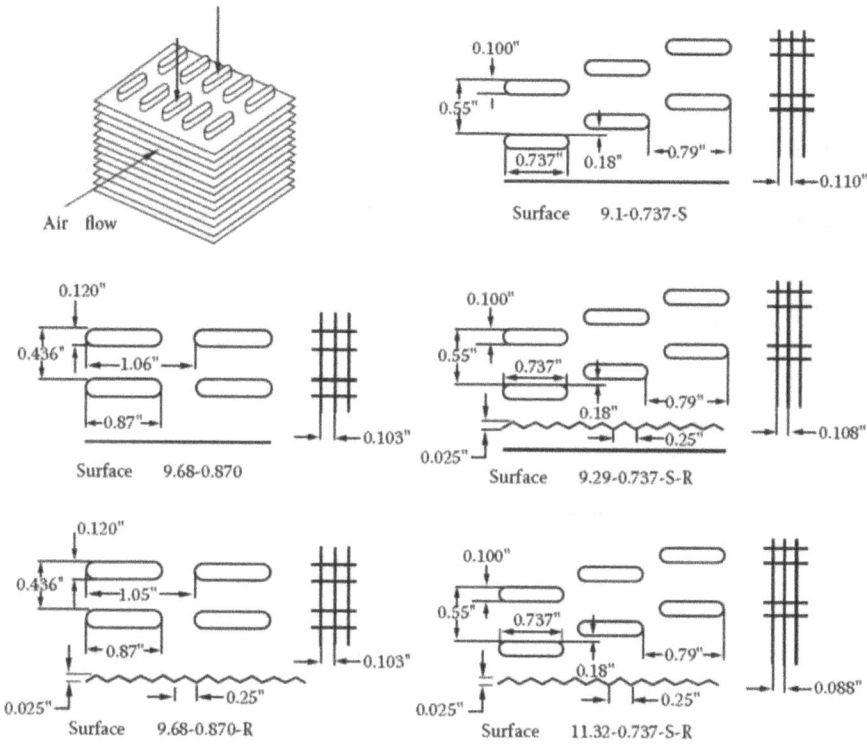

Figure 5.32 Various flattened tube-plate fin compact surfaces for which test data are presented in Figure 5.33. (Kakac et al. [2012].)

presents five different surface geometries of fin-tube configuration and four different plate-fin types. Figure 5.32 shows the geometrical details of various flattened tube plate-fin surfaces. Figure 5.33 shows the heat transfer and friction factor for flow across a finned flat-tube matrix for the surfaces shown in Figure 5.32 and Table 5.12. Figure 5.34 shows the friction factor and heat transfer data for four plain plate-fin heat transfer matrices presented in Table 5.12. The heat transfer and friction factor data for some of the configurations are presented in Figures 5.35–5.37. These curves show the correlation using three dimensionless groups given as:

Stanton number, $St = \frac{h}{G c_p}$

Prandtl number, $Pr = \frac{\mu c_p}{k}$

Reynolds number, $Re = \frac{G D_h}{\mu}$

where, mass velocity (G) is given by

$$G = \rho u_{max} = \frac{\dot{m}}{A_{min}}$$

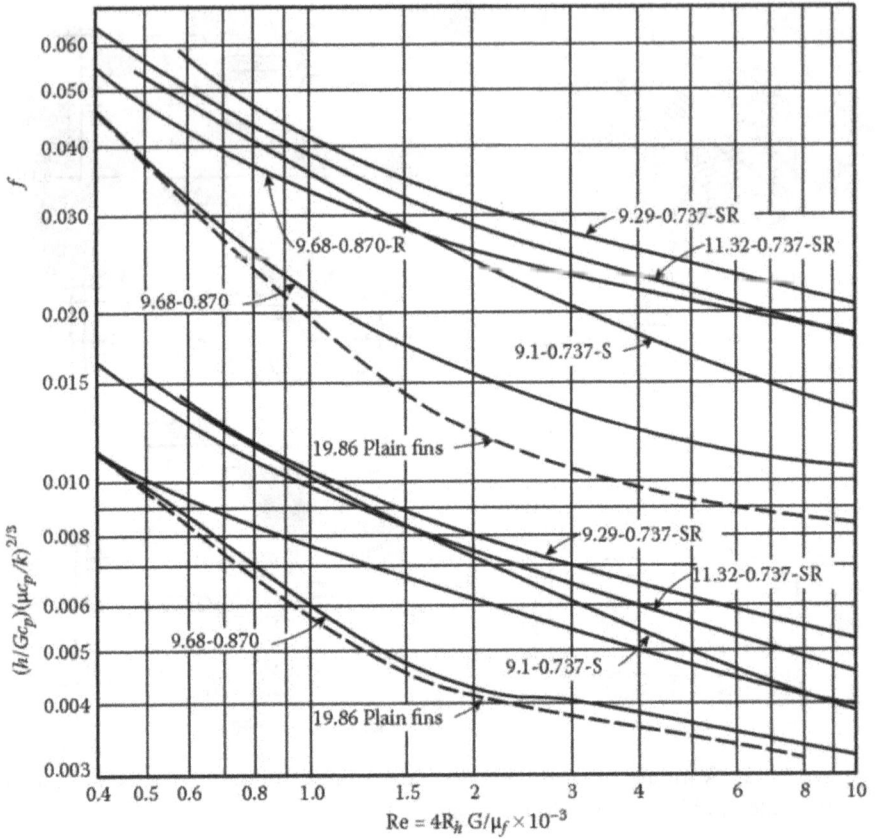

Figure 5.33 Heat transfer and friction factor for flow across a finned flat-tube matrix for the surfaces shown in Figure 5.32 and Table 5.12. (Kakac et al. [2012].)

where, $\dot{m}$ is the total mass flow rate of fluid and A_{min} is the free-flow cross-sectional area.

The hydraulic diameter is defined as

$$D_h = 4\frac{LA_{min}}{A}$$

where, LA_{min} is the flow passage volume, A is the total heat transfer area and L is the flow length of the heat exchanger matrix. The overall heat transfer coefficient based on the gas-side surface area in a gas-liquid heat exchanger can be written as

$$\frac{1}{U_0} = \frac{A_t}{A_i}\frac{1}{h_i} + A_t R_w + \frac{1}{\eta_o h_o}$$

Here, the fouling resistance has been neglected. The total external air-side surface area is expressed as

$$A_t = A_u + A_f$$

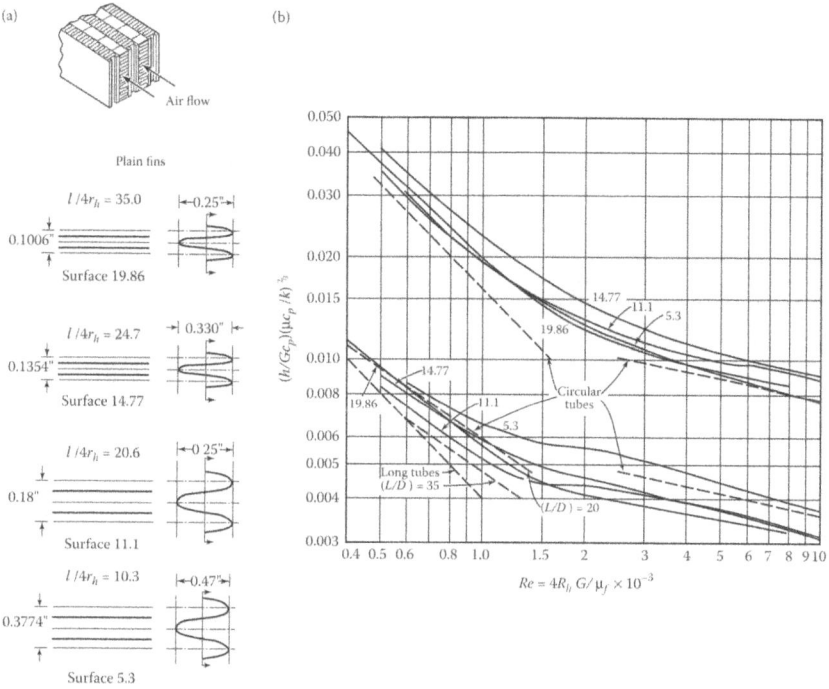

Figure 5.34   Heat transfer and friction factor for four plain plate-fin heat transfer matrices of Table 5.12. (Kakac et al. [2012].)

The fin efficiency, η_f can be calculated using appropriate graphs and equations available in most heat transfer books.

5.22.2   PRESSURE DROP FOR FINNED-TUBE EXCHANGERS

The friction factor f calculation for finned-tube banks was discussed in Section 5.11. The total pressure drop between the pressures at the inlet and outlet is given by

$$\Delta p = \frac{G^2}{2\rho_i}\left[f\frac{A_t}{A_{min}}\frac{\rho_i}{\rho} + \left(1+\sigma^2\right)\left(\frac{\rho_i}{\rho} - 1\right)\right] \tag{5.73}$$

where,

$$\sigma = \frac{A_{min}}{A_{fr}} = \frac{\text{Minimum free-flow area}}{\text{Frontal area}}$$

$$\frac{A_t}{A_{min}} = \frac{4L}{D_h} = \frac{\text{Total heat transfer area}}{\text{Minimum flow area}}$$

$$G = \frac{\rho u_\infty A_{fr}}{A_{min}} = \frac{\rho u_\infty}{\sigma}$$

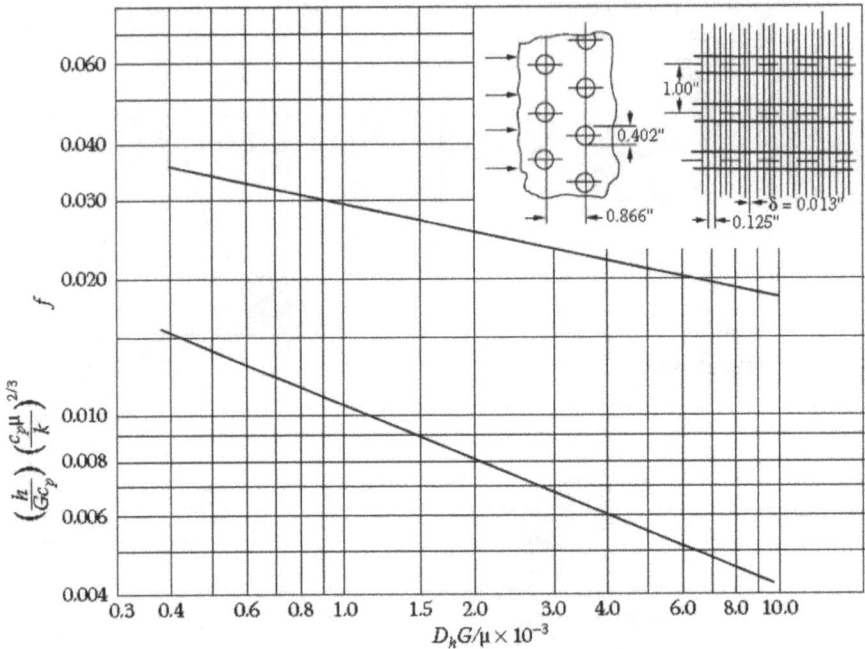

Figure 5.35 Heat transfer and friction factor for a circular tube continuous fin heat exchanger. Surface 8.0-3/8 T: tube OD = 1.02 cm; fin pitch = 3.15/cm; fin thickness = 0.033 cm; fin area/total area = 0.839; air-passage hydraulic diameter = 0.3633 cm; free-flow area/frontal area, $\sigma = 0.534$; heat transfer area/total volume = 587 m^2/m^3. (Kakac et al. [2012].)

Here, ρ is the average density evaluated at the average temperature between the inlet and outlet. It can be estimated by averaging the fluid-specific volume at the inlet and outlet, which can be expressed as

$$\frac{1}{\rho} = \frac{1}{2}\left(\frac{1}{\rho_i} + \frac{1}{\rho_o}\right)$$

The second term in the above pressure drop calculation equation accounts for the acceleration or deceleration of flow. This term is negligible for liquids in which the density is almost constant.

5.22.3 PRESSURE DROP FOR PLATE-FIN EXCHANGERS

The total pressure drop for flow across the plate-fin heat exchanger matrix is expressed as

$$\Delta p = \frac{G^2}{2\rho_i}\left[\left(k_c + 1 - \sigma^2\right) + 2\left(\frac{\rho_i}{\rho} - 1\right) + f\frac{A}{A_{min}}\frac{\rho_i}{\rho} - \frac{\rho_i}{\rho_o}\left(1 - k_e - \sigma^2\right)\right] \quad (5.74)$$

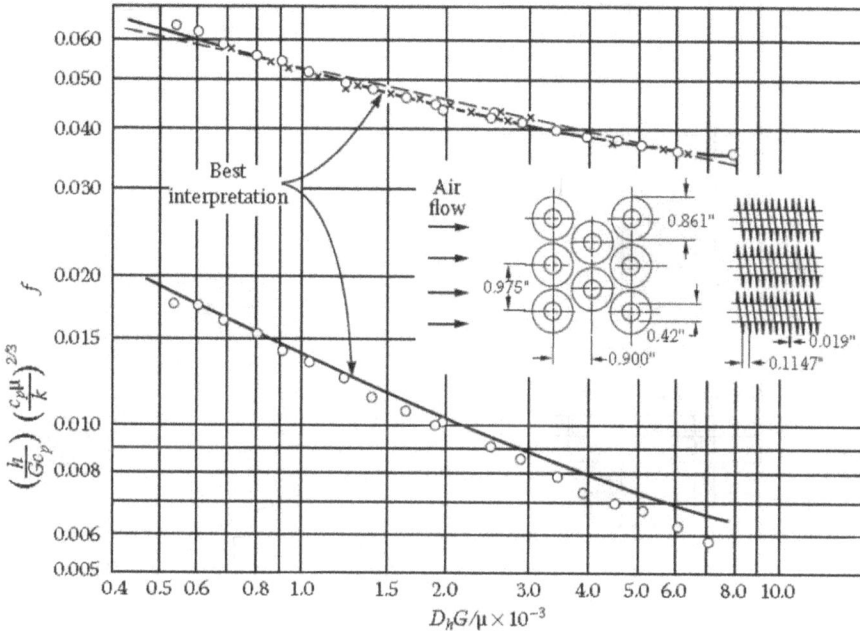

Figure 5.36 Heat transfer and friction factor for flow across a circular finned-tube matrix. Surface CF-8.72 (c): tube OD = 1.07 cm; fin pitch = 3.43/cm; fin thickness = 0.048 cm; fin area/total area = 0.876; air passage hydraulic diameter, $d_h = 0.443$ cm; free-flow area/frontal area, $\sigma = 0.494$; heat transfer area/total volume = 446 m^2/m^3. (Kakac et al. [2012].)

Typical values of contraction loss coefficient, k_c and enlargement loss coefficient, k_e are given in Figure 5.38.

The first term in the RHS of equation 5.74 is attributed to the entrance effect, the second term is due to the flow acceleration effect, the third term is due to core friction and the last term is due to the exit effect. Mostly the frictional pressure drop dominates. The entrance and exit losses are important for short cores.

5.23 OPTIMIZATION OF HEAT EXCHANGERS

Heat exchanger design involves different optimization criteria depending on the type of applications i.e. minimum initial cost, minimum weight, minimum operating cost etc. Accordingly, a performance measure can be defined quantitatively, which has to be either maximized or minimized. This is known as an objective function. A design has to meet certain requirements i.e. required heat transfer, allowable pressure drop, limitation on height or width or length of the heat exchanger. These requirements are known as constraints for the optimization of heat exchangers. A heat exchanger has many geometric variables i.e. types of surfaces and their sizes. Similarly, the

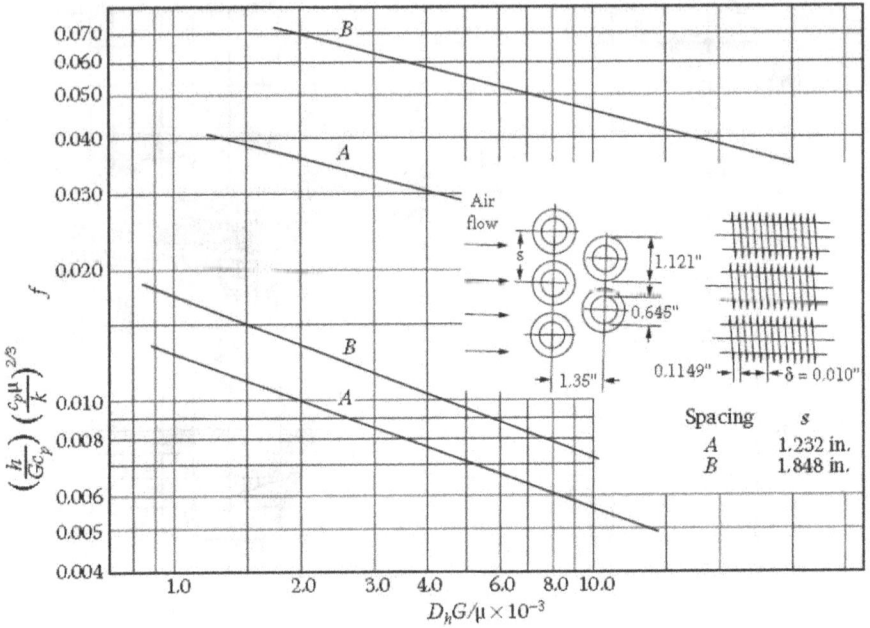

Figure 5.37 Heat transfer and friction factor for flow across a finned-tube matrix. Surface CF-8.7-5/8 J: tube OD = 1.638 cm; fin pitch = 3.43/cm; fin thickness = 0.0254 cm; fin area/total area = 0.862; air passage hydraulic diameter, $d_h = 0.5477$ cm (A), 1.1673 cm (B); free-flow area/frontal area, $\sigma = 0.443$ (A), 0.628 (B); heat transfer area/total volume = 323.8 m^2/m^3 (A), 215.6 m^2/m^3 (B). (Kakac et al. [2012].)

operating variables i.e. mass flow rates and temperature can be varied. As a result, a large number of design variables are used for heat exchanger designs. The aim of optimization of heat exchangers is to effectively adjust these design variables within imposed constraints and achieve an optimum objective function. Figure 5.31 shows the flow chart for the optimization of a heat exchanger. It consists of three preliminary steps i.e. problem formulation, heat exchanger design and optimization. The heat exchanger design and optimization stage can be carried out using a computational tool/package.

The designer starts with one set of heat exchanger surface and geometrical dimensions. The heat transfer and pressure drop are evaluated based on the heat exchanger design procedure. The output of the heat exchanger is fed to the optimization program, which evaluates the objective function and constraint. Subsequently, the design variables are changed and reevaluated by the heat exchanger program. The iteration is continued till the objective function is optimized.

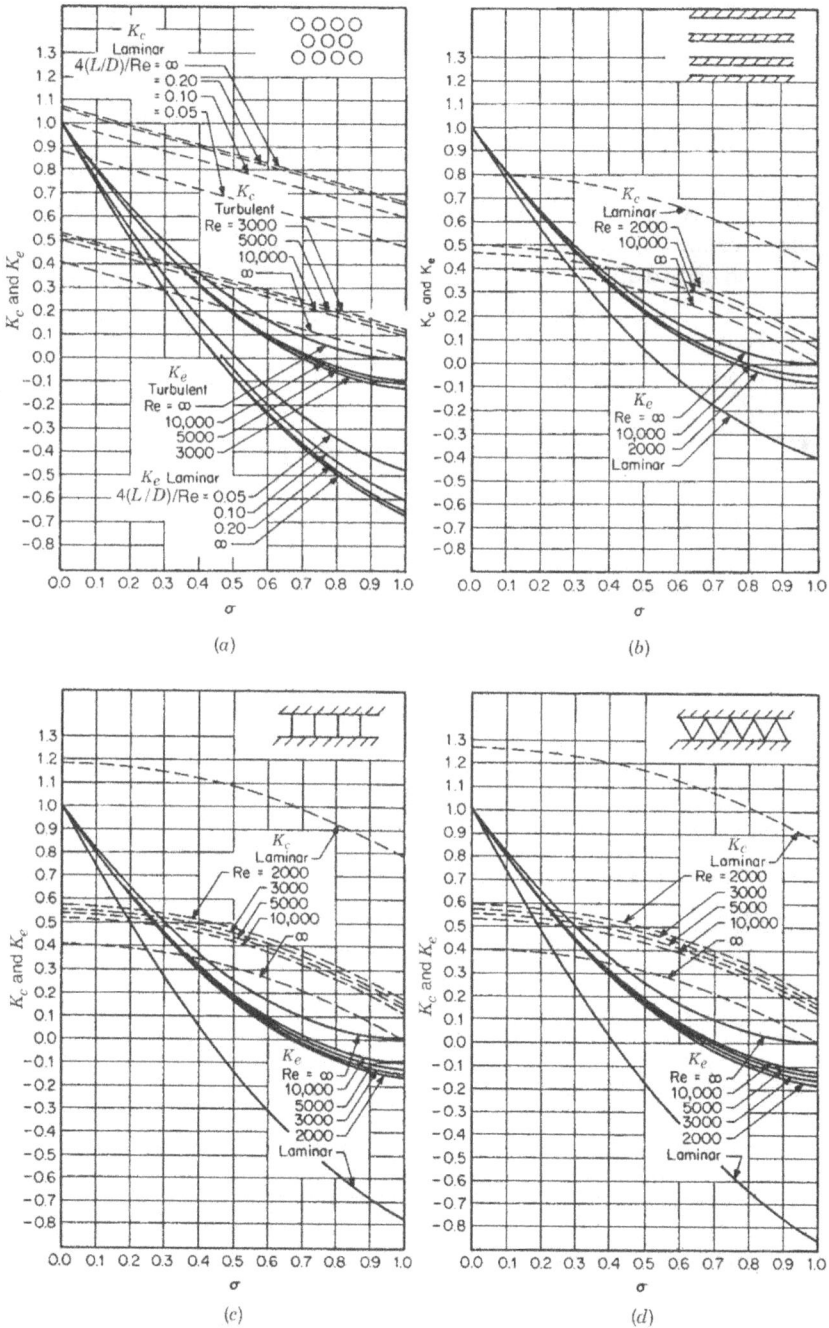

Figure 5.38 Entrance and exit pressure loss coefficients for (a) a multiple circular tube core, (b) multiple-tube flat-tube core, (c) multiple square tube core and (d) multiple triangular tube core with abrupt contraction (entrance) and abrupt expansion (exit). (Shah and Sekulic [1985].)

Example: Suppose a design engineer has to design a radiator such that the coolant at the radiator inlet (top tank) does not exceed 120 °C at 100 kPa gauge radiator cap pressure. Consider the following conditions: engine heat rejection rate, $q = 35$ kW, air flow rate is 0.75 kg/s, air inlet temperature $= 53$ °C and coolant flow rate is equal to 1.4 kg/s. The specific heat for the air and coolant is 1009 J/kg–K and 3664 J/kg–K, respectively. Let's assume that $UA = 1180$ W/K for the initial design of heat exchanger. What are the radiator top tank temperature and outlet temperature of the coolant? Comment on the results. Consider the radiator with one fluid mixed and another fluid unmixed.

SOLUTION

$$C_{air} = C_c = (\dot{m}c_p)_{air} = 0.75 \times 1009 = 756.75 \text{ W/K}$$

$$C_{liq} = C_h = (\dot{m}c_p)_{coolant} = 1.4 \times 3664 = 5129.6 \text{ W/K}$$

$$C^* = \frac{C_{air}}{C_{liq}} = \frac{756.75}{5129.6} = 0.148$$

$$NTU = \frac{UA}{C_{min}} = \frac{1180}{756.75} = 1.559$$

$$\varepsilon = 1 - exp\left[-\frac{1 - exp(-C^*NTU)}{C^*}\right]$$

$$= 1 - exp\left[-\frac{1 - exp(-0.148 \times 1.559)}{0.148}\right]$$

$$= 1 - exp\left[-\frac{1 - (0.7939)}{0.148}\right]$$

$$= 1 - exp[-1.4047] = 0.7545$$

$$T_{h,i} = T_{c,i} + \frac{q}{\varepsilon C_{min}}$$

$$= 53 + \frac{35000}{0.7545 \times 756.75} = 114.29 \text{ °C}$$

It is less than 120 °C, so the design is safe. If the temperature $T_{h,i} > 120$ °C, we would have to either increase A or U.

$$T_{h,o} = T_{h,i} - \frac{q}{C_h} = 114.29 - \frac{35,000}{5,129.6} = 107.46 \text{ °C}$$

Example: Find the surface area of a 1-2 TEMA E-shell and tube heat exchanger using the following data. Water enters the shell at 20 °C at a rate of 1.4 kg/s. Engine oil flows at a rate of 1.0 kg/s with inlet and outlet temperatures equal to 150 °C and 90 °C, respectively. Assume the overall heat transfer coefficient, $U = 225$ W/m²–K. The specific heat of water and oil is equal to 4.19 and 1.67 J/g–K, respectively.

SOLUTION

The heat capacity of shell liquid (water) is

$$C_s = (\dot{m}c_p)_s = 1.4 \times 4.19 \times 10^3 = 5866 \text{ W/K}$$

The heat capacity of tube fluid (oil) is

$$C_t = (\dot{m}c_p)_t = 1.0 \times 1.67 \times 10^3 = 1670 \text{ W/K}$$

The heat transfer rate from oil is

$$q = C_t(T_{t,i} - T_{t,o}) = 1670(150 - 90) = 100.2 \times 10^3 \text{ W}$$

Using the energy balance equation, we write

$$T_{s,o} = T_{s,i} + \frac{q}{C_s} = 20 + \frac{100.2 \times 10^3}{5866} = 37.1^\circ C$$

$$C^* = \frac{C_{[t]}}{C_s} = \frac{1670}{5866} = 0.2867$$

$$\varepsilon = \frac{T_{t,i} - T_{t,o}}{T_{t,i} - T_{s,i}} = \frac{150 - 90}{150 - 20} = 0.4615$$

$$\text{NTU} = \frac{1}{\sqrt{1+C^{*2}}} \ln \left[\frac{2 - \varepsilon \left(1 + C^* - \sqrt{1+C^{*2}}\right)}{2 - \varepsilon \left(1 + C^* + \sqrt{1+C^{*2}}\right)} \right]$$

$$A = \frac{C_{min}}{U} \times \text{NTU} = \frac{1670}{225} \times 0.6828 = 5.068 \text{ m}^2$$

Example: Calculate the shell-side pressure drop of a shell and tube heat exchanger using the following data:

Cross-flow area near the shell centerline (A_s): 0.0444 m^2
Flow area through the window zone (A_w): 0.01261 m^2
Number of effective tube row baffle section (N_c): 9
Number of effective rows crossed in one window zone (N_{cw}): 3.868
Number of baffles (N_b): 14
Oil flow rate $(\dot{m}_s)$: 36.3 kg/s
Ideal tube bank friction factor (f_c): 0.23
Oil density (ρ_s): 849 kg/m^3
Correction factor for baffle-to-shell and tube-to-baffle leakage streams (R_l): 0.59
Correction factor for baffle to shell bypass streams (R_b): 0.69
Correction factor for unequal baffle spacing on inlet and exit baffle section (R_s): 0.81
Shell-side Reynolds number (Re_s): 242
Tube outside diameter (d_o): 19 mm
Tube wall thickness (t): 1.2 mm
Tube pitch (P_t): 25 mm (square layout)
Tube shell-side lubricating oil is cooled from 70 °C to 65 °C

SOLUTION

$$\Delta p_s = R_l \left[(N_b - 1)\Delta p_{bi} + N_b \Delta p_{wi} \right] + 2\Delta p_{bi} \left(1 + \frac{N_{cw}}{N_c} \right) R_b R_s$$

$$\Delta p_{bi} = 4 f_i \frac{G_s^2}{2\rho_s} \left(\frac{\mu_{s,w}}{\mu_s} \right)^{0.14} \times N_c$$

$$G_s = \frac{\dot{m}}{A_s} = \frac{36.3}{0.0444} = 817.56 \text{ kg/m}^2\text{-s}$$

Assuming $\frac{\mu_{s,w}}{\mu_s} \simeq 1$ small (because of the small temperature difference). Pressure in the tube bank in one baffle (Δp_{bi}) and pressure drop in one window section (Δp_{wi}) are calculated as

$$\Delta p_{bi} = 4 \times 0.23 \times \frac{817.56^2}{2 \times 849} \times 1^{0.14} \times 9 = 3259.35 \text{ Pa}$$

$$\Delta p_{wi} = \frac{\dot{m}_s^2 (2 + 0.6 N_{cw})}{2\rho_s A_s A_w}$$

$$= \frac{36.3^2 \times (2 + 0.6 \times 3.868)}{2 \times 849 \times 0.0444 \times 0.01261} = 5784.42 \text{ Pa}$$

$$\Delta p_s = 0.59 \left[(14 - 1)3259.35 + 14 \times 5784.42 \right]$$

$$+ 2 \times 3259.35 \left(1 + \frac{3.868}{9} \right) \times 0.69 \times 0.81$$

$$= 17249 + 47779.3 + 5209.11$$

$$= 70237.41 \text{ Pa}$$

Example: Consider a single shell and single-tube pass heat exchanger with the following data:

Outside diameter of the tube (d_o): 19 mm
Inside diameter of the tube (d_i): 16 mm
Square pitch of the tube (P_t): 0.0254 m
Baffle spacing (B): 0.5 m
Shell fluid: Distilled water
Shell flow rate $(\dot{m}_s)$: 50 kg/s
Shell fluid temperature at the inlet $(T_{h,1})$: 32 °C
Shell fluid temperature at the outlet $(T_{h,2})$: 25 °C
Tube fluid: Raw water
Tube flow rate $(\dot{m}_t)$: 150 kg/s
Tube fluid temperature at inlet $(T_{c,1})$: 20 °C
Thermal conductivity of tube material (k): 42.3 W/m–K
Total fouling resistance (R_f): 0.000176 m²–K/W
Flow velocity through each tube (u_m): 2 m/s

Properties of tube-side fluid	Properties of shell-side fluid
$c_{p,t} = 4182$ J/kg–K	$c_{p,s} = 4179$ J/kg–K
$k_t = 0.598$ W/m^2–K	$k_s = 0.612$ W/m^2–K
$Pr_t = 7.01$	$Pr_s = 5.75$
$\rho_t = 998.2$ kg/m^3	$\rho_s = 995.9$ kg/m^3
$\mu_t = 10.02 \times 10^{-4}$ N-s/m^2	$\mu_s = 8.15 \times 10^{-4}$ N-s/m^2

(a) Calculate the shell-side heat transfer coefficient by the Kern method.
(b) Calculate the shell-side heat transfer coefficient by the Bell–Delaware method.
(c) Calculate the tube-side heat transfer coefficient.
(d) Calculate the overall heat transfer coefficient of the clean surface.
(e) Calculate the overall heat transfer coefficient of the fouled surface.
(f) Calculate the net heat transfer requirement.
(g) Calculate the net heat transfer area required considering the clean heat exchanger.
(h) Calculate the net heat transfer area required considering the fouled heat exchanger.

SOLUTION

Total mass flow rate through a tube $(\dot{m}_t) = N_t \rho_t u_m A_c$

Thus,

$$\text{Number of tubes, } N_t = \frac{\dot{m}_t}{\rho_t u_m A_c}$$

$$= \frac{4 \times 150}{998.2 \times \pi \times (0.016)^2 \times 2} = 373.88 \simeq 374$$

$$\text{Shell diameter, } D_s = \left[\frac{N_t (C_L)(PR)^2 d_o^2}{0.785(CTP)} \right]^{1/2}$$

$$CTP = 0.93$$

$$CL = 1.0$$

$$PR = \frac{P_t}{d_o} = \frac{0.0254}{0.019} = 1.336$$

$$D_s = \left[\frac{374 \times 1 \times 1.336^2 \times 0.019^2}{0.785 \times 0.93} \right]^{1/2}$$

$$= 0.5745 \text{ m} = 575 \text{ mm}$$

(a) Kern method

Cross-flow area at the shell diameter, $A_s = B(D_s - N_{tc}d_o)$

where, $N_{tc} = \frac{D_s}{P_t} = \frac{575}{25.4} = 22.637$

$$A_s = 0.5(0.575 - 22.637 \times 0.019) = 0.0724 \text{ m}^2$$

Equivalent diameter, $D_e = \dfrac{P_t^2 - \pi d_o^2/4}{\pi d_o/4}$

$$= \frac{0.0254^2 - \pi \times 0.019^2/4}{\Pi \times 0.019/4}$$

$$= 0.0242 \text{ m}$$

$$Re = \frac{\dot{m}_s D_e}{A_s \mu_s} = \frac{50 \times 0.0242}{0.0724 \times 8.15 \times 10^{-4}}$$

$$= 20506.38$$

$$\frac{h_o D_e}{k} = 0.36 Re^{0.55} Pr^{1/3} \left(\frac{\mu_b}{\mu_w}\right)^{0.14}$$

Assuming $\mu_b = \mu_w$ due to a small temperature difference.

$$= 0.36 \times 20506.38^{0.55} \times 5.75^{1/3}$$

$$= 0.36 \times 235.25 \times 1.7904 = 151.628$$

$$h_o = \frac{151.628 \times 0.612}{0.0242} = 3834.55 \text{ W/m}^2\text{-K}$$

(b) Bell-Delaware method

Let's assume the combined effect of different correction factors i.e. j_c, j_l, j_b, j_s, j_r is 60%.

Hence, $h_o = 0.6 \times h_{id}$

where, $h_{id} = j_i C_{ps} \left(\dfrac{\dot{m}_s}{A_s}\right) \left(\dfrac{k_s}{C_{ps}\mu_s}\right)^{2/3} \left(\dfrac{\mu_s}{\mu_{s,w}}\right)^{0.14}$

$$j_i = a_1 Re_s^{a_2} \left(\frac{1.3}{P_t/d_o}\right)^a$$

$$a = \frac{a_3}{1 + 0.14 Re_s^{a_4}}$$

From Table 5.11, $a_1 = 0.37$, $a_2 = -0.395$, $a_3 = 1.187$, $a_4 = 0.37$

$$Re_s = \frac{\dot{m}_s d_o}{A_s \mu_s} = \frac{0.019 \times 50}{0.0724 \times 8.15 \times 10^{-4}} = 16100$$

$$a = \frac{1.187}{1 + 0.14(16100)^{0.37}} = 0.196$$

$$\frac{1.3 d_o}{P_t} = \frac{1.3 \times 0.019}{0.0254} = 0.9724$$

$$\left(\frac{1.3 d_o}{P_t}\right)^a = 0.9724^{0.196} = 0.994 \simeq 1$$

Hence, $j_1 = 0.37 \times Re_s^{-0.395} = 0.37 \times 16100^{-0.395} = 0.0081$

$$h_{id} = 0.0081 \times 4179 \left(\frac{50}{0.0724}\right) \left(\frac{0.612}{4179 \times 8.15 \times 10^{-4}}\right)^{2/3}$$
$$= 7443.9 \ \text{W/m}^2\text{--K}$$

$$h_o = 0.6 \times h_{id} = 0.6 \times 7443.9 = 4466.3 \ \text{W/m}^2 - K$$

(c)
$$Nu_t = \frac{(f/2)Re_b Pr_b}{1.07 + 12.7(f/2)^{1/2}(Pr_b^{2/3} - 1)}$$

$$Re_t = \frac{\rho_t \mu_m d_t}{\mu_t} = \frac{998.2 \times 2 \times 0.016}{10.02 \times 10^{-4}} = 31878.6$$

$$f = (1.58 \times ln(Re_b) - 3.28)^{-2}$$
$$= (1.58 \times ln(31878.6) - 3.28)^{-2} = 0.0058$$

$$f/2 = 0.0058/2 = 0.0029$$

$$Nu_t = \frac{0.0029 \times 31878.6 \times 7.01}{1.07 + 12.7 \times 0.0029^{1/2}(7.01^{2/3} - 1)}$$
$$= 224.16$$

$$h_t = \frac{Nu_t k}{d_i} = \frac{224.16 \times 0.598}{0.016} = 8377.98 \ \text{W/m}^2\text{--K}$$

(d) Overall heat transfer coefficient for a clean surface is

$$\frac{1}{U_c} = \frac{1}{h_o} + \frac{1}{h_i} \frac{d_o}{d_i} + \frac{r_o \ ln(r_o/r_i)}{k}$$

$$= \frac{1}{3835.55} + \frac{1}{8378} \frac{0.019}{0.016} + \frac{9.5 \times 10^{-3} \times ln(19/16)}{42.3}$$

$$U_c = 2445.5 \ \text{W/m}^2\text{--K}$$

(e) Overall heat transfer coefficient of the fouled surface is

$$\frac{1}{U_f} = \frac{1}{U_c} + R_f = \frac{1}{2445.5} + 0.000176$$

$$U_f = 1709.7 \text{ W/m}^2\text{-K}$$

(f)

$$Q = (\dot{m}c_p)_h(T_{h,1} - T_{h,2})$$
$$= 50 \times 4179 \times (32 - 25) = 1462650 \text{ W}$$

$$T_{c,2} = \frac{Q}{(\dot{m}c_p)_c} + T_{c,1}$$
$$= \frac{1462650}{150 \times 4182} + 20 = 22.33°\text{C}$$

$$\text{LMTD} = \frac{(32 - 22.3) - (25 - 20)}{\ln\left(\frac{32-22.3}{25.20}\right)} = 7.09 °\text{C}$$

(g) Heat transfer area for a clean surface is

$$A_c = \frac{Q}{U_c \Delta T_m} = \frac{1462650}{2445.5 \times 7.09} = 84.36 \text{ m}^2$$

(h) Heat transfer area for a fouled surface is

$$A_s = \frac{Q}{U_f \Delta T_m} = \frac{1462650}{1709.7 \times 7.09} = 120.66 \text{ m}^2$$

Example: Consider a compact heat exchanger using matrix surface type $8.0 - 3/8$ T. The matrix is 0.6 m long, and air at 1 atm and 400 K flows across the matrix at a velocity of 24.8 m/s. Calculate the heat transfer coefficient and friction pressure drop for the air side.

Use the following properties of air at 1 atm and 400 K.

$$\rho = 0.8825 \text{ kg/m}^3, \quad \mu = 2.29 \times 10^{-5} \text{ kg/m-s}, \quad c_p = 1013 \text{ J/kg-K}, \quad Pr = 0.719$$

SOLUTION

From Figure 5.35 for this surface,

$$\frac{A_{min}}{A_{fr}} = \sigma = 0.534$$

$$D_h = 0.3633 \text{ cm}$$

$$G = \frac{\rho u_\infty A_{fr}}{A_{min}} = \frac{\rho u_\infty}{\sigma} = \frac{0.8829 \times 24.8}{0.534} == 40.98 \text{ kg/m}^2\text{-s}$$

$$Re = \frac{GD_h}{\mu} = \frac{40.98 \times 0.3633 \times 10^{-2}}{2.29 \times 10^{-5}} = 6501.3$$

From Figure 5.35 for $Re = 6501.3$,

$$\frac{h}{Gc_p} Pr^{2/3} = 0.005$$

$$h = \frac{0.005 \times 40.98 \times 1013}{(0.719)^{2/3}} = 258.6 \text{ W/m}^2\text{–K}$$

For $Re = 6501.3$, $f = 0.02$ (from Figure 5.35)

$$\frac{A_t}{A_{min}} = \frac{4L}{D_h} = \frac{4 \times 0.6}{0.3633 \times 10^{-2}} = 660.6$$

$$\Delta p_f = f \frac{G^2}{2 \rho_a} \frac{A_t}{A_{min}} = 0.02 \times \frac{40.98^2}{2 \times 0.8825} \times 660.6$$

$$= 12570.94 \text{ Pa}$$

PROBLEMS

1. A shell and tube heat exchanger with one-shell pass and multiples of two-tube passes is constructed from 0.0254 m OD tube to cool 693 kg/s of a 95% ethyl alcohol ($C_p = 3810$ J/kg-K)from 66 °C to 42 °C, using 6.3 kg/s of water available at 10 °C ($C_p = 4187$ J/kg-K). In the heat exchanger, 72 tubes will be used. Assume that the overall heat transfer coefficient based on the outer-tube area is 568 W/m²-K. Calculate the surface area and the length of the heat exchanger.

2. A two-pass tube baffled single-pass shell; shell and tube heat exchanger is used as an oil cooler. Cooling water flows through the tubes at 20 °C at a flow rate of 4.082 kg/s. Engine oil enters the shell side at a flow rate of 10 kg/s. The inlet and outlet temperatures of oil are 90 °C and 60 °C, respectively. Determine the surface area of the heat exchanger by the e–NTU method if the overall heat transfer coefficient based on the outside tube area is 262 W/m²-K. The specific heats of water and oil are 4179 J/kg-K and 2118 J/kg-K, respectively.

3. A heat exchanger is to be designed to heat raw water by the use of condensed water at 67 °C and 0.2 bar, which will flow in the shell side with a mass flow rate of 50,000 kg/h. The heat will be transferred to 30,000 kg/h of city water coming from a supply at 17 °C (C_p : 4184 J/kg-K). The fouling resistance equal to 0.000176 m²K/W is assumed. The maximum pressure drop on the shell side is 5.0 psi. The water outlet temperature should not be less than 40 °C. Calculate the heat exchanger length and pressure drops in both streams. The TEMA standard shell and tube heat exchanger with the following geometrical parameters is selected.

Shell internal diameter: 0.39 m
Number of tubes: 124

Outer tube diameter: 19 mm
Inner tube diameter: 16 mm
Thermal conductivity of tube material: 60 W/m^2-K
Pitch size: 0.024 m
Number of tube passes: 2

4. A single-pass shell and tube heat exchanger (condenser) heats 946 m^3/h of water from 10 °C to 38 °C . The heat exchanger uses a plain steel tube (k: 45 W/m-°C) with an internal diameter 0.0266 m and an outside diameter 0.0333 m. Steam is supplied at 107 °C with the steam-side heat transfer coefficient equal to 68143 W/m^2-°C. The water-side heat transfer coefficient is equal to 1937 W/m. The mass velocity through the tube is equal to 8258.3 kg/m^2-s. Calculate the number of tubes required and the length of the tube.

5. Air at 2 atm and 500 K with a velocity of $U = 20$ m/s flows across a compact heat exchanger matrix having surface type I1.32-0737-S-R. The length of the matrix is 0.8 m. Calculate the heat transfer coefficient and the frictional pressure drop. Use the following properties of air at 500 K and 2 atm: $\rho = 1.41$ kg/m^3; $C_p = 1030$ J/kg-K; $\mu = 2.69x10^{-5}$ kg/m-s; $Pr = 0.718$.
   Air at 2 atm and 500 K with a velocity of $U = 20$ m/s flows across a compact heat exchanger matrix having surface type 9.29–0.737–S–R. Calculate the heat transfer coefficient and the frictional pressure drop. The length of the matrix is 0.8 m.

REFERENCES

K. J. Bell. *Heat Transfer Equipment Design*, chapter Delaware method for shell design, pages 145–166. Hemisphere Publishing, 1988a.

K. J. Bell. *Heat Transfer Equipment Design*, chapter Overall design methodology for shell and tube heat exchangers, pages 131–144. Hemisphere Publishing, 1988b.

S. Kakac, H. Liu, and A. Pramuanjaroenkij. *Heat Exchangers: Selection, Rating, and Thermal Design, Third Edition*. CRC Press, 2012. ISBN 978-1-4398-4991-0.

W. M. Kays and K. L. London. *Compact Heat Exchangers*. Mc-Graw Hill, third edition, 1984.

R. K. Shah and D. P. Sekulic. *Handbook of Heat Transfer Applications*, chapter Heat exchangers, pages 17.1–17.169. McGraw-Hill, 1985.

J. Taborek. *Handbook of Heat Exchanger Design*, chapter Shell and tube heat exchangers: single phase flow, pages 3.3.3–1–3.3.11–5. Begell House, 1991.

TEMA. *Standards of the Tubular Exchanger Manufacturers Association*, ninth edition. Tubular Exchanger Manufacturers Association, Inc., 2007.

6 Piping Flow

6.1 INTRODUCTION

A piping network is an integral part of the majority of energy systems. It consists of pumps, turbines, heat exchangers, pipes, valves and other auxiliary devices. There are two aspects of a piping system design: (1) to obtain the performance characteristics of a given piping network and (2) to decide the sizing of the system i.e. the diameter, length, arrangement, material of the pipes etc. Most of the devices present in a piping system are described by the manufacturers' performance curves for specified fluids. The procedure for computer implementation of piping network design is presented in the following sections.

6.2 ENERGY EQUATIONS

The energy equation forms the basis of piping system design, which is a modified Bernouli's equation considering the energy losses, energy consumption and energy output of constituent devices. The general equation based on the conservation of energy is given as

$$\frac{P_1}{\rho g} + \frac{U_1^2}{2g} + Z_1 = \frac{P_2}{\rho g} + \frac{U_2^2}{2g} + Z_2 + \sum_{i=1}^{I} \frac{W_{si}}{g} - \sum_{j=1}^{J} \frac{W_{sj}}{g} + \sum_{k=1}^{K} h_{f_k} + \sum_{l=1}^{L} h_{f_l}$$

Here, W_{si} is the work output per unit mass for turbine, W_{sj} is the work per unit mass for pumps, h_{f_k} is the minor loss due to valves, fittings etc. and h_{f_l} is the major loss due to friction in pipes. The head loss due to friction of fluid in a pipe is expressed as

$$h_f = \frac{f_{D-W} L V^2}{2gD} \quad \text{(Darcy–Weisback)}$$

The friction factor for laminar flow $Re_D < 2300$ of circular pipe is given as

$$f = \frac{64}{Re_D}$$

$$h_f = \frac{4 f_F L V^2}{2gD} \quad \text{(Fanning)}$$

where, f_{D-W} is the Darcy friction factor, f_F is the Fanning friction factor, L is the length of pipe and D is the diameter of pipe. It can be noted that $4 f_F = f_{D-W}$ and f_{D-W} can be obtained from the Moody diagram. However, it is not comfortable to implement Moody diagram in computer. Therefore, different empirical expressions are used. Some sample expressions are presented below.

DOI: 10.1201/9781003049272-6

The implicit algebraic equation by Celebrook is

$$\frac{1}{\sqrt{f_{D-W}}} = ln\left(\frac{\varepsilon}{3.7D} + \frac{2.51}{Re_D\sqrt{f_{D-W}}}\right)^{-2}$$

where ε is absolute wall roughness and Re_D is Reynolds number based on the pipe diameter $(\rho U D/\mu)$. Here, ρ is density, U is velocity and μ is dynamic viscosity. The explicit algebraic equation by Haaland is given as

$$f_{D-W} = \frac{0.3086}{\left\{ln\left[\frac{6.9}{Re_D} + \left(\frac{\varepsilon}{3.7D}\right)^{1.11}\right]\right\}^2}$$

The equation by Churchill–Churchill is valid for laminar, transition and turbulent flow, as given below:

$$f_{D-W} = 8\left[\left(\frac{8}{Re_D}\right)^{12} + \frac{1}{(A+B)^{1.5}}\right]^{1/12}$$

where,

$$A = \left\{2.457\,ln\left[\frac{1}{(7/Re_D)^{0.9} + (0.27\,\varepsilon/D)}\right]\right\}^{16}, \qquad B = \left(\frac{37,530}{Re_D}\right)^{16}$$

6.2.1 MINOR LOSSES

The head loss in valves and fittings is known as minor loss in piping systems, which takes place due to obstructions to flow, changes in flow path, changes in cross section and shape of the flow path etc. One method to characterize the head loss is using equivalent length ratio, L/D, which will cause the same pressure drop as the obstruction under the same flow condition in the pipeline. The equivalent length of pipe is added to the actual length of straight pipe, and the head loss in straight pipe is calculated using

$$h_L = \frac{fLV^2}{2gD}$$

Another way to calculate the head loss due to valve and fittings is using dimensionless coefficient, K. The head loss due to valve and fittings is then given as

$$h_f = \frac{KU^2}{2g}$$

where K is the minor loss coefficient. Minor loss is attributed to the loss due to the presence of bends, tees and other fittings. The values of the head loss coefficient for various types of fittings, valves and bends can be found in references Cengel et al. (2010) and White et al. (2022).

6.2.2 GRAPHICS SYMBOL CONVENTIONS

Different graphics symbol conventions for heating, ventilating and air conditioning purposes are used in water distribution systems, chilled water lines, solar heating, natural gas lines, compressed air lines, heat pump systems, cryogenic applications etc for graphical representation of different components in a piping circuit.

6.2.3 GENERAL CONSIDERATIONS

A simple piping system can be classified into three categories.

Category	Given Conditions	Solution	Comment
I	$Q, L, D, \rho, \varepsilon, \mu, k$	h_f	Direct
II	$h_f, L, D, \rho, \varepsilon, \mu, k$	Q	Iterative
III	$h_f, Q, L, \rho, \varepsilon, \mu, k$	D	Iterative

Among these, category III shows design problems, since Q, h_f, L, ρ, ε, μ, k are given and the diameter of the pipe is solution variable.

6.2.4 RESISTANCE ANALOGY

The flow in a pipe leads to a pressure drop ΔP. Pressure drop ΔP can be related to flow rate Q by resistance analogy given as:

$$\Delta P = R_{hyd} Q$$

where, R_{hyd} is the hydraulic resistance of the pipe. For example, the hydraulic resistance for laminar flow in a pipe based on the Poiseuille flow assumption is given as

$$R_{hyd} = \frac{128\mu L}{\pi D^4}$$

Similarly, the hydraulic resistance value can also be calculated for turbulent flow in the piping circuit. The piping network can be arranged in either series or parallel configuration (Figure 6.1). Using the additivity of pressure drop in a series coupling, the equivalent resistance R can be written as

$$R = R_1 + R_2 + R_3 + R_4$$

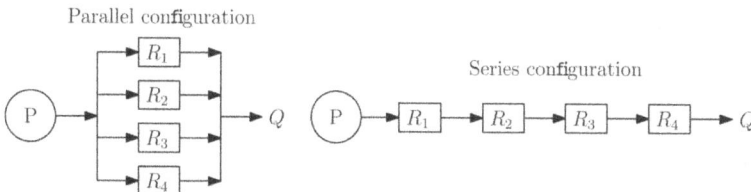

Figure 6.1 A typical series or parallel configuration of piping systems.

Similarly, in a parallel coupling, the law of additivity of inverse hydraulic resistance is applicable i.e.

$$\frac{1}{R} = \left[\frac{1}{R_1} + \frac{1}{R_2} + \frac{1}{R_3} + \frac{1}{R_4} \right]^{-1}$$

These laws are applicable to N i.e. number of pipes in either series or parallel configuration. The procedure for complex piping system design is presented in the following section.

6.2.5 CLASSIFICATION OF PUMPS

The basic system of pump classification is based on the principle by which energy is added to the fluid. On this basis, pumps can be classified into two major categories: (1) dynamic and (2) displacement. In the dynamic category, energy is continuously added to increase the fluid velocities such that the subsequent velocity reduction within the pump produces a pressure increase. The dynamic pump can be further subdivided as a centrifugal pump and other special effect pumps such as a jet pump, electromagnetic pump etc. In a displacement pump, energy is periodically added by the application of force to one or more than one movable boundaries resulting in a direct increase of pressure. They are broadly divided into reciprocating and rotary types depending on the nature of the movement of the pressure-producing members.

6.2.6 PUMP SELECTION

Figure 6.2 shows a pump capacity chart, which can help in preliminary selection by reviewing the wide range of pump casing sizes for a specific impeller speed. This chart helps to narrow down the choice of pumps that can satisfy the system requirements. Figure 6.3 shows a typical pump performance chart for a given model, casing size and impeller rotational speed. The performance curve is a plot of the total head versus flow rate for a specific impeller diameter and rotational speed. The plot starts at zero flow and the corresponding head point is known as the "shut-off head" of the pump. The head decreases starting from this point till a minimum head, which is called the "run-out point" representing the maximum flow of the pump. Beyond this point, the pump cannot operate.

The pump's efficiency varies throughout the operating range of a pump, which is required to calculate the motor power. The best efficiency point, BEP is the highest efficiency point of the pump. The pumping efficiency should be optimized while selecting a pump. Several performance charts at different speeds should be examined such that one model satisfies the efficiency requirement compared to other models. The lowest pump speed should be preferred from multiple available options, as a lower speed will reduce the wear and tear of the rotating parts. Consequently, the pump will have a longer life.

The horsepower curve indicated in the chart gives the power required to operate the pump within a certain range. For example, all points to the left of 1.1 kW curve will be attainable with a 1.5 kW motor (see Figure 6.3). It may be noted that the

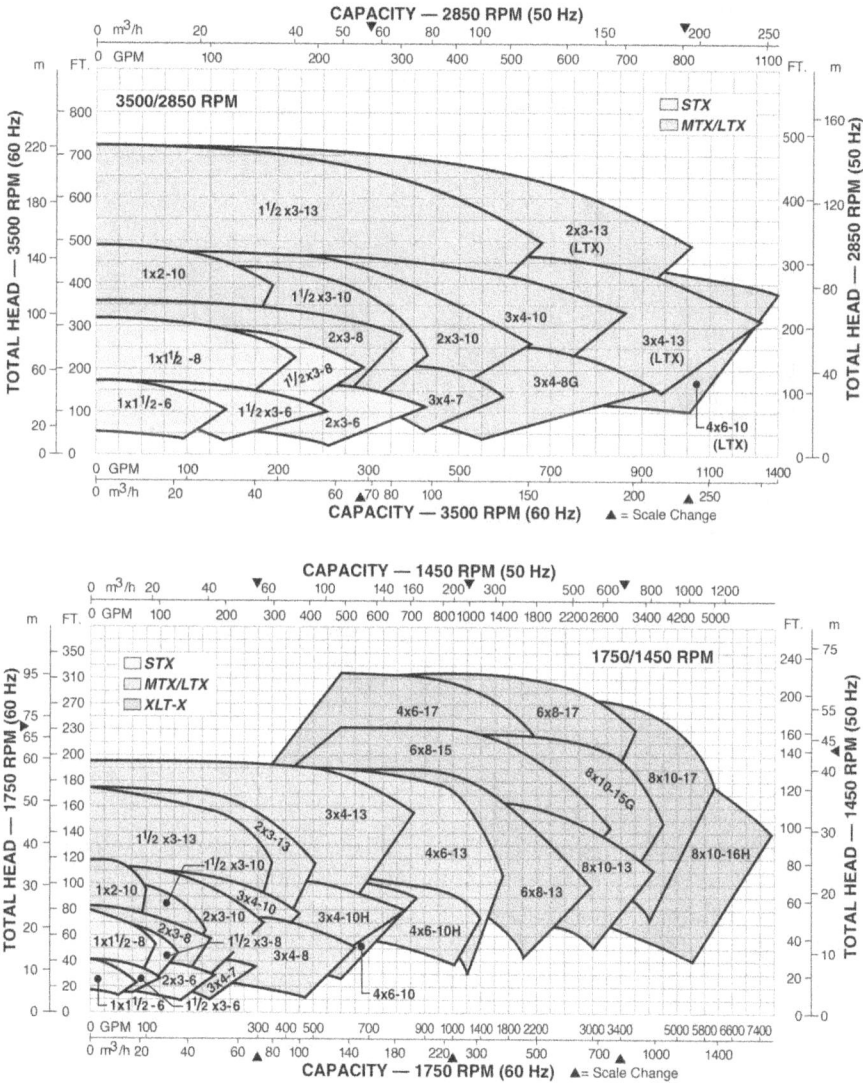

Figure 6.2 A typical pump capacity chart. (Adapted from Goulds Pumps.)

horsepower curve shown in the performance curves are valid for water only. The horsepower can be calculated using the total head, flow and efficiency at the operating point.

The Net Positive Suction Head (NPSH) requirement shown in the pump curve specifies the minimum requirement of suction head for the pump to operate at its design capacity. The NPSH requirement becomes higher as the flow rate increases,

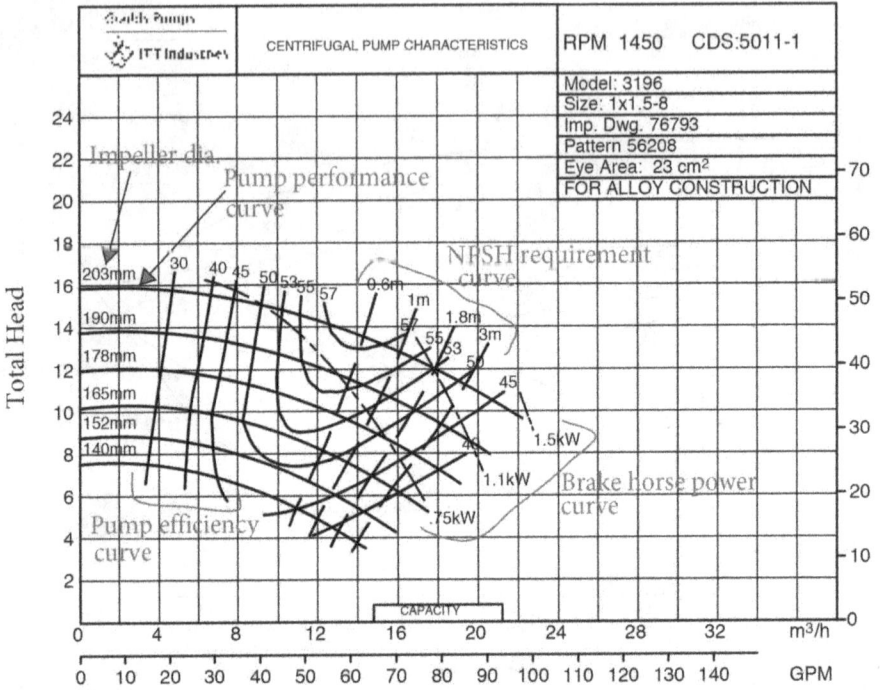

Figure 6.3 Typical pump performance curve. (Adapted from Goulds Pumps, STX $1 \times 1\frac{1}{2} - 8$ AA at 1450 RPM.)

and lower as the flow rate decreases. This essentially means that more pressure head is required at the pump suction for high flows than low flows.

Sometimes the operating point may be located between two curves on the performance chart. In this situation, the impeller size required can be calculated by linear interpolation. Suppose the operating point is located between 165 mm and 178 mm impeller diameter. The correct size of the impeller for this situation can be calculated using

$$D_{op} = 165 + \left(\frac{178 - 165}{\Delta H_{178} - \Delta H_{165}} \right) \left(\Delta H_{op} - \Delta H_{165} \right)$$

where, D_{op} is the required impeller diameter, ΔH_{op} is the pump total head at the operating point, ΔH_{165} is the pump total head at the intersection of 165 mm impeller curve and flow rate and ΔH_{178} is the pump total head at the intersection of 178 mm impeller curve and flow rate.

Whenever possible, it is a good practice to select a pump with an impeller that can be either increased or decreased in size permitting a future requirement of change in head and capacity. As a guide, one should select a pump with an impeller size no greater than between 1/3 and 2/3 of the impeller range for that casing with an

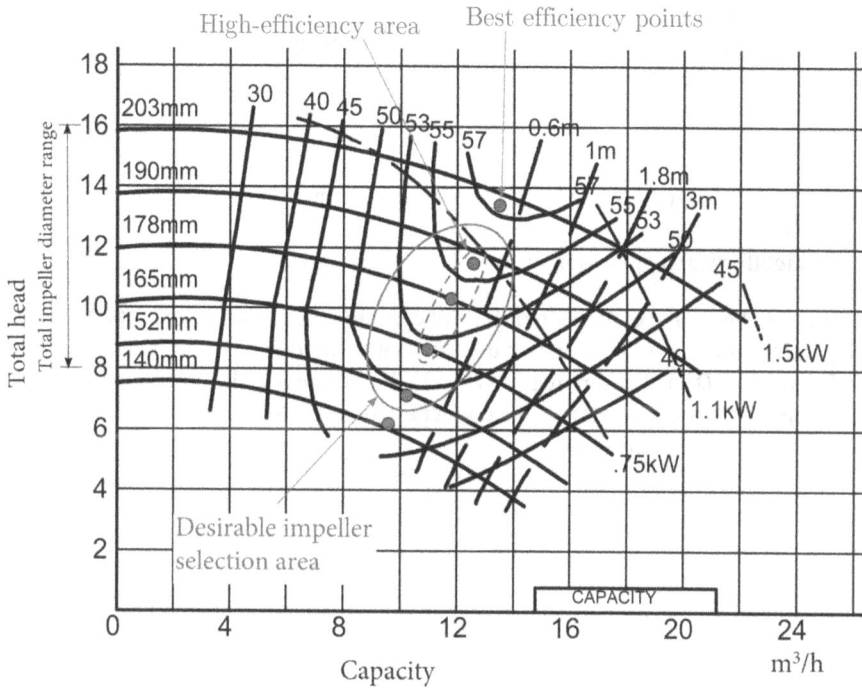

Figure 6.4 Typical pump performance curve. (Adapted from Goulds Pumps, STX $1 \times 1\frac{1}{2} - 8$ AA at 1450 RPM.)

operating point in the high-efficiency area (see Figure 6.4). It is also important not to opt for too far right or left from the BEPs. The general guideline is to locate the operating point between 110% and 80% of the BEP flow rate with an operating point in the desirable impeller selection area.

6.3 PUMP PERFORMANCE USING DIMENSIONAL ANALYSIS

Generally, the pump performance characteristics are available for water at some specific speed and impeller diameter. However, *water is not always the working fluid* and *the operational speed* is different than that given in the pump curve for many applications. Three approaches allow for obtaining information from the manufacturer performance curve at different operating conditions: (1) dimensional analysis, (2) *correction factors* for very viscous fluids and (3) curve fitting.

6.3.1 DIMENSIONAL ANALYSIS

Based on the dimensional analysis, a turbomachine or pump can be described using the following important non-dimensional parameters:

$$\Pi_1 = \frac{Q}{ND^3}, \; \Pi_2 = \frac{H}{N^2D^2}, \; \Pi_3 = \frac{P}{\rho N^3 D^5}, \; \Pi_4 = \frac{\mu}{\rho ND^2}, \; \Pi_5 = \eta$$

Here, Q is the flow rate, N is the rotational speed of impeller D is the diameter of impeller,, H is the developed head, η is the efficiency, μ is the viscosity, ρ is the density of the working fluid and P is the power. The following points should be kept in mind while using the non-dimensional numbers for predicting the pump performance.

1. Π_5 is redundant as Π_1, Π_2 and Π_3 can be combined to determine Π_5 i.e. $\Pi_5 = \frac{\Pi_1 \Pi_2}{\Pi_3}$.
2. Straightforward application of similarity concept should be avoided.

Example: Using the expression of Π_1, we expect Q (flow rate) to be proportional to D^3, and from Π_3, we expect power to be proportional to D^5. However, it is not true. In Π_1 and Π_3, D^3 and D^5 are really AD and AD^3, where A is the cross-sectional area at pump inlet and outlet. Hence, the correct relationship when the *impeller diameter changes in the same casing* is

$$\frac{Q_2}{Q_1} = \frac{D_2}{D_1}$$

Similarly, the change in head can be related as

$$\frac{H_2}{H_1} = \left(\frac{D_2}{D_1}\right)^2$$

Hence, the change in power with respect to the change in impeller diameter can be expressed as

$$\frac{P_2}{P_1} = \left(\frac{D_2}{D_1}\right)^3$$

Therefore, the blind use of non-dimensional numbers relating power to D is most correct.

If only speed changes, for a particular pump, we have

$$\frac{Q_2}{Q_1} = \frac{N_2}{N_1}, \frac{H_2}{H_1} = \left(\frac{N_2}{N_1}\right)^2, \; and \; \frac{P_2}{P_1} = \left(\frac{N_2}{N_1}\right)^3$$

It may be noted that the above dimensionless rule assumes that the two operating points that are being compared are at the same efficiency.

6.3.2 SPECIFIC SPEED

Specific speed is another important non-dimensional number, which is defined as the speed of an ideal pump geometrically similar to the actual pump, which when running at this speed will raise a unit of volume in a unit of time through a unit of the head. It can be determined from the above non-dimensional numbers as follows:

$$\Pi_1 = \frac{Q}{ND^3} \quad \Rightarrow \quad D = \left(\frac{Q}{\Pi_1 N}\right)^{1/3}$$

$$\Pi_2 = \frac{H}{N^2 D^2} \qquad \Rightarrow \qquad \Pi_2 = \frac{H}{N^2}\left(\frac{\Pi_1 N}{Q}\right)^{2/3}$$

$$\Rightarrow \qquad \Pi_5 = \frac{\Pi_2}{(\Pi_1)^{2/3}} \equiv \frac{H(N)^{2/3}}{N^2 Q^{2/3}} \Rightarrow \Pi_5 = \frac{H}{N^{4/3} Q^{2/3}}$$

$$\Rightarrow \qquad N^{4/3} = \frac{H}{\Pi_5 Q^{2/3}} \Rightarrow N = \frac{H^{3/4}}{\Pi_5 Q^{1/2}}$$

Note that we have written $\Pi_5^{3/4}$ as Π_5 in this equation as it is a non-dimensional number and its power is not going to influence the general relationship. We can rewrite the new non-dimensional number, Π_5 as

$$\Rightarrow \qquad \Pi_5' = \frac{N\sqrt{Q}}{H^{3/4}}$$

Here, Π_5' is known as the specific speed.

6.4 PUMP CURVE FOR VISCOUS FLUID

In general, the similarity principle is applied for relatively low-viscosity fluid. Hence, Π_4 *is relatively small* and the remaining Π groups are *nearly invariant* with respect to Π_4. If the viscosity is large, the similarity principle cannot be applied to extend the manufacturer's data.

The performance characteristics of a large viscous fluid for a pump can be known from that of water by applying suitable correction as below (Gulich (1999a, 1999b).

$$Q_{vis} = C_Q Q_W, \qquad H_{vis} = C_H H_W, \qquad n_{vis} = C_n n_W$$

Here, subscript W indicates water, and subscript *vis* indicates viscous fluid. The co-efficients, C_Q, C_H and C_n are correction factors. The correction factors are published in the form of curves by the Hydraulic Institute (www.pumps.org). The correction factor charts are also available on www.fluidedesign.com. Fluids that are viscous can significantly affect the performance of centrifugal pumps. Hence, these correction factors can be *used as approximations* only. The exact pump performance for a viscous fluid should be obtained by actual tests on the pump with the viscous fluid.

6.4.1 PROCEDURE TO OBTAIN THE CORRECTION FACTOR AND PUMP CURVE FOR VISCOUS FLUID

The following procedure can be followed to obtain the correction factor for the pump curve of viscous fluid from that of water.

1. Locate the point of maximum efficiency on the $H - Q$ curve for water (see Figure 6.5). If this is Q_{MW}, then obtain $0.6 \times Q_{MW}$, $0.8 \times Q_{MW}$ and $1.2\, Q_{MW}$.

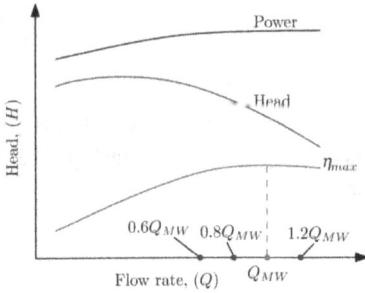

Figure 6.5 A representative pump performance curve for water used to extend for very viscous fluid.

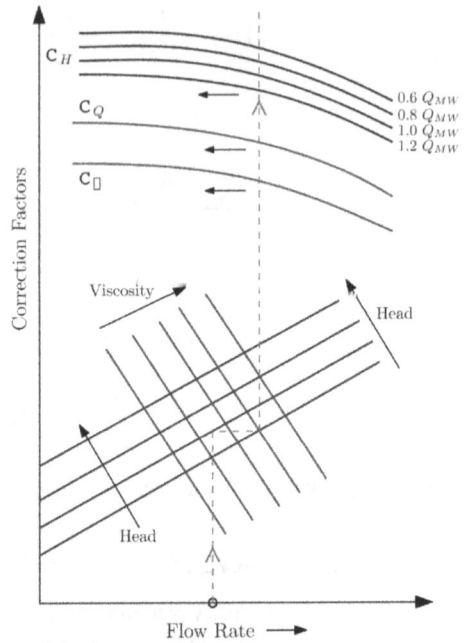

Figure 6.6 A typical correction factor curve for obtaining the pump curve of viscous fluid from that of water.

2. Go to the correction factor curve (Figure 6.6) at capacity corresponding to n_{max} and go forward to the desired head developed and then horizontal to the desired viscosity. From that point, proceed upward to obtain C_H, C_Q and C_n.
3. Read C_Q, C_H and C_n for all four capacities. Note that C_Q and C_n are determined once at maximum efficiency point only.
4. Multiply H_W by C_H to get H_{vis}.
5. Multiply η_W by C_n to get n_{vis}.
6. Multiply Q_W by C_Q to get Q_{vis}.
7. The power required can be calculated using Q_{vis}, H_{vis} and n_{vis}.
8. Plot the corrected values and draw a smooth curve through them.

6.5 EFFECTIVE PUMP PERFORMANCE CURVE

Pumps are operated either in series or parallel for *added head* or *flow rate* requirements, respectively. *Check valves* are usually provided on the suction (inlet) side of the pump to prevent back flow and *shut-off valves* are provided in the discharge (outlet) lines for complete isolation when needed. The losses due to these valves are usually neglected for pump selection.

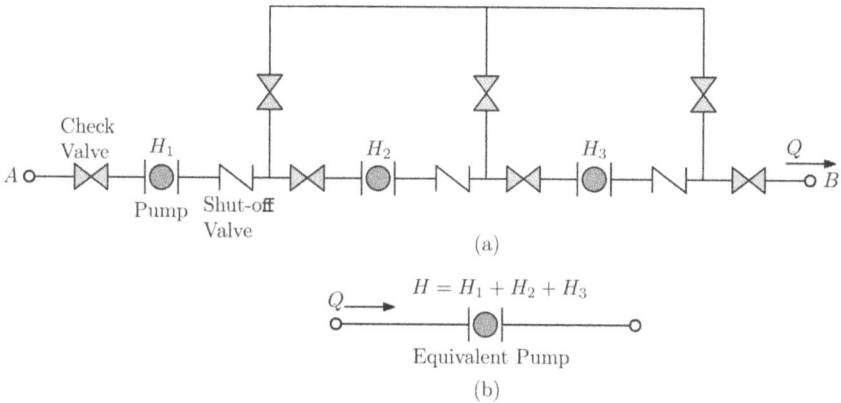

(a)

$$H = H_1 + H_2 + H_3$$

Equivalent Pump

(b)

Figure 6.7 A typical pump system arranged in series (a) and its equivalent pump representation (b).

Analysis of pumps in series or parallel arrangement is facilitated by an *effective pump performance curve*. This effective pump performance curve provides a single head-capacity relationship equivalent to that of all pumps in the network. Figures 6.7 and 6.8 show typical equivalent pump representation in series and parallel arrangement, respectively. Figure 6.9 shows the procedure for the generation of an effective pump performance curve when the pumps are arranged either in series or parallel. For pumps in series, heads are added at a constant flow rate. For pumps in parallel, flow rates are added at a constant head. For a dissimilar pump in parallel, the pump with the lower *shut-off head* cannot be brought into operation until the head of the larger pump is decreased below this lower shut-off head. Shut-off head is the maximum head generated by a pump with zero flow. Otherwise, the more powerful pump will block the output of the lower shut-off head pump.

Figure 6.9 shows the equivalent pump curve for similar pumps (A) and dissimilar pumps (A and B) in both series and parallel arrangements. For series arrangement, H for each pump is added at a fixed value of Q. For parallel arrangement, Q is added for a fixed value of H. For dissimilar pumps in parallel, the head of the lower shut-off head is only added for a flow rate greater than its shut-off head.

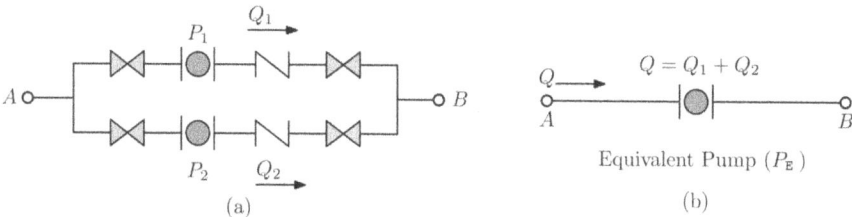

(a)

$$Q = Q_1 + Q_2$$

Equivalent Pump (P_E)

(b)

Figure 6.8 A typical pumping system arranged in parallel (a) with its equivalent pump representation (b).

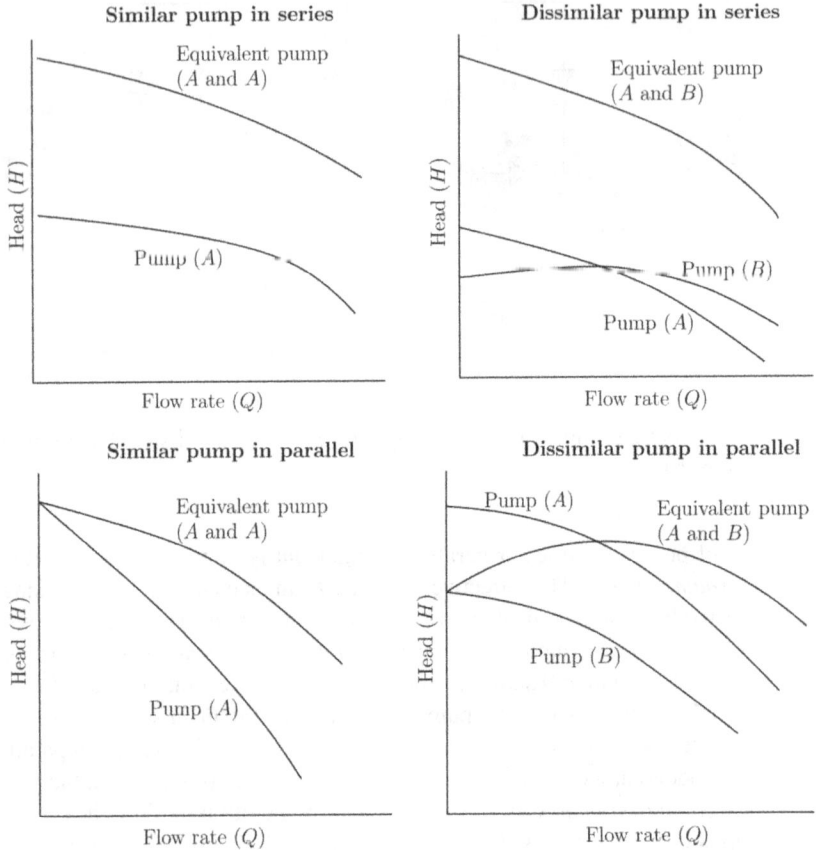

Figure 6.9 Construction of an effective pump performance curve for pumps in series and parallel.

6.5.1 COMPUTER IMPLEMENTATION

The pump performance curve in suitable form can also be used for design purpose using a computer. The procedure for computer implementation of pump performance curve is outlined below.

6.5.1.1 Pumps in Series

For pumps in series, a pump curve for each pump can be obtained by curve fitting as

$$H_i = A_{1i} + A_{2i}Q + A_{3i}Q^2 \qquad \text{(Pump curve)}$$

Subsequently, for the equivalent pump curve, the head developed by each pump is added as follows:

$$H = H_1 + H_2 + H_3 + \cdots$$

Here, H_1, H_2 and H_3 are the pump curves for pumps 1, 2 and 3, respectively. Thus, we get

$$H = (A_{11} + A_{12} + \cdots) + (A_{21} + A_{22} + \cdots)Q + (A_{31} + A_{32} + \cdots)Q^2$$

In a generalized form, for pumps in series, we can write

$$H = \sum_{\substack{i=0,1,\ldots,M \\ j=0,1,\ldots,N}} A_{ij} Q^i$$

Here, i indicates the order and j indicates the pump number for coefficient A_{ij}.

6.5.1.2 Pumps in Parallel

For pumps in parallel, the flow rate of each pump is expressed as a function of the head developed.

$$Q_i = B_{1i} + B_{2i}H + B_{3i}H^2$$

The flow rate developed by each pump is added for pumps in parallel.

$$Q = Q_1 + Q_2 + \cdots$$

Here, Q_1, Q_2,..., corresponds to flow rate of pump 1 and pump 2, respectively.

$$Q = (B_{11} + B_{12} + \cdots) + (B_{12} + B_{22} + \cdots)H + (B_{31} + B_{32} + \cdots)H^2$$

In a generalized form, we can write the equivalent pump performance as

$$Q = \sum_{\substack{i=0,1,\ldots,M \\ j=0,1,\ldots,N}} B_{ij} H^i$$

Here, B_{ij} is the coefficients of the equivalent pump curve of pumps in parallel, with i indicating the order and j indicating the pump number.

6.6   SYSTEM CHARACTERISTICS

For a given piping system, the head versus flow rate curve are known as system characteristics. This curve can be determined from the *basic energy equation*. The system characteristics have two uses: (1) pump selection or specification for a desired system flow rate and (2) determination of the operating point for a given pump-system combination.

The first objective is accomplished with a pump curve by selecting the appropriate pump for a specified flow rate. The second objective i.e. the determination of the operating point is illustrated in Figure 6.10. The intersection of the system characteristic and pump curve indicates the operating point. Figure 6.10 shows the H–Q for two piping systems (A and B). The system curves A and B intersect the pump curve at point C and D, receptively. Here, points C and D are the operating points for systems A and B, receptively.

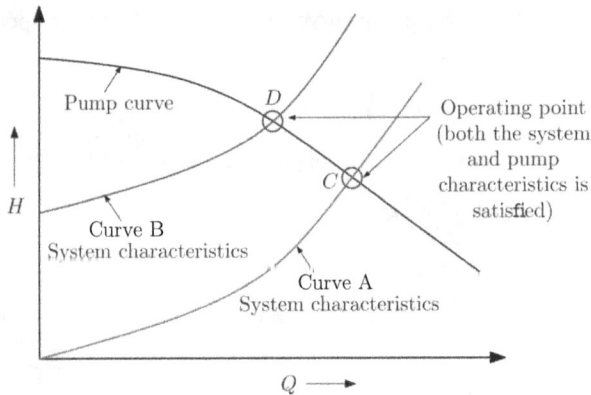

Figure 6.10 Operating point for two arrangements i.e. system characteristic. System characteristic A corresponds to a pipe network between two reservoirs at the same surface elevation. System characteristic B corresponds to pipe networks between two reservoirs at different surface elevations.

6.7   PUMP PLACEMENT

Pump placement is an important factor in the design of piping systems. The issue of cavitation and recirculation needs to be considered for the proper placement of pumps.

6.7.1   CAVITATION

When the static pressure of the fluid is reduced below its vapor pressure, pockets of vapor form and cavitation is said to take place. The growth and collapse of these pockets can cause pressure fluctuation leading to mechanical damage of the pump's components. Therefore, the pump placement in a piping circuit should be appropriately decided to avoid cavitation.

6.7.2   NET POSITIVE SUCTION HEAD

It is the difference between the total pressure at the pump suction (inlet) and the vapor pressure of the fluid $(P_s - P_v)$. The required NPSH $(NPSH_R)$ is determined on the basis of the head across the pump for a given speed and flow rate. A *decrease of 3%* in the head across the pump as the suction pressure is decreased is considered the evidence of cavitation. It is assumed that mechanical efficiency, head increase and power required are essentially constant under the variable suction condition as long as the cavitation is not present. The cavitation problem can be avoided by locating the pump where the available NPSH $(NPSH_A)$ is greater than the required $NPSH_R$ specified by the pump performance curve.

   A typical pump arrangement is shown in Figure 6.11. The suction pressure at the inlet can be expressed as $P_s = P_{atm} + \rho Hg - P_f$, where P_f is the pressure loss due to

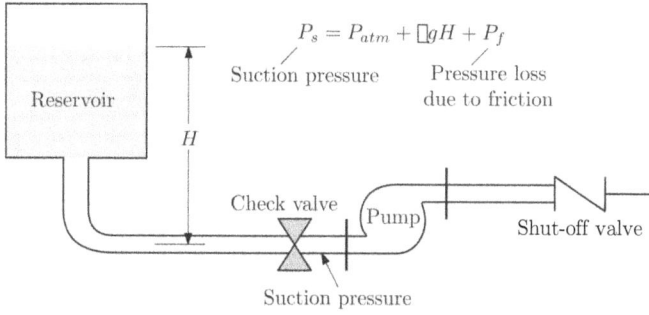

Figure 6.11 A typical pump arrangement and calculation of $NPSH_A$.

friction. The available suction pressure is given as

$$NPSP_A = P_{atm} + \rho Hg - P_f - P_v$$

where, P_v is the vapor pressure of the working fluid. Thus, the net positive suction head is expressed as

$$NPSH_A = \frac{P_{atm}}{\rho g} + H - \frac{P_f}{\rho g} - \frac{P_v}{\rho g}$$

It may be noted that the lower the pump placement, the higher is the $NPSH_A$. Therefore, it is preferred to place the pump at the lowest elevation to meet the cavitation requirement.

6.7.3 RECIRCULATION PROBLEM

Recirculation within the pump inlet and at rotor discharge may be observed at flow rates below the *BEP*. At the *BEP*, velocities are high enough to preclude or minimize recirculation. However, as the flow rate and hence the fluid velocities decrease recirculating regions develop (see Figure 6.12). Suction-specific speed is calculated to predict the recirculation.

6.8 SUCTION-SPECIFIC SPEED

Suction specific speed (N_{ss}) is a parameter that identifies the recirculation characteristic. It is constant for a given pump and function of suction inlet design.

$$N_{ss} = \frac{N\sqrt{Q}}{(NPSH_R)^{3/4}}(n_{max})$$

The recirculation problem is minimized by selecting a pump with N_{ss} less than 9000 to 11000. This value is dependent on the fluid type (9000 for water and 11000 for hydrocarbon). The recirculation problem is more significant for *large pumps* than small pumps. If a pump operates near *BEP*, recirculation may not be a problem.

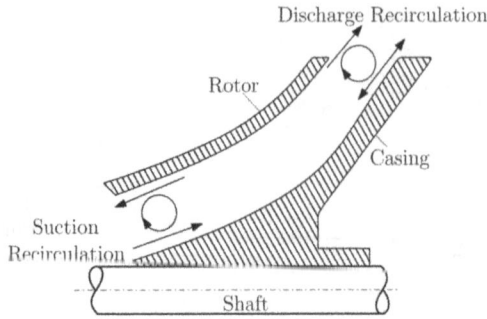

Figure 6.12 Schematic showing suction and discharge recirculation at low flow rate.

It may be noted that $(NPSH)_R$ should be low to avoid cavitation. However, low $(NPSH)_R$ leads to high N_{ss} and may lead to recirculation problem Karassik et al. (2001). Hence, the requirement of *NPSH* have a conflicting requirement to avoid cavitation and recirculation problem.

6.9   NET POSITIVE SUCTION HEAD AVAILABLE

The Net Positive Suction Head Available (NPSH$_A$) is the total energy per unit weight or head at the suction flange of the pump less the vapor pressure of the fluid. It is required to avoid cavitation and can be determined from the pressure head at the suction flange of the pump. The pressure at which a liquid boils is called vapor pressure, which is associated with a specific temperature. The boiling temperature drops with the decrease in environment pressure. The pressure near the impeller eye is lower than the pressure at the pump suction flange. The point of lowest pressure is near the eye of the impeller on the underside of the vane (see Figure 6.25), where bubbles can form. These bubbles get rapidly compressed while traveling from the start of the impeller vane to its tip. The rapid compression of bubbles can cause small pieces of metal to be dislodged from the surface. The collapse of the bubbles near the tip of the vane also causes noise and vibration. The decrease in pressure from points A to D is attributed to friction loss, turbulence and entrance loss due to the right angle turn of horizontal flow to outward radial flow in the impeller. The usual recommendation is to ensure that NPSH$_A$ is equal to 15% or more above NPSH$_R$.

6.10   UNCERTAINTY EFFECT ON PUMP SELECTION

Thermal system design specifications have uncertainty due to either use of a standard design database (pump curves) containing its own uncertainty or the use of dimensional analysis. Therefore, there is a need to determine the uncertainty limits of the system/component parameters due to the uncertainty of the design database. This is important for the proper operation of the pumping systems.

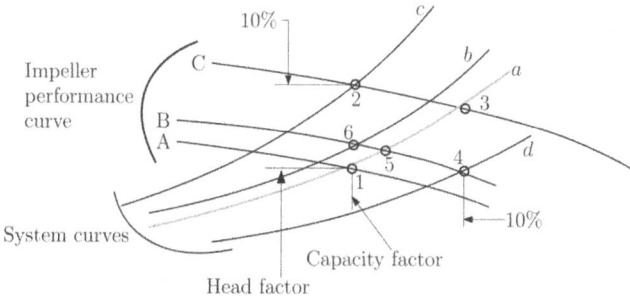

Figure 6.13 The procedure for specification of excessive pump capacity to provide factor of safety.

However, the uncertainty qualification in thermal system design is not well developed. Therefore, thermal system design procedure is generally based on adhoc *safety factor*. This approach leads to *overspecification* and adverse consequences. The procedure for overspecification of a pumping system is demonstrated in Figure 6.13.

Before finalizing the impeller size, one may have to apply an extra capacity/flow factor depending on the requirement. Figure 6.13 demonstrates the systematic procedure for providing the safety factor. Let point 1 on the impeller curve A is the operating point before applying the capacity factor. For safety margin on the total head, if we apply a head factor of 10%, we will have to select impeller C. The new operating point for this curve will be point 2. This means that the new system curve will be c instead of a. If the pressure drop calculation is correct and we are operating with curve a, the operating point will shift from point 2 to point 3 on impeller curve C. If we need to get back to the flow corresponding to point 1 for process reason, then throttling a valve at the pump discharge will be required. This will change the operating curve to match curve c and bring the operating point back to point 2.

We have to select impeller curve, B if we apply a capacity factor for safety margin on flow rate say 10%. The system curve will be d in this situation. If the original flow estimate is correct, the operating point will shift from point 4 to 5. We have to throttle back so that we shift to point 6 to get back the original flow. Now we will operate on a new system curve b.

The overspecification of the pump has the following drawbacks:

1. It consumes an excess amount of energy.
2. Operating point will be lower than the best efficiency point.
3. Flow control is difficult due to the excessive nonlinearity of the valve in the near closed position. It may also lead to flow-induced vibration. To alleviate this problem, rotor diameter is reduced in industry, which can lead to maintenance problem.

It may be noted that the pump overspecification is in response to uncertainties in the pump selection process.

6.11   UNCERTAINTY ANALYSIS PROCEDURE

Experimentalists use the procedure for calculation of uncertainty of derived quantities from the experiment. The above procedure can be modified for the calculation of uncertainty in piping system design parameters. Let's assume that the following data reduction equation is used to report the experimental results.

$$R = f(X_1, X_2, \ldots, X_j)$$

where, R represents experimental results and X_j are measured quantities.

The uncertainties of experimented results are calculated using the following expression:

$$U_R = \left[\left(\frac{\partial R}{\partial X_1} U_{X_1} \right)^2 + \left(\frac{\partial R}{\partial X_2} U_{X_2} \right)^2 + \cdots + \left(\frac{\partial R}{\partial X_j} U_{X_j} \right)^2 \right]^{1/2}$$

Here, U_R is the uncertainty in result, $\frac{\partial R}{\partial X_j}$ is the sensitivity of the results to the change in measured variable and U_{Xj} is the error or uncertainty of the measured variable. We can relate the uncertainty calculation for experiments to design of the piping system. The above uncertainty equation can also be rewritten as

$$\left(\frac{U_R}{R} \right)^2 = \left(\frac{X_1}{R} \frac{\partial R}{\partial X_1} \right)^2 \left(\frac{U_{X_1}}{X_1} \right)^2 + \left(\frac{X_2}{R} \frac{\partial R}{\partial X_2} \right)^2 \left(\frac{U_{X_2}}{X_2} \right)^2 + \cdots + \left(\frac{X_J}{R} \frac{\partial R}{\partial X_J} \right)^2 \left(\frac{U_{X_j}}{X_j} \right)^2$$

Here, $\left(\frac{X_j}{R} \frac{\partial R}{\partial X_j} \right)$ is the normalized sensitivity coefficient. This parameter identifies the input parameter to which the computed parameter is most sensitive. We can identify which design inputs need to be known *with the most fidelity* from the calculation of this parameter. Uncertainty contribution of each design inputs is also given by

$$\text{Relative contribution:} \quad \left(U_{X_i} \frac{\partial R}{\partial X_i} \right)^2 \Big/ U_R^2$$

6.11.1   PIPING NETWORK DESIGN

The previous uncertainty equation can be written for the calculation of uncertainty in piping network design as

$$U_{Q_j}^2 = \sum_{i=1}^{Pipes} \left\{ \left(\frac{\partial Q_j}{\partial D_i} \right)^2 U_{D_i}^2 + \left(\frac{\partial Q_j}{\partial L_i} \right)^2 U_{L_i}^2 + \left(\frac{\partial Q_j}{\partial K_i} \right)^2 U_{K_i}^2 + \left(\frac{\partial Q_j}{\partial \dot{m}_i} \right)^2 U_{\dot{m}_i}^2 \right.$$

$$\left. + \left(\frac{\partial Q_j}{\partial f_i} \right)^2 U_{f_i}^2 + \left(\frac{\partial Q_j}{\partial \varepsilon_i} \right)^2 U_{\varepsilon_i}^2 + \left(\frac{\partial Q_j}{\partial \mu} \right)^2 U_\mu^2 + \left(\frac{\partial Q_j}{\partial \rho} \right)^2 U_\rho^2 \right\}^{1/2}$$

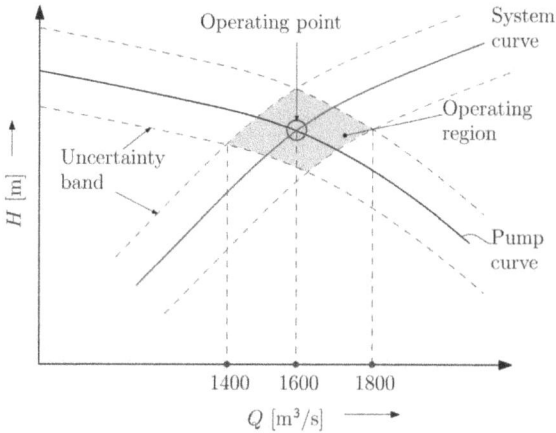

Figure 6.14 The uncertainty band of a pump curve and system curve.

Here, U_{Di}, U_{Li}, U_{Ki}, $U_{\dot{m}_i}$, U_{fi}, $U_{\varepsilon i}$, U_μ and U_ρ correspond to the error or uncertainty in diameter, length, minor loss coefficient, flow rate, friction factor, roughness value, viscosity and density, respectively. Here, Q_j is the flow rate in each line j of the piping network. Uncertainties in the specifications of each pipeline influence the flow rate uncertainty in other pipelines.

Figure 6.14 shows the uncertainty band of a typical pump curve and system characteristic curve. In Figure 6.14, the nominal operating flow rate is 1600 m³/s. If all the uncertainties combine to degrade the system performance, the flow rate will be 1400 m³/s. If all the uncertainties combine to enhance the system performance, the flow rate is 1800 m³/s. If a pump is required to guarantee 1400 m³/s flow rate, its entire operating region must be to the right of 1400 m³/s. This way the excess capacity of pump selection can be tolerated. The sensitivity coefficients can be determined by finite-difference schemes if the analytical partial derivative calculation is cumbersome.

Figure 6.15 shows typical normalized sensitivity coefficients and relative contributions for a piping circuit problem. The largest normalized sensitivity coefficient is associated with the diameter. This may be related to the power law relation of the flow rate with respect to the pipe diameter. The next contribution is due to the viscosity, which may be related to a large error in viscosity due to temperature fluctuation etc.

6.12 PIPING SYSTEM DESIGN

Most energy systems are composed of piping, pumps, compressor, turbine (primary movers) and heat exchangers. In this section, we discuss the systematic approach for the design of complex series-parallel piping networks.

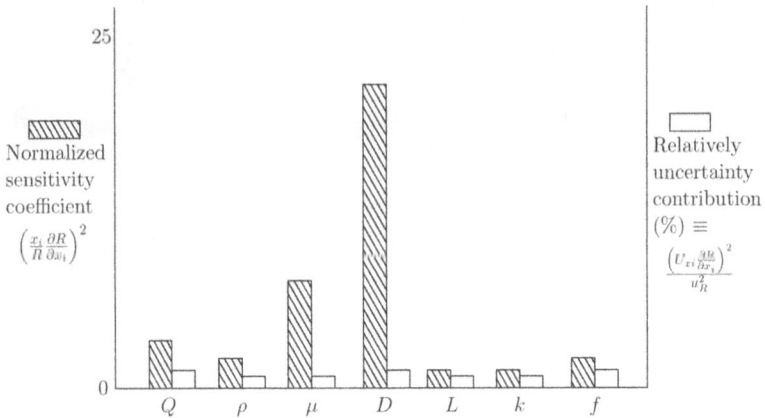

Figure 6.15 A histogram showing the uncertainty estimate of a typical piping system.

6.12.1 HARDY CROSS METHOD

The Hardy Cross method can readily be adapted to computers, and number of software firms have generalized Hardy Cross programs for sale. The basis for the Hardy Cross analysis technique is: (1) conservation of mass at *a node* and (2) uniqueness of pressure at a given point in the *loop*.

A *loop* is defined as a series of pipes forming a closed path. A *node* is defined as a point where two or more lines are joined (see Figure 6.16).

(1) Conservation of mass at a node says that the node cannot accumulate mass. For node α, based on mass conservation, we have

$$\sum_{\beta=1} Q_{\alpha\beta} = 0$$

where, β is the no of lines connecting the node, α.

(2) The pressure at a node must be single valued i.e. the sum of the pressure drops around a loop must be zero. For the ith loop, we can write

$$\sum_{j=1} h_{f_{ij}} = 0$$

Here, j is the number of lines in the loop.

6.12.2 HAZEN–WILLIAMS COEFFICIENT

Since flow rate rather than velocity is usually of primary interest, the head loss expression can be written as

$$h_f = \frac{fLV^2}{2gD} \quad \text{(Darcy–Weisback)}$$

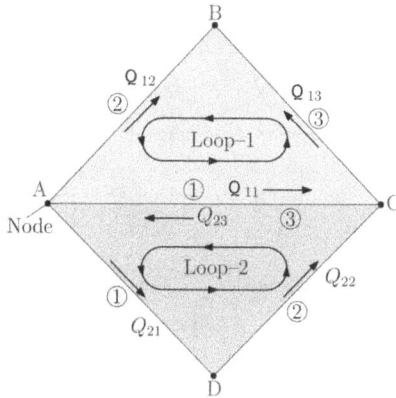

Figure 6.16 Representation of a piping system.

$$h_f = \frac{fL}{2gD}\frac{Q^2}{A^2} = \frac{16fLQ^2}{2\Pi^2 gD^5} \quad \text{(For circular pipes)}$$

$$= K_1\frac{fLQ^2}{D^5} \quad (\text{where } K_1 = 8/\Pi^2 g)$$

Hazen and Williams suggested that for a general fluid the head loss can be written as

$$h_f = KQ^n = \frac{k_1 L}{C^{1.852}D^{4.8704}}Q^{1.852}$$

where, C is a dimensionless number that is indicative of the roughness of the pipe. Hazen and Williams coefficient k_1 is a constant depending on the dimensions of Q i.e. if $Q =(\text{ft}^3/\text{s})$, $k_1 = 4.727$ and if $Q =(\text{m}^3/\text{s})$, $k_1 = 10.466$. Different values of C are given as follows:

Extremely smooth and straight pipes: $C = 140$
Cast iron pipes with some years of service: $C = 140$
Cast iron pipes in bad condition: $C = 80$

6.12.3   BASIC IDEA

First, the conservation of mass at each node is established without consideration of uniqueness of pressure. Then, the uniqueness of pressure is used to calculate the correction factor for each loop.

6.12.4   CORRECTION FACTOR

Figure 6.16 shows a typical representation of a piping system. The piping network is divided into two loops, and each pipe in each loop is assigned a number. The flow rate in each pipe is then identified as $Q_{(ij)}$, where i is the loop number and j is the pipe number.

Note: The pipes that are common between the two loops have flow rates which are related as for example

$$Q_{11} = -Q_{23}$$

i.e. when the flow is in the *counter clockwise* direction around a loop, that flow toward the node is positive.

We know from the consistency that a negative Q_{ij} yields a negative $h_{f_{ij}}$ and similarly, from sign convection, Q_{ij} can be either negative or positive. Therefore, we can write

$$h_f = \begin{cases} KQ^n & Q \geq 0 \\ -K(-Q)^n & Q < 0 \end{cases}$$

Taking a derivative of this expression, we can write

$$\frac{dh_f}{dQ} = \begin{cases} nKQ^{n-1} & Q \geq 0 \\ nK(-Q)^{n-1} & Q < 0 \end{cases}$$

Using the Taylor series to expand the head loss about Q, we have

$$h_f(Q + \Delta Q) = h_f(Q) + \frac{dh_f}{dQ}\Delta Q + \frac{d^2 h_f}{dQ^2}\frac{\Delta Q^2}{2} + \cdots$$

Neglecting higher-order terms,

$$h_f(Q + \Delta Q) = \begin{cases} K[Q^n + nQ^{n-1}\Delta Q] & Q \geq 0 \\ -K[(-Q)^n - n(-Q)^{n-1}\Delta Q] & Q < 0 \end{cases}$$

With reference to Figure 6.16 in loop 1, $Q_{11}^0 > 0$, $Q_{13}^0 > 0$ and $Q_{12}^0 < 0$. Here, the superscript "0" denotes the first guess. Using the initial guesses Q_{ij}^0 for loop 1, we have

$$h_{f_{11}}^0 + h_{f_{13}}^0 - h_{f_{12}}^0 \neq 0$$

But, it is desired that

$$\pm h_{f_{11}}^1 \pm h_{f_{13}}^1 \pm h_{f_{12}}^1 = 0$$

Here, "1" stands for the next iteration. Substituting the expression of $h_f(Q + \Delta Q)$, we have

$$K_{11}\left[(Q_{11}^0)^n + n(Q_{11}^0)^{n-1}\Delta Q_1\right] + K_{13}\left[(Q_{13}^0)^n + n(Q_{13}^0)^{n-1}\Delta Q_1\right]$$

$$-K_{12}\left[(-Q_{12}^0)^n - n(-Q_{12}^0)^{n-1}\Delta Q_1\right] = 0$$

Solving for ΔQ_1, we obtain

$$\Delta Q_1 = -\frac{K_{11}(Q_{11}^0)^n + K_{13}(Q_{13}^0)^n - K_{12}(-Q_{12}^0)^n}{K_{11}n(Q_{11}^0)^{n-1} + K_{13}n(Q_{13}^0)^{n-1} + K_{12}n(-Q_{12}^0)^{n-1}}$$

or using absolute values,

$$\Delta Q_1 = -\frac{K_{11}Q_{11}^0|Q_{11}^0|^{n-1} + K_{12}Q_{12}^0|Q_{12}^0|^{n-1} + K_{13}Q_{13}^0|Q_{13}^0|^{n-1}}{n(K_{11}|Q_{11}^0|^{n-1} + K_{12}|Q_{12}^0|^{n-1} + K_{13}|Q_{13}^0|^{n-1})}$$

Note: Use of the absolute value signs allows the correct sense of the sign of $h_{f_{ij}}$ to be maintained.

Using the summation notation for loop 1, we can write

$$\Delta Q_1 = \frac{-\sum_{j=1}^{3} K_{1j}Q_{1j}^0|Q_{1j}^0|^{n-1}}{n\sum_{j=1}^{3} K_{1j}|Q_{1j}^0|^{n-1}}$$

For any loop i with J lines, we can write the expression for correction as below:

$$\Delta Q_i = -\frac{\sum_{j=1}^{J} K_{ij}Q_{ij}^0|Q_{ij}^0|^{n-1}}{n\sum_{j=1}^{J} K_{ij}|Q_{ij}^0|^{n-1}}$$

6.12.5 IMPLEMENTATION PROCEDURE

The following procedure can be adopted for the implementation of Hardy Cross method in a piping circuit.

1. Subdivide the network into a number of loops. Be sure that all pipes are included in at least one loop.
2. Determine the zeroth estimate for the flow rate $Q_{\alpha\beta}^0$. If s be the number of nodes and r be the total number of lines, write a node equation for each node. As $r > s$, there will be more unknowns than the equation. If we assume $(r - s)$ values of $Q_{\alpha\beta}^0$, then the system should reduce to s linear algebraic equations with s unknowns. If the resulting set is linearly independent, then a solution for all other $Q_{\alpha\beta}^0$ can be obtained. Linear independence is tested by taking the determinant of the coefficient matrix. If the system of equations turns out to be linearly dependent, then one additional $Q_{\alpha\beta}^0$ value must be assumed before a set of values $\{Q_{\alpha\beta}^0\}$ can be established.
3. Determine the correction factor ΔQ_1 for each loop.
4. Obtain a new value for the flow rate in each line $Q_{ij}^I = Q_{ij}^0 + \Delta Q_i$.
5. Repeat step 3 and step 4 till all the corrections are equal to zero.

Example: Obtain the flow rates in each of the lines of the network (Figure 6.17). Assume the Hazen–Williams coefficient, $C = 130$.

Solution

Step 1: Figure 6.18 shows the numbering of nodes, loops and lines.

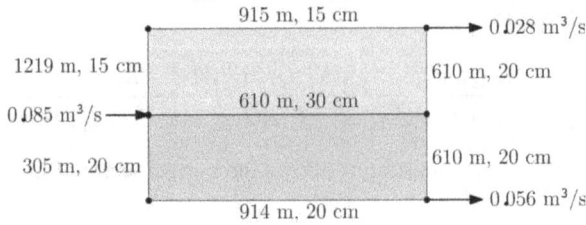

Figure 6.17 A piping network. Length in meter and diameter in centimeter along with the inlet and outlet flow rates are provided in the sketch.

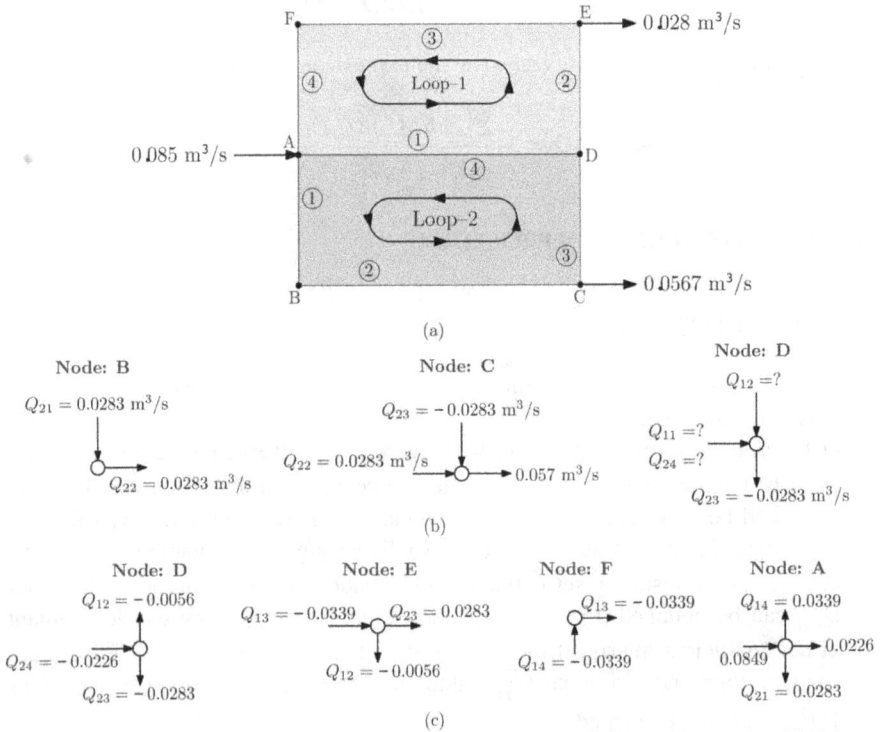

Figure 6.18 Schematic showing different steps of implementation. Step 1: (a) the numbering of nodes, loops and lines. Step 2: (b) application of node equation at nodes B, C and D and (c) application of node equations at nodes D, E, F and A.

Step 2: Determine the zeroth estimate.

Here, $S = 6$ and $r = 7$

We have $r - s = 1$. Therefore, minimum one number of $Q^0_{\alpha\beta}$ should be assumed. In fact, it is suggested that $(r - s + 1)$ values of $Q^0_{\alpha\beta}$ must be assumed in general i.e. we have to assume two values of $Q^0_{\alpha\beta}$. To start the process, let's assume $Q_{21} = 0.0283$ m^3/s. Figure 6.18b shows the application of node equation for nodes

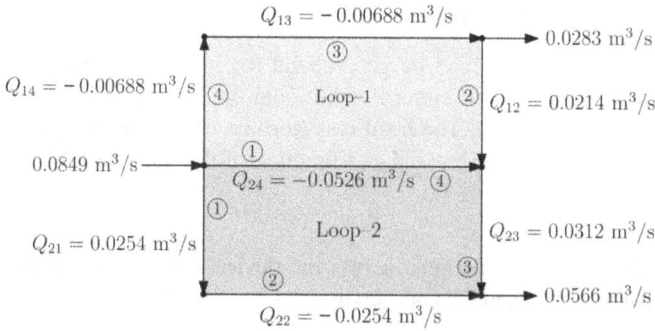

$Q_{13} = -0.00688$ m^3/s

0.0283 m^3/s

$Q_{14} = -0.00688$ m^3/s ④

Loop-1

② $Q_{12} = 0.0214$ m^3/s

0.0849 m^3/s

$Q_{24} = -0.0526$ m^3/s ④

$Q_{21} = 0.0254$ m^3/s

Loop-2

$Q_{23} = 0.0312$ m^3/s

0.0566 m^3/s

$Q_{22} = -0.0254$ m^3/s

Figure 6.19 Converged solution.

B, C and D. For the application of the node equation of D, we need a second assumption. Now, we make second assumption:

$$Q_{11} = 0.0226 \ m^3/s \quad \text{or} \quad Q_{24} = -0.02266 \ m^3/s$$

Thus, we have all the flow rates for node D. The applications of node equations for nodes D, E, F and A are shown in Figure 6.18c.

Steps 3 and 4: Determine the correction factor ΔQ_i for each loop.

$$\Delta Q_1 = -\frac{\sum_{j=1}^{4} K_{ij} Q_{ij}^0 |Q_{ij}^0|^{(1.852-1)}}{1.852 \sum_{j=1}^{4} K_{ij} |Q_{ij}^0|^{(1.852-1)}} = 0.0179 \ \text{m}^3/\text{s}$$

where,

$$K = \frac{K,L}{C^{1.852}\Delta^{4.8704}}$$

$$\Delta Q_2 = 0.004457 \ \text{m}^3/\text{s}$$

$$\therefore \quad Q_{13}^1 = -0.034 + 0.0179 = -0.0161 \ \text{m}^3/\text{s}$$

Flow rates for lines that are contained in more than a single loop are corrected by considering the correction factors for common loops.

$$Q_{11}^1 = Q_{11}^0 + \Delta Q_1 - \Delta Q_2 = 0.0226 + 0.0179 - (-0.004457) = 0.045 \ \text{m}^3/\text{s}$$

$$Q_{12}^1 = 0.0123 \ \text{m}^3/\text{s}, \quad Q_{14}^1 = -0.0161 \ \text{m}^3/\text{s}$$

Similarly,

$$Q_{21}^1 = 0.0239 \ \text{m}^3/\text{s}, \quad Q_{22}^1 = 0.0239 \ \text{m}^3/\text{s}$$

$$Q_{23}^1 = -0.0328 \ \text{m}^3/\text{s}, \quad Q_{24}^1 = -0.045 \ \text{m}^3/\text{s}$$

$$Q_{24}^1 = Q_{24}^1 + \Delta Q_2 - \Delta Q_1 = -0.0226 - 0.00446 - 0.0179 = -0.045 \ \text{m}^3/\text{s}$$

The converged solution will be obtained after the fifth iteration, which is shown in Figure 6.19.

6.13   GENERALIZED HARDY CROSS ANALYSIS

The Hardy Cross approach can be generalized for including minor loss and lines containing devices that result either in additional pressure drop i.e. heat exchanger or turbine (see Figure 6.20). The head loss (turbine or heat exchanger) or pressure increase in a pump can be expressed as a function of flow rate i.e.

$$h_{fD_{ij}} = g_{ij}(Q)$$

In general, the change in the head across the device will depend on the flow rate. This expression of the head loss or increase can be represented as

$$h_{fD_{ij}} = A_{ij} + \sum_{m}^{M} B_{ijm} Q^m$$

where, m represents the degree of polynomial.

Example: The head loss through a pipe fitting is typically described as

$$h_{fD_{ij}} = C_{ij}\frac{V_{ij}^2}{2g} = C_{ij}\frac{Q_{ij}^2}{2gA^2}$$

Hence, the coefficients of the equivalent polynomial expression are $M = 2$, $A_{ij} = 0$, $B_{ij1} = 0$, and $B_{ij2} = C_{ij}\frac{1}{2gA^2}$. For a centrifugal pump, the head versus flow rate can be represented as

$$h_{fD_{ij}} = -(A_{ij} - B_{ij1}Q)$$

Note, $h_{fD_{ij}}$ is less than zero because a pump represents a *negative head loss* i.e. increase in head. A_{ij} and B_{ijm} are obtained from the curve fitting of experimental data i.e. pump performance curve. Incorporation of this polynomial representation in the Hardy Cross method discussed in the previous section can be modified as

$$\Delta Q_i = -\frac{\sum_{j=1}^{J}\left[K_{ij}Q_{ij}|Q_{ij}|^{n-1} + SGN(Q_{ij})A_{ij} + \sum_{m=1}^{M}B_{ijm}Q_{ij}|Q_{ij}|^{m-1}\right]}{\sum_{j=1}^{J}\left(K_{ij}n|Q_{ij}|^{n-1} + \sum_{m=1}^{M}B_{ijm}m|Q_{ij}|^{m-1}\right)}$$

where,

$$SGN(Q_{ij}) = 1 \text{ when } Q_{ij} > 0$$
$$SGN(Q_{ij}) = -1 \text{ when } Q_{ij} < 0$$

Example: Let's investigate the effect of adding the following device in line 2 of loop 2 of Example 6.1.

(a) A pump with $h_{fD} = -(15.235 - 4.24Q)$ m
(b) A heat exchanger with $h_{fD} = (19022Q^2)$ m
(c) A very large pump with $h_{fD} = -304.79$ m

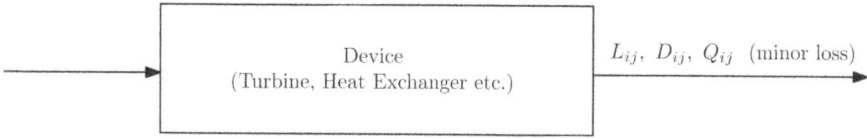

Figure 6.20 A sketch showing the minor loss and device representation for implementation of the Hardy Cross method.

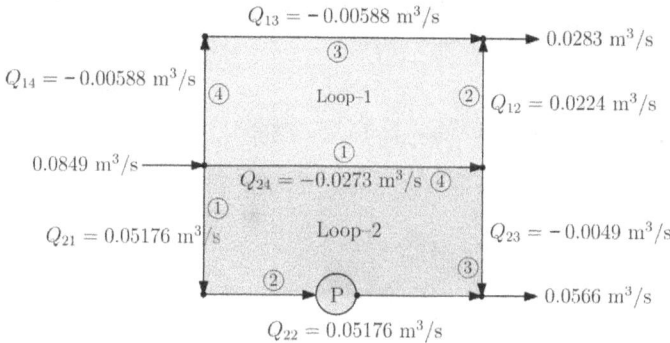

Figure 6.21 Converged solution when a pump is added to the piping circuit.

Solution

(a) Here, $A_{22} = -15.235$, $B_{22\,1} = 4.24$ and $B_{22\,2} = 0.0$. Using the same initial estimate as in the previous example, we get the solution shown in Figure 6.21. Note, the addition of the pump results in an increase in the flow rate through lines 1 and 2 of loop 2. Loop 1 is affected very little.

(b) $h_{fD} = 19022Q^2$
    Here, $A_{22} = 0$, $B_{22\,1} = 0$ and $B_{22\,2} = 19022$
    The converged solution is given in Figure 6.22.
    Note: The addition of the heat exchanger reduces the flow rate in lines 1 and 2 of loop 2. Loop 1 is not affected much due to the addition of heat exchanger.

(c) $h_{fD} = -304.79m$
    Here, $A_{22} = -304.79$, $B_{22\,1} = B_{22\,2} = 0$
    The converged solution is shown in Figure 6.23.
    Note that this large pump completely changes the flow rate distributions. The size of the head increase dominates loop 2 and forces a useless circulation in the loop, i.e. Pipe (11) or Pipe (24) now flows in a direction opposite from that of the previous cases.

Note: When specific Hazen–Williams coefficients are not known or when a different fluid is used, we have to use the Moody diagram and compute the head loss for a

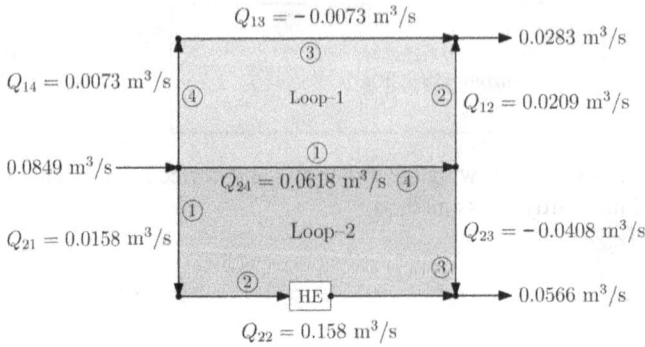

Figure 6.22 Converged solution when a heat exchanger is added to the flow circuit.

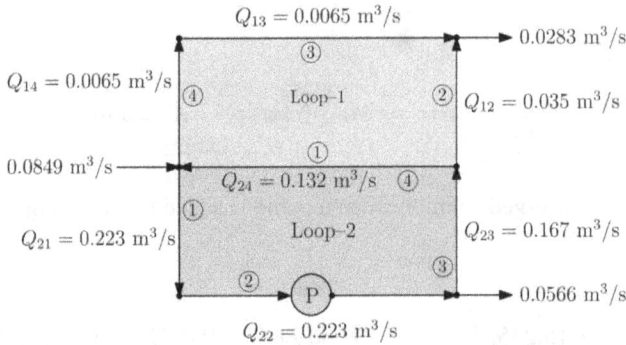

Figure 6.23 Converged solution when a very large pump is added.

variety of flow rates. A curve fit is used to obtain the K and n values. For fluids in process networks, the exponent n may very from pipe to pipe. These circumstances lead to the most general Hazen–Williams Hardy Cross expression as

$$\Delta Q_i = -\frac{\sum_{j=1}^{J}\left[K_{ij}Q_{ij}|Q_{ij}|^{n_{ij}-1} + SGN(Q_{ij})A_{ij} + \sum_{m=1}^{M}B_{ijm}Q_{ij}|Q_{ij}|\right]^{m-1}}{\sum_{j=1}^{J}\left(K_{ij}n_{ij}|Q_{ij}|^{n_{ij}-1} + \sum_{m=1}^{M}B_{ijm}m|Q_{ij}|^{m-1}\right)}$$

6.13.1 BLOCK DIAGRAM

A block diagram or flow chart showing the implementation of generalized Hardy Cross method is shown in Figure 6.24. In the first step, the parameters associated with the number of loops, number of lines in each loop, the length, diameter and nature of line/pipe materials in each loop are provided as input to the program. The specifications of device i.e. A_{ij} and B_{ij} in a line are also provided. If the device is common to more than one loop, then the device coefficients for each line of each loop are provided. The initial flow rate for each line is estimated. The corrections to the

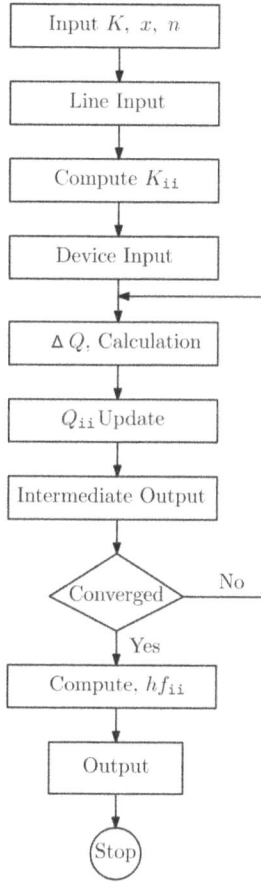

Figure 6.24 A block diagram showing the implementation of the Hardy Cross method for piping circuit.

flow rate is calculated for each loop, and the new flow rate for the line is calculated. If a line is common to two loops, the correction sign is properly taken care of. The intermediate point of output is provided. The iteration is continued till convergence. The convergence is attained when the absolute value of all flow rate corrections is less than a specified tolerance.

Example: A piping circuit with a pump is shown in Figure 6.26. The pump characteristic used in circuit is given as

$$h_{fD} = -A + 0.4Q$$

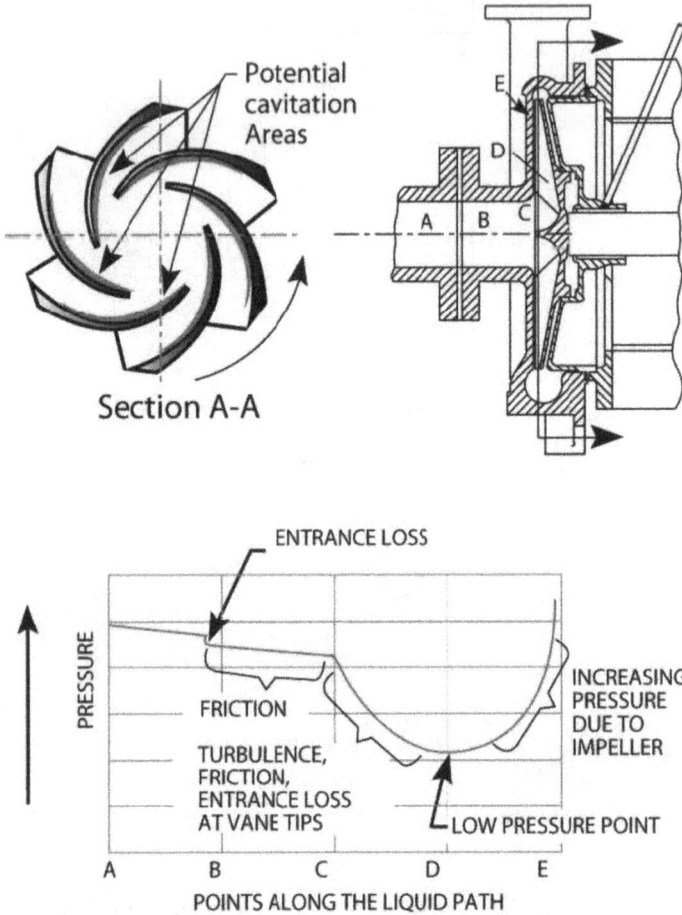

Figure 6.25 Pressure variation within a pump. (`https://www.pumpfundamentals.com/images/cavitation.jpg`, accessed on 30 Nov. 2020.)

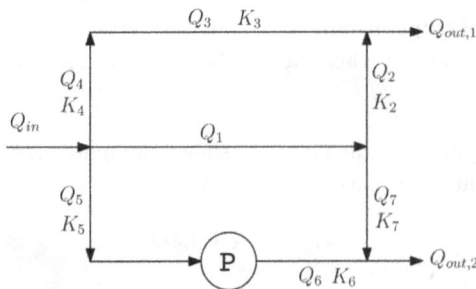

Figure 6.26 Schematic of a sample piping circuit.

where, A is a constant and equal to the shut-off head of the pump and Q is the flow rate. What is the value of constant A? Use the following data for the piping circuit:

$$Q_1 = 0.963 \text{ m}^3/\text{s} \quad Q_2 = 0.792 \text{ m}^3/\text{s} \quad Q_3 = 0.208 \text{ m}^3/\text{s} \quad Q_4 = 0.208 \text{ m}^3/\text{s}$$

$$Q_5 = 1.828 \text{ m}^3/\text{s} \quad Q_6 = 1.828 \text{ m}^3/\text{s} \quad Q_7 = 0.172 \text{ m}^3/\text{s} \quad Q_8 = 3.0 \text{ m}^3/\text{s}$$

$$Q_{out,1} = 1.0 \text{ m}^3/\text{s} \quad Q_{out,2} = 2.0 \text{ m}^3/\text{s}$$

$$K_1 = 1.148 \quad K_2 = 8.269 \quad K_3 = 50.36 \quad K_4 = 67.146 \quad K_5 = 4.134$$

$$K_6 = 12.403 \quad K_7 = 8.269 \quad K_8 = 1.148$$

Solution

The head loss with a loop having a pump should be equal to zero.

$$\sum h_{fD}(\text{Loop 2}) = 0$$

$$k_5 Q_5^{1.852} - A + 0.4 Q_6 + k_6 Q_6^{1.852} - k_7 Q_7^{1.852} - k_1 Q_1^{1.852} = 0$$

$$A = k_5 Q_5^{1.852} + 0.4 Q_6 + k_6 Q_6^{1.852} - k_7 Q_7^{1.852} - k_1 Q_1^{1.852}$$

$$= 4.134(1.828)^{1.852} + 0.4 \times 1.828 + 12.403(1.828)^{1.852} - 8.269(0.172)^{1.852}$$

$$- 1.148(0.963)^{1.852}$$

$$= 12.634 + 0.731 + 37.905 - 0.317 - 1.07$$

$$= 49.883.$$

Example: A pump and pipe system moves water from a reservoir to a water tower. The water level in the tower is 30 m higher than the reservoir. The pipe system uses a 200 mm diameter pipe. The pipe running from the reservoir to the pump is 5 m. The pipe running from the pump to the water tower is 250 m long. For your calculations, assume that the friction factor in the pipe is a constant, $f = 0.024$ (note that you would need to check this assumption later in actual applications). The only minor loss is at the entrance to the pipe system ($K = 20$). The pump characteristic curve is given by $h_p = 100 - 750 Q^2$ where, h_p is the head added by the pump and Q is the flow rate in m^3/s.

Compute the flow rate in the pump/pipe system and the head added by the pump (i.e. its operating point).

Solution

For the piping system,

$$(z_B - z_A) + h_L(Q) = 30 + \left(f \frac{L_1}{D^5} + f \frac{L_2}{D^5} + \frac{K}{D^4} \right) \left(\frac{8}{g\pi^2} \right) Q^2$$

$$= 30 + \left(0.024 \frac{5}{0.2^5} + 0.024 \frac{250}{0.2^5} + \frac{2}{0.2^4} \right) \left(\frac{8}{g\pi^2} \right) Q^2$$

$$= 30 + (375 + 18750 + 1250) \left(\frac{8}{g\pi^2} \right) Q^2$$

$$= 30 + 1683.5 Q^2$$

For the pump, $h_p(Q) = 100 - 750Q^2$.

The operating point occurs at

$$100 - 750Q^2 = 30 + 1683.5Q^2$$

$$70 = 2433.5Q^2$$

$$Q = \sqrt{\frac{70}{2433.5}} = 0.170 \text{ m}^3/\text{s}$$

The head added by the pump is $h_p = 100 - 750(0.170)^2 = 78.33$ m.

Example: A pump delivers 5.66 m³/s against a head of 122 m at 4500 RPM. The impeller diameter is 2.1 m. The positive suction head including velocity head is equal to 3 m. The pump is too large for a laboratory testing. Therefore, a small model of the pump needs to be tested for predicting its operational characteristics. The model pump has an impeller diameter equal to 0.46 m. It is to be tested at a reduced head of 97.5 m. At what speed, capacity and suction head should the test be conducted?

Solution

$$N_1 = N \frac{D}{D_1} \sqrt{\frac{H_1}{H}} = 450 \left(\frac{2.1}{0.46} \right) \sqrt{\frac{97.5}{122}} = 1836.5 \text{ rpm.}$$

$$Q_1 = Q \left(\frac{D_1}{D} \right)^2 \sqrt{\frac{H_1}{H}} = 5.66 \left(\frac{0.46}{2.1} \right)^2 \sqrt{\frac{97.5}{122}} = 0.2427 \text{ m}^3/\text{s.}$$

$$\text{Cavitation factor, } \sigma = \frac{H_b - H_s}{H} = \frac{10 - 3}{122} = 0.057$$

Hence, $H_{s1} = H_b - \sigma H = 10 - 0.057 \times 97.5 = 4.44$ m

Hence, the model should be tested with a positive suction head of 4.44 m.

PROBLEMS

1. The piping network shown below uses smooth cast iron pipes (C:130). The diameter and length of pipes in the network are pipe 1: dia. = 0.31 m, length = 609.6 m; pipe 2: dia. = 0.203 m, length = 609.6 m; pipe 3: dia. = 0.1524 m, length = 914.4 m; pipe 4: dia. = 0.1524 m, length = 1219.2 m; pipe 5: dia. = 0.203 m, length = 304.8 m; pipe 6: dia. = 0.203 m, length = 914.4 m; pipe 7: dia. = 0.203 m, length = 609.6 m. Perform the Hardy Cross analysis and calculate the flow rate in each line after one iteration. Note that $K_1 = 4.721$ when Q is in cfs. Use the following initial flow rates: Q_1 (top loop) = 1.811 cfs, Q_1 (bottom loop) = −1.811 cfs, $Q_2 = 0.686$ cfs, $Q_3 = -0.314$ cfs, $Q_4 = -0.314$ cfs, Q_5 cfs, $Q_6 = 0.875$ cfs, $Q_7 = -1.125$ cfs.

   Take 1 cfs = 0.02832 m³/s.

2. Find out the flow rate in each line of the following piping circuit. Here, K is the Hazen–Williams coefficient with $n = 2$.

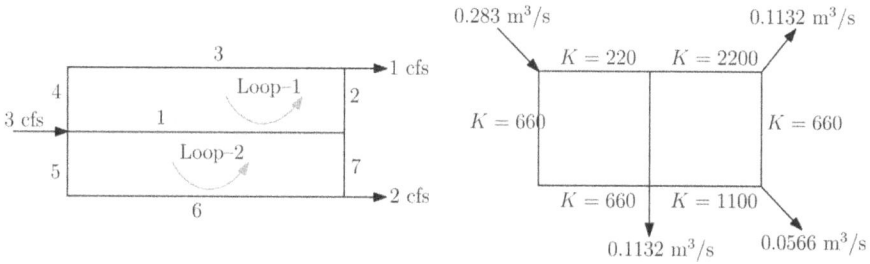

3. A piping circuit with a pump is shown in Figure 6.26. The pump characteristic used in the circuit is given as $h_{fD} = 0.4Q - A$ where, A is a constant equal to the shut-off head of the pump and Q is the flow rate. Calculate the value of constant A using the following data:

Q_1	Q_2	Q_3	Q_4	Q_5	Q_6	Q_7	Q_{in}	$Q_{out,1}$	$Q_{out,2}$
0.963	0.792	0.208	0.0208	1.828	1.828	0.172	3.0	1.0	2.0

K_1	K_2	K_3	K_4	K_5	K_6	K_7	K_8
1.148	8.269	50.36	67.146	4.134	12.403	8.269	1.148

Note: All the flow rates (Q) are in m^3/s.

REFERENCES

Cengel, Y. and Cimbala, J., *Fluid Mechanics: Theory and Applications*, McGraw-Hill Education, New York (2010).

Crane Co. *Flow of Fluids Through Valves, Fittings, and Pipe.* Crane Company. Technical paper. Crane Company, 2013. ISBN 9781400527120. https://books.google.co.in/books?id=8kDYAAAAMAAJ.

Frank M. White, H Xue, *Fluid Mechanics*, McGraw-Hill (2022)

J. Gülich. Pumping highly viscous fluids with centrifugal pumps — Part 1. *World Pumps*, 1999(395):30–34, 1999a. doi: 10.1016/s0262-1762(00)87528-8.

J. Gülich. Pumping highly viscous fluids with centrifugal pumps — Part 2. *World Pumps*, 1999(396):39–42, 1999b. doi: 10.1016/s0262-1762(00)87492-1.

I. Karassik, J. Messina, P. Cooper, and C. Heald. *Pump Handbook.* McGraw-Hill Education, 2001. ISBN 9780071500111. https://books.google.co.in/books?id=d29fFdOkiSgC.

7 Artificial Intelligence for Thermal Systems

7.1 INTRODUCTION

Artificial intelligence (AI) is defined as a branch of computer science which investigates symbolic and non-algorithmic reasoning processes for use in machine inference. It has two general common traits i.e. it consists of several computer methods and can also reproduce a non-quantitative human thought process. AI consists of several branches, namely, expert systems (ESs), artificial neural networks (ANNs), genetic algorithms (GAs), fuzzy logic (FL), natural language processing (NLP), computer vision (CV) etc. These computer-based algorithms are also widely classified in the domain of soft computing. Algorithms in soft computing are mostly based on simplistic models of human intelligence and evolutionary experience. They generally have very simple computational steps, often accompanied by a large number of repeated computational cycles. This is very much in contrast to hard computing, which generally deals with numerical solutions to differential equations such as conservation laws. Up to the very recent past, thermal science problems have largely been treated by traditional hard-computing approaches, along with experiments carried out for the purpose of validation or development of correlations. However, thermal problems are becoming increasingly more complex, and there is a need for dealing with process dynamics, optimization and control. Unfortunately, the traditional approaches are simply not robust enough to handle such increased complexity, and new methodologies are definitely needed for this purpose.

This chapter introduces the concept of AI for thermal system design and analysis. Specific emphasis is given on the discussion of ANN technique, followed by sample examples.

7.2 EXPERT SYSTEM

An ES is an AI application aimed at the resolution of a specific class of problems. It is a well-organized and well-cross-referenced task list with computer as a work tool. An ES is an application consisting of a series of procedures aimed at the solution of a specific class of problems built around the general expert knowledge that can be collected for these problems. ESs allow computers to "make decisions" by interpreting data and selecting from a list of alternatives. ESs take computers a step beyond straightforward programming. It is based on a technique called rule-based inference, in which pre-established rule systems are used to process the data. It associates with a rule for every suggestion or piece of advice a specialist expert would give on how to solve the problem. The ES is better described as knowledge-based system, which is a computer program that emulates the decision-making ability of a human expert.

DOI: 10.1201/9781003049272-7

An ES makes extensive use of specialized knowledge to solve problems at the level of a human expert. It is an intelligent computer program that uses knowledge and inference procedures to solve problems that are difficult enough to require significant human expertise for the solution. ESs are capable of computational, qualitative, descriptive and explanatory functions. The main idea is to create programs where knowledge and reasoning techniques are introduced, such that it can generate answers similar to those that would be provided by a highly experienced human being. In effect, a user can access the human experts knowledge and experience through the user interface of the computer. The user of an ES asks questions and receives answers and explanations presented in various forms, such as text, video, sound, photo, figure etc.

7.2.1 ADVANTAGES OF EXPERT SYSTEMS

The salient features and advantages of an ES can be summarized as follows:

1. *Increased Availability:* An ES can be considered as a mass production of expertise because it can be made available on any suitable computer hardware.
2. *Reduced Cost:* The cost per user of an ES for expertise is less.
3. *Life Span:* The human experts can retire, quit or die. However, an ES is permanent
4. *Multiple Expertise:* The knowledge of multiple experts is available in an ES at the same time. An ES can provide the second opinion to a human expert or break a tie in case of disagreements between multiple human experts. An ES can have fast response and is more readily available than a human expert.
5. *Intelligent Database:* An ES can access a database in an intelligent manner.

7.2.2 DISADVANTAGES OF EXPERT SYSTEMS

An ES can have the following disadvantages:

1. *Incorrect Answer:* An ES can make an error which can be very costly.
2. *Answer outside Its Domain:* An ES may provide a solution outside its field of expertise. It may have limited knowledge in a particular domain. As a result, it may give misleading or incorrect answers. A human expert, in contrast, will know the limits of his/her abilities and may not try to solve problems outside his/her expertise.
3. *Common Sense Knowledge:* It can be difficult to represent common sense knowledge in an ES.

7.2.3 STRUCTURE OF EXPERT SYSTEMS

Figure 7.1 shows a representative architecture of an ES. Knowledge base, inference mechanism and explanation mechanism are the main components of an ES. The ES developer receives knowledge from the experts and prepares it in the form of rules and charts, which are stored in a knowledge base. The inference mechanism receives

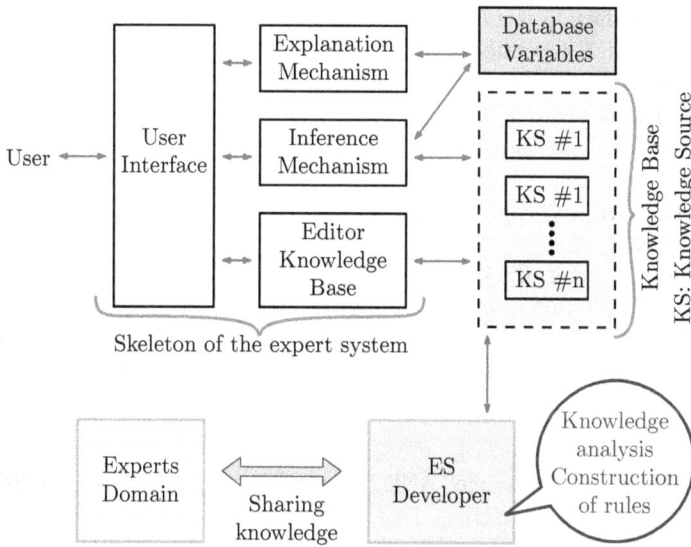

Figure 7.1 Architecture of a knowledge-based expert system with its components. (Gonciarz [2014].)

input from the user interface, processes the input using a knowledge base and explains it to the user. Several terminologies used in the description and operation of an ES are as follows:

1. *Knowledge Base:* A knowledge base contains facts and rules that are necessary to solve problems related to a specific field. It contains description of all elements under consideration with a list of their mutual relations, and a list of rules (including mathematical formulae) according to which these elements operate.
2. *Editor Knowledge Base:* The editor can modify the knowledge contained in an ES. This helps in extension or improvement of the ES with time.
3. *Inference Engine:* An inference engine is the software that uses the data represented in the knowledge base to reach a conclusion for particular cases. It describes the strategy to be used in solving a problem i.e. it guides from query to solution.
4. *Rules:* A rule is a way of formalizing declarative knowledge. A rule is a statement of relationships and not a series of instructions. For example, a turbine can be declared as follows: IF a machine is rotating AND IF it has blades AND IF a fluid expands in it, THEN the machine is a turbine.
5. *Facts:* Facts are specific expressions that describe a particular situation. These expressions can be numerical, logical, symbolic or probabilistic. For example, IF (pressure high-frequency oscillations) is TRUE AND (shaft high-frequency vibrations) is TRUE, THEN probability (compressor stall) = 75 percent.

6. *Induction:* Induction is a logical procedure that proceeds from effects backward to their causes. For example, high-frequency vibrations detected on a gas turbine shaft may be caused by compressor stall, rotor unbalance in the compressor or turbine, fluid dynamic instabilities in the flow, irregular combustion, mechanical misalignment of the shaft, failure in the lubrication system or bearings wear. Pure induction does not always produce a solution. Other data need to be available that can be included in the inductive process to eliminate some of the probable causes from the list.

7. *Backward Chaining:* Backward chaining is a procedure that attempts to validate or deny a proposition (goal) by searching through a list of conditional rules and facts to see if they univocally determine the possibility of reaching the goal. Let's consider an example to ascertain if the machine under consideration is a Francis turbine.

Here are the rules:
Rule 1: "Francis" IF "turbine" AND "hydraulic" AND "specific speed lower than 2 and higher than 0.3"
Rule 2: "turbine" IF "rotating" AND "vaned"
Rule 3: "hydraulic"IF "fluid = water"

Fact 1: "specific speed = rpm" $* \dfrac{(volume\ flowrate)^{0.5}}{(g*total\ head)^{3/4}}$

Here are the facts:
Fact 2: "rpm = 300"
Fact 3: "volume flowrate = 30 m^3/s"
Fact 4: "total head = 500 m"
Fact 5: "fluid = water"
Fact 6: "number of blades = 13"

Goal: "given the knowledge base, determine if the machine is a Francis"

Result: "it is not". Because the specific speed (2.805) criteria is not satisfied.

7.2.4 AN EXAMPLE FOR FEED WATER PUMP SELECTION

Here, we consider an example of feed water pump selection for a power plant. The objective is to devise an automatic procedure for the choice and specification of a feed water pump. The procedure should be able to handle any kind of pump geometry (axial, mixed, centrifugal) and cover the entire range of possible operations. The following tasks list may be constructed by consulting a pump design textbook or by asking an expert:

- Read the design data: Flow rate (Q) and Head (H).
- Read constraints (if applicable): Net Positive Suction Head (NPSH), speed, etc.
- Is rpm given?
  if Yes: Then compute: Specific speed, N_s.
  if No: Then is NPSH given?

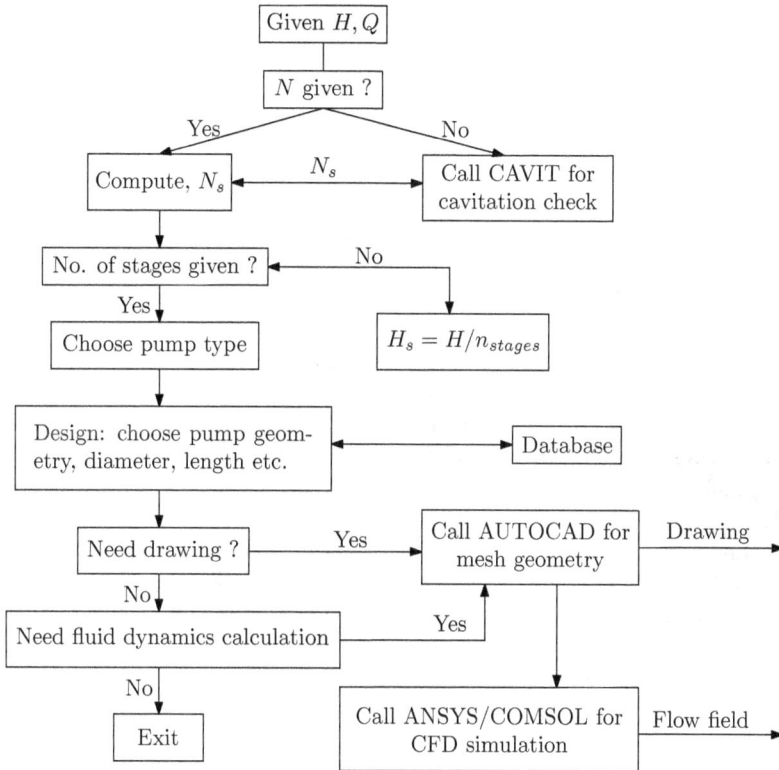

Figure 7.2 Decision tree for selection and design of a feed water pump. (Paoletti and Sciubba [1997].)

    if Yes: Compute iteratively suction head and derive maximum rpm.
    if No: Assume rpm, compute NPSH, check N_s. Iterate.
- Check if multi-staging is required.
- Check if multiple pumps in parallel are required.
- Compute overall dimensions of the pump.
- Check component library for the nearest defined pump type from the database.
- Produce a technical specification sheet.

The decision tree for the selection and design of the feed water pump is shown in Figure 7.2. Here, CAVIT, AUTOCAD and ANSYS/COMSOL are codes/routines of the knowledge base for cavitation check, meshing and computational fluid dynamics (CFD) calculation, respectively. The CAVIT code is specific to the pump problem. However, AUTOCAD and ANSYS/COMSOL are useful for different types of thermal design problems.

7.3   ARTIFICIAL NEURAL NETWORK (ANN) OVERVIEW

ANN is the leading methodology for the solution of general thermal problems. There are several reasons for this. ANN has the ability to recognize accurately the inherent relationship between any set of input and output without a physical model. This ability is essentially independent of the complexity of the underlying relation such as nonlinearity, multiple variables, noise, certain input and output data. ANN is also inherently fault tolerant due to a large number of processing units in the network. The learning ability of ANN also allows it to adapt to changes in its parameters, which enables the ANN to deal with time-dependent dynamic modeling and adaptive control. The ANN also has ability to incorporate elements of other soft-computing methodologies such as FL and GA to further improve its capability for dealing with additional complexity in thermal problems.

One of the limitations of ANN is the requirement of input-output data sets in the learning process to train neural networks. However, it is not a serious shortcoming as a large amount of experimental data sets are available for various thermal systems and device performances. In addition, experimental data obtained under specific dynamic conditions can also be used to train dynamic ANNs. The neural network can be trained in real time when the experimental data are being acquired. This feature is useful in the development of dynamic adaptive-control schemes.

7.3.1   STRUCTURE OF ANNS

The ANN is an electrical analog of biological neural networks. Biological nerve cells, called neurons, receive signals from the neighboring neurons or receptors through dendrites. These neurons process the received electrical pulses at the cell body and transmit signals through a large and thick nerve fiber, called as axon. The electrical model of a typical biological neuron consists of a linear activator, followed by a nonlinear inhibiting function. The linear activation function yields the sum of the weighted input excitation. The nonlinear inhibiting function attempts to capture the signal levels of the sum. An ANN is a collection of such electrical neurons connected in different topologies. The schematic explaining the operation of a neuron is shown in Figure 7.3.

Here, x_i, y, w_i and θ are known as input, output, weight and bias, respectively. The neuron consists of two parts i.e. net function (Σ) and activation function (f). The net function determines how inputs are combined inside the neuron. The output of the neuron is determined by the activation function.

A fully interconnected ANN consists of a number of processing units known as nodes or artificial neurons, organized in layers. There are three groups of node layers i.e. the input layer, one or more hidden layers and an output layer. Each layer is occupied by a number of nodes. All the nodes of each hidden layer are connected to all the nodes of the previous and following layers by means of inter-node synaptic connectors or simply connectors. Each of the connectors, which mimic the biological neural synapsis, is characterized by a synaptic weight. The nodes of the input layer are used to designate the parameter space of the problem under consideration. The

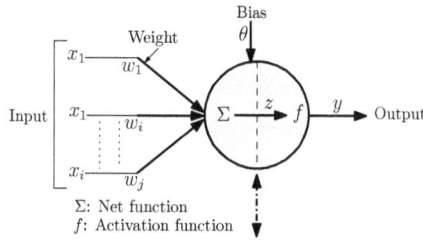

Bias
θ

Weight

x_1——
w_1

Input x_1----:w_i Σ → z → f → y → Output

x_i——
w_j

Σ: Net function
f: Activation function

Figure 7.3   Schematic representation of a basic neuron.

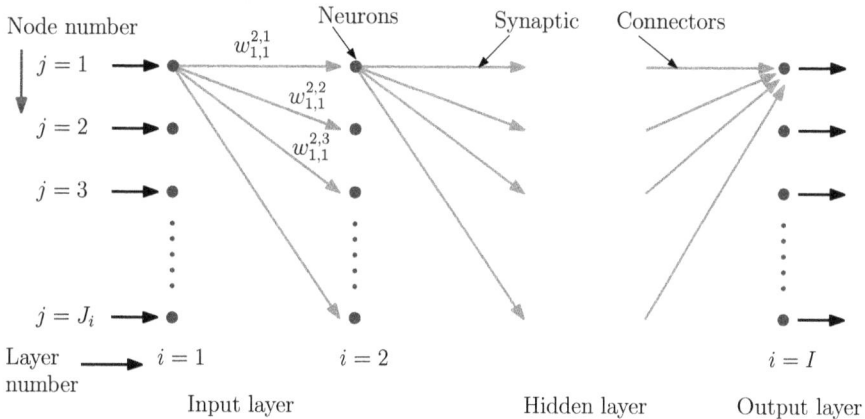

Node number   Neurons   Synaptic   Connectors

$w_{1,1}^{2,1}$

$j = 1$

$w_{1,1}^{2,2}$

$j = 2$
$w_{1,1}^{2,3}$

$j = 3$

$j = J_i$

Layer ——→ $i = 1$ $i = 2$ $i = I$
number
          Input layer   Hidden layer   Output layer

Figure 7.4   Schematic of a fully connected multilayer ANN.

output layer nodes correspond to the unknowns of the problem. The parameters in the input layer and output layer need not be all independent.

At each hidden-layer node, the node input consists of a sum of all the node outputs from the nodes in the previous layer modified by the individual interconnector weights (w) and a local node bias (θ). The bias represents the propensity of the combined input to trigger a response at the node. The weights are simply weighting functions that determine the relative importance of the signals from all the nodes in the previous layer. At each hidden node, the node output is determined by an activation function, which determines whether the particular node is to activate or not. Information, which starts at the input layer moves forward toward the output layer by the connector and node operations. Such a network is known as a fully connected feed-forward network.

Figure 7.4 shows the structure or configuration of a network, consisting of the input, hidden and output layers with node and layer designations. Here, i refers to the layer number with $i = 1$ is the input layer, and $i = I$ is the output layer with I being the total number of layers. The node number in any layer is denoted as j. Since node numbers are likely to vary from layer to layer, the maximum j value is designated by J_i, depending on the layer number, and J_I is thus the number of unknowns in the output layer. Each node is designated by (i, j). A somewhat different designation is used for all the connectors since there are two nodes involved. The node on the left is designated by subscripts, while the right node in the forward direction is designated by superscripts. The synaptic weight $w_{1,1}^{2,3}$ refers to the connector from Node $(1,1)$ to Node $(2,3)$. The nodal input to node (i, j) is written as

$$x_{i,j)} = \theta_{i,j} + \sum_{k-1}^{j_i-1} w_{i-1,k}^{i,j} y_{i-1,k} \tag{7.1}$$

where, $\theta_{i,j}$ is the nodal bias at (i, j), and $y_{i-1,k}$ is the nodal output at $(i-1,k)$ of the previous layer. This equation indicates that each signal coming from the previous layer is tampered by the weight in the same connector before they are added, and modified by the local node bias to form the input to the local node (i, j). The information to be processed represents the combined influence of all nodes from the previous layer. The node output, $y_{i,j}$, is driven by the input, $x_{i,j}$, through the activation function or threshold function given as

$$y_{i,j} = \phi_{i,j}(x_{i,j}) \tag{7.2}$$

This expression plays the role of the biological neuron i.e. whether it should fire or not on the basis of the strength of the input signal. When the input signal is weak, the artificial neuron simply produces a small output. On the other hand, when the input signal exceeds a certain threshold, the artificial neuron fires and then sends a strong signal to all the connectors and then to all the nodes in the next layer. Several relevant activation functions have been proposed i.e. the step function, the logistic sigmoid function, the hyperbolic tangent, the Gaussian, the wavelet etc. The activation function can also be changed from one hidden layer to another. The most popular and preferred activation function is the continuous version of the step function, known as the logistic sigmoid function. This function possesses continuous derivatives to avoid computational difficulties. It is also highly nonlinear, which is beneficial in dealing with highly nonlinear input-output relations. It is generally written as

$$\phi_{i,j}(\xi) = \left(1 - e^{\xi/c}\right)^{-1}, \quad i > 1 \tag{7.3}$$

$$= \xi, \quad i = 1 \tag{7.4}$$

Here, the constant c determines the steepness of the function. It may be noted that the node output $y_{i,j}$ represented by the sigmoid function always lies between 0 and 1 for all $x_{i,j}$. Therefore, it is desirable to normalize the network input and output data with the largest and the smallest of each of the data sets used in the ANN analysis.

7.3.2   TRAINING OF ANNs

Training is the most important step of ANNs for modeling a thermal system. This section describes the training process of ANNs. When the information reaches from input to the output layer, errors can be determined by comparing the calculated feed-forward data with the output data to determine the error at each of the output nodes. These errors are then used to adjust all the node biases and connector weights in the entire network to minimize the errors by means of a learning or training procedure. Figure 7.5 shows the flow chart of ANN training process. The most popular training procedure for fully connected feed-forward networks is known as the supervised backpropagation learning scheme based on the steepest-gradient error correction process. The weights and biases are adjusted layer by layer from the output layer toward the input layer during training. The whole process of feeding forward with backward learning is then repeated until a satisfactory error level is reached or becomes stationary. Levenberg–Marquardt and scaled conjugate gradient etc. are the other learning algorithms.

For a given chosen network architecture of layers and nodes, the very first step in the training process is to assign initial values to all the synaptic weights and biases in the network. The values may be either positive or negative and are usually taken to be less than unity in absolute values. The second step is to complete all the node input and output calculations based on the equations presented in the previous section. The backpropagation procedure starts with an error function expressed as

$$\delta_{I,j} = (t_{I,j} - y_{I,j})y_{I,j}(1 - y_{I,j}) \tag{7.5}$$

where, $t_{I,j}$ is the normalized output target for the j-th node of the last output layer. This equation is simply a finite-difference approximation of the derivative of the sigmoid function. Once all $\delta_{I,j}$ are calculated, the computation moves back to the previous layer, $I-1$ subsequently. The target outputs for this layer do not exist. Therefore, a surrogate error is utilized and calculated instead for the hidden layer $I-1$, which is given by

$$\delta_{I-1,k} = y_{I-1,k}(1 - y_{I-1,k})\sum_{j=1}^{j_i} \delta_{I,j}w_{I-1,k}^{I,j} \tag{7.6}$$

Similar calculations are continued from layer to layer in the backward direction until Layer 2. After all the errors ($\delta_{i,j}$) are known, the changes in the weights and biases can be determined by the generalized delta rule as

$$\Delta w_{i-1,k}^{i,j} = \lambda \delta_{i,j}y_{i-1,k} \tag{7.7}$$

for all $i < I$, from which all the adjustments in the weights and biases can be determined. The quantity λ is known as the learning rate that is used to scale down the degree of change made to the connectors and nodes. The larger the learning or training rate, the faster the network learns. However, there may be a chance that the ANN may not reach the desired outcome due to oscillatory error behaviors. Its value is normally determined by numerical experimentation, and a commonly arrived value is in the range of 0.4–0.5. The error-correction rate is also modulated by addition of a momentum term based on the old weight and bias changes in the previous learning

Start

↓

Identify inputs and outputs
for the thermal system

↓

Collection of input and output
from experiment or simulation

↓

Normalization of training data between 0.15 and 0.85

↓

Identify the network structure
depending on application

↓

Training of the network by changing

- Learning rate

- Momentum factor

- Weights

- Number of hidden layers

- Number of neurons in hidden layer

↓

Validation of network

↓

Error goal reached   No

Yes ↓

Select the best network architecture with
optimum network parameters

↓

Network is ready for
performance prediction

↓

End

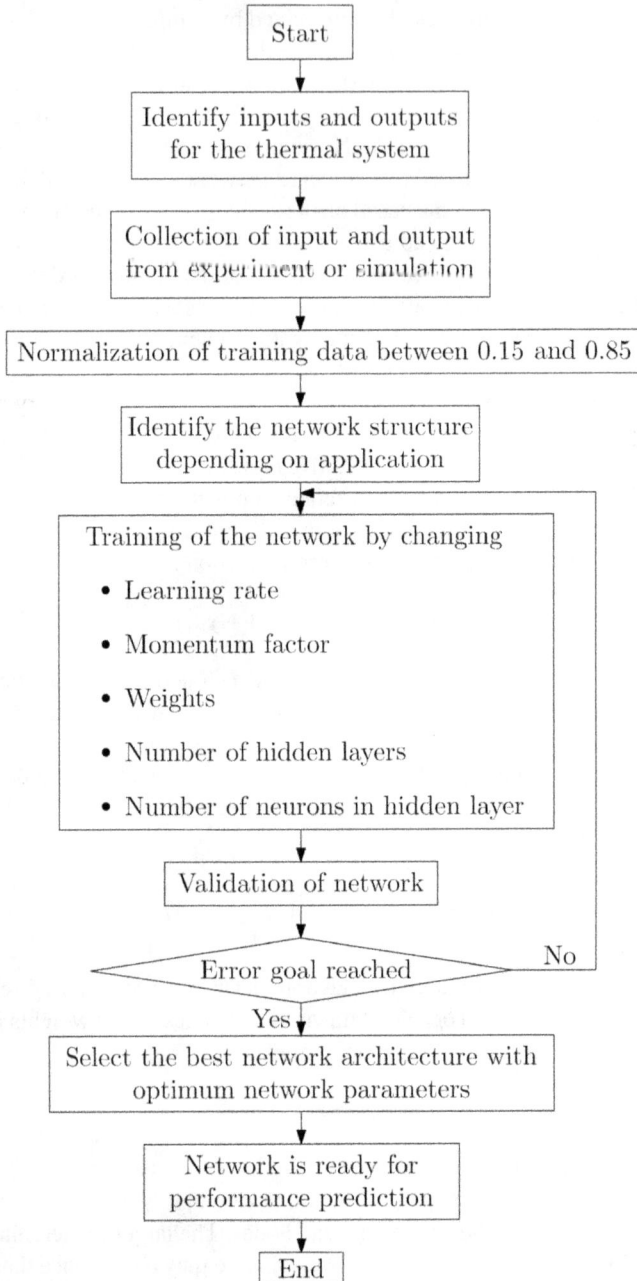

Figure 7.5   Flow chart for the ANN training process. (Mohanraj et al. [2015].)

iteration, which can be written as

$$\Delta w^{i,j}_{i-1,k}(l) = \lambda \delta_{i,j} y_{i-1,k} + \beta \Delta w^{i,j}_{i-1,k}(l-1) \tag{7.8}$$

$$\Delta \theta^{i,j}_{i-1,k}(l) = \lambda \delta_{i,j} + \beta \Delta \theta^{i,j}_{i-1,k}(l-1) \tag{7.9}$$

where, β is the momentum factor and l is the iteration counter.

A cycle of training consists of computing a new set of weights and biases successively for all the experimental runs in the training data. The calculations are then repeated over many cycles while recording an overall error quantity for a specific run within each cycle, given by

$$E_r = \frac{1}{2} \sum_{j=1}^{J_l} \left(t_{l,j} - y_{l,j} \right)^2 \tag{7.10}$$

The weights and biases are continually updated throughout the runs and cycles. The training is terminated when the last cycle error falls below a prescribed threshold or becomes stationary. The final sets of weights and biases can now be used for prediction purposes, and the corresponding ANN becomes a model of the input-output relation of the thermal system.

The overall ANN analysis involves few deterministic and algebraic steps repeated many times on the computer. It involves a relatively large number of free parameters and choices i.e. the number of hidden layers, the number of nodes in each layer, the initial weights and biases, the learning rate, the minimum number of training data sets and the choice of input parameters. The computational steps are simple. However, overall effort depends on the total number of nodes in the network, as a large number of nodes tends to slow down the training process. One flexibility of the ANN methodology is that both numbers of the hidden layer and the corresponding nodes can be increased at will from one training cycle to another training cycle if the cycle errors do not decrease as expected. On the other hand, it may be noted that too many nodes may suffer from the localizing effect of specific data points similar to selecting the degree of polynomial during curve fitting.

The issue of assigning initial weights and biases is also difficult in a new application. Without past information or data, the current practice is simply to generate a set of initial data from a random number generator of bounded numbers. The choice of the training rate can be between 0.4 and 0.5 during the starting point. The sigmoid activation function possesses asymptotic limits of 0 and 1 and may cause difficulties when these limits are approached. Therefore, the usual practice is to normalize all physical variables in an arbitrarily restricted range such as 0.15–0.85 to limit the computational efforts. It is desirable to include as many training data sets as possible as the experimental data set can have error and uncertainty. It is also important to set aside about one-quarter of the entire data set to serve as the testing data set to evaluate the accuracy of the ANN results.

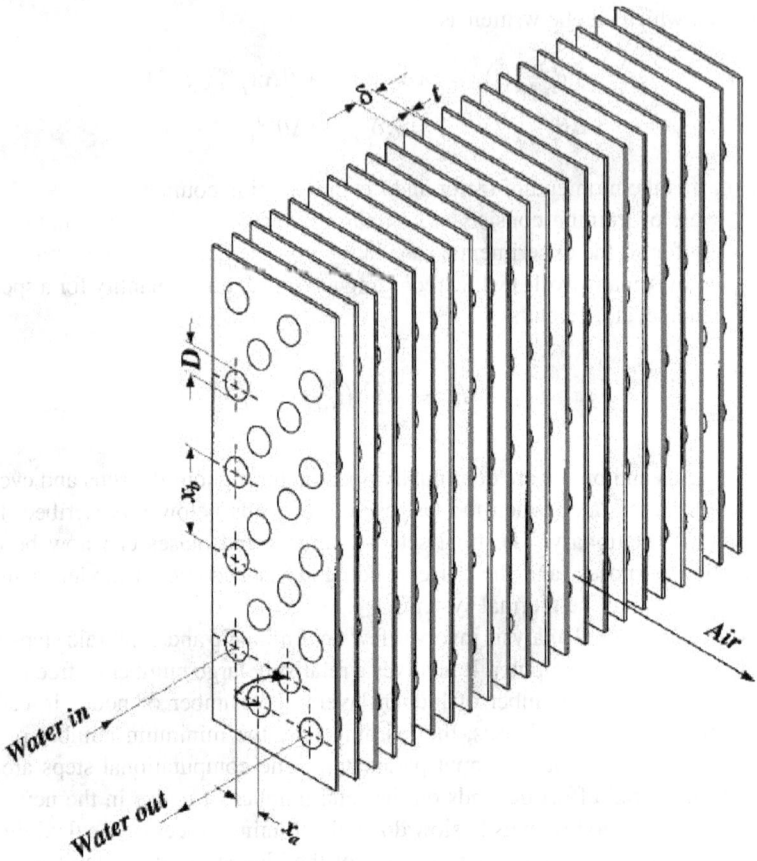

Figure 7.6   Schematic of a multi-column fin tube heat exchanger. (Pacheco Vega [2001].)

7.4   ANNs FOR HEAT EXCHANGER ANALYSIS

Heat exchangers involve significant geometrical complexity and operating conditions involving complex physics. In this section, we will evaluate the capability of ANN in modeling a heat exchanger. The compact multi-row, multi-column, fin-tube heat exchanger shown in Figure 7.6 is considered a sample. This heat exchanger utilizes chilled water flow inside the tubes for air-cooling purpose. Mc-quiston [1978a,b] studied this heat exchanger in great detail by extensive careful experimental measurements and developed correlations in terms of the Colburn j-factors. The chilled water temperature can cause the air temperature to fall below its dew point leading to condensation on the fin surfaces. The data set contained all three fin-surface conditions i.e. dry surface, surface with dropwise condensation and surface with film condensation. Dropwise and film condensation cases were differentiated by subjective visualization. The film spacing was an added important parameter because of the condensation phenomena, which was also treated as an input parameter. Only high Reynolds number turbulent flow conditions were considered on the water side. The following correlations were proposed:

$$j_s = 0.0014 + 0.2618 Re_D^{-0.4} \left(\frac{A}{A_{tb}} \right)^{-0.15} f_s \tag{7.11}$$

$$j_t = 0.0014 + 0.2618 Re_D^{-0.4} \left(\frac{A}{A_{tb}} \right)^{-0.15} f_t \tag{7.12}$$

Here, A is the total air-side heat transfer area, A_{tb} is the surface area of tubes without fins, f_s and f_t are the functions of the fin geometry and Re_D is the Reynolds number based on the tube diameter.

For dry surface,

$$f_s = 1.0 \tag{7.13}$$

$$f_t = 1.0 \tag{7.14}$$

For dropwise condensation,

$$f_s = \left(0.90 + 4.3 \times 10^{-5} Re_D^{1.25} \right) \left(\frac{\delta}{\delta - t} \right)^{-1} \tag{7.15}$$

$$f_t = \left(0.80 + 4.0 \times 10^{-5} Re_D^{1.25} \right) \left(\frac{\delta}{\delta - t} \right)^{4} \tag{7.16}$$

where, δ and t are the fin spacing and thickness, respectively.

For filmwise condensation,

$$f_s = 0.84 + 4.0 \times 10^{-5} Re_D^{1.25} \tag{7.17}$$

$$f_t = \left(0.95 + 4.0 \times 10^{-5} Re_D^{1.25} \right) \left(\frac{\delta}{\delta - t} \right)^{2} \tag{7.18}$$

The Colburn factors are defined as follows:

$$j_s = \frac{h_a}{Gc c_{pa}} Pr_a^{2/3} ; \quad j_t = \frac{h_{ta}}{Gc} Sc_a^{2/3}$$

$$Re_D = \frac{GcD}{\mu_a} ; \quad Re_\delta = \frac{Gc\delta}{\mu_a} ; \quad \frac{A}{A_{tb}} = \frac{4}{\pi} \frac{x_a}{D_h} \frac{x_b}{D} \sigma_f$$

Here, Gc is the air mass velocity based on the free flow area, h is the heat transfer coefficient, h_t is the total heat transfer coefficient, σ_f is the ratio of the free flow cross-sectional area to the frontal area, Re_δ is the Reynolds number based on fin spacing, j_s is the Colburn j-factor for the sensible heat and j_t is for the total heat. Figure 7.6 shows the geometrical arrangement of the tube denoted by x_a and x_b. Later, Gray and Webb [1986] proposed a new correlation using data from additional sources. For a dry surface,

$$j_s = 0.14 Re_D^{-0.328} \left(\frac{x_b}{x_a} \right)^{-0.502} \left(\frac{\delta - t}{D} \right)^{0.0312} \tag{7.19}$$

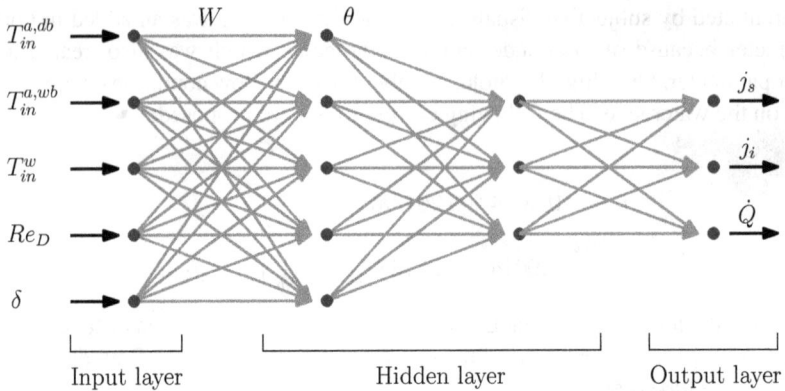

Figure 7.7   A 5-5-3-3 ANN for analysis of a multi-column fin tube heat exchanger. (Pacheco Vega [2001].)

Pacheco-Vega et al. [2000] carried out an ANN analysis of a fin tube heat exchanger and compared it with that of the correlations. They used a fully connected feed-forward network of 5-5-3-3 configuration (Figure 7.7) and backpropagation learning algorithm. The five input nodes correspond to the air-inlet, dry-bulb and wet-bulb temperatures, the chilled water inlet temperature, the airflow Reynolds number and the fin spacing. The three output nodes correspond to j_s for sensible heat transfer j_t for total heat transfer and Q for the total heat-transfer rate. They used 91 data sets for dry-surface conditions, 117 data sets for dropwise condensation and 119 data sets for film condensations. They also trained entire 327 data sets for training to compare the ANNs trained with separate data sets involving different physics with the ANN trained with the complete data set. Table 7.1 shows the results in the rms percentage deviations of the prediction by ANNs and correlations. A low level of error in total heat transfer is observed which is comparable to the expected experimental uncertainties. The ANNs give better predictions for dry surfaces than those for wet surfaces, possibly due to the complex physics of wet surfaces. When the ANNs are trained with the entire data sets by disregarding surface conditions, all deviations tend to increase. This may be attributed to more complexity and variability in physics involved.

7.5   ANNs FOR A THERMOPHYSICAL PROPERTY DATABASE

As discussed earlier in Chapter 4, the selection of fluid is an important component of several thermal system designs. There are a large number of fluid types for selection, and also many more types of fluid are also being added to the database continuously. In addition, the properties of fluid are functions of the physical state i.e. temperature, pressure, composition etc. Several correlations are available for predicting the physical properties of fluid as a function of the operational parameters. However, errors in using these correlations are high. Therefore, it's a challenge for a design engineer to

Table 7.1

Comparison of Percentage Errors in Predictions between the ANN and Standard Power Law Correlations of Heat Exchangers

Surface	Method	j_s	j_t	Q_i
	McQuiston	14.57	14.57	6.07
Dry	Gray and Webb	11.62	11.62	4.95
	ANN	1.002	1.002	0.928
	McQuiston	8.5	7.5	–
Dropwise	Gray and Webb	–	–	–
	ANN	3.32	3.87	1.446
	McQuiston	9.01	14.98	–
Filmwise	Gray and Webb	–	–	–
	ANN	2.58	3.15	1.96
Combined	ANN	4.58	5.05	2.69

Source: Yang et al. [2008].

select the best working fluid. The availability of several soft computing techniques has opened a new tool to easily accommodate large data sets of working fluids at various operating conditions for design.

For example, nanofluid of several types is being considered a promising working fluid for several applications. The prediction of nanofluid properties is discussed in this section. Nanofluids are suspension of small diameter particles (less than 100 nm) i.e. TiO_2, Al_2O_3, SiC, SiO_2, CuO, etc. dispersed in base fluid i.e. water, deionized water, ethylene glycol (EG), ethanol, polyalphaolefin, transformer oil, mixture of EG and water, mixture of propylene glycol and water, R11 refrigerant and so on. The thermal conductivity of nanofluids has been observed to be higher than base fluid even at very low concentrations. Therefore, nanofluid is actively considered as a coolant in several applications. Accurate prediction of properties e.g. viscosity and thermal conductivity is difficult because of complexities due to hydrodynamic and particle–particle interactions of nanoparticles in dispersion. Several empirical models have been proposed in literature. However, the limitation of these models is the requirement of a sufficient amount of data for calibration and validation purposes that makes these models computationally incompetent. The advantages of ANN methodology compared to conceptual models are its high speed, simplicity and large capacity with a lower computational requirement. Let's consider one example of nanofluid property modeling using ANN.

Heidari et al. [2016] developed a feed-forward backpropagation multilayer perceptron ANN for predicting nanofluid viscosity in broad ranges of operating parameters. They evaluated several configurations and the network performance was optimized by changing the number of hidden layers, number of neurons in the hidden layer and network training algorithm in order to obtain the best network for

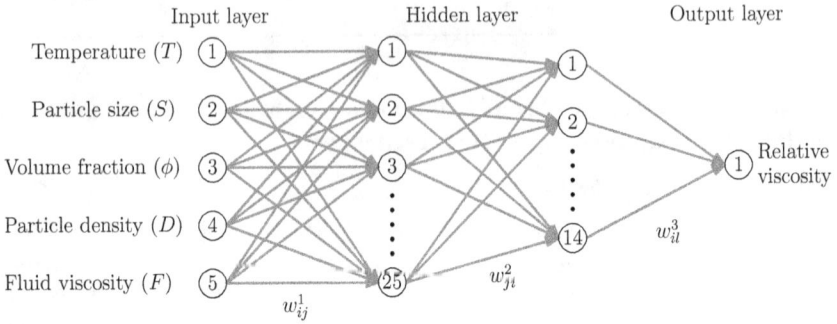

Figure 7.8 Structure of ANN network for modeling the property of nanofluid. (Heidari et al. [2016].)

the prediction of nanofluid relative viscosity. The schematic of the ANN arcitechture used is shown in Figure 7.8. The model input parameters are temperature (T), nanoparticle size (S), density (D), volume fraction (Φ) and base fluid viscosity (F). They used 1490 data points for different nanofluids with different base fluids (water, ethanol, EG, transformer oil, R-11 refrigerant, toluene etc.) and nanoparticles ($TiO_2, Al_2O_3, SiC, CUo, Fe_3O_4$ etc.) for a wide range of different parameters. The ANN network was compared with the correlation developed by Meybodi et al. [2015] based on the data set of viscosity of water-based Al_2O_3, TiO_2, SiO_2 and CuO nanofluids. The mathematical form of the correlation is

Relative viscosity =

$$\frac{135.54064976 - 343.82413843(e^{\phi/s}) + 290.11804759(e^{\phi/s})^2 - 78.993120761(e^{\phi/s})^3}{0.91161630781 + 32.33014233\frac{Ln(s)}{T} - 11.732514460\frac{(Ln(s))^2}{T}}$$

(7.20)

where, S, ϕ and T are size of nanoparticles in nm, volumetric concentration of nanoparticle in percent and temperature of the system in Kelvin, respectively. Several statistical criteria used to evaluate the performance of ANN i.e. average absolute relative deviation (AARD, %), coefficient of determination (R^2), mean square error (MSE) are defined as follows:

$$AARD(\%) = \frac{1}{N}\sum_{i=1}^{N}\left|\frac{y_i - y_p}{y_i}\right| \times 100$$

(7.21)

$$R^2 = 1 - \frac{\sum_{i=1}^{N}(y_i - y_p)^2}{\sum_{i=1}^{N}(y_i - \bar{y})^2}$$

(7.22)

$$MSE = \frac{1}{N}\sum_{i=1}^{N}(y_i - y_p)^2$$

(7.23)

where, $y_i, y_p, \bar{y}$ and N are experimental data, predicted data, average value of experimental data and number of data points, respectively.

Table 7.2

ANN Network Performance for Different Input Type Combinations

No. of Test	Combination	R^2 of All Data	No. of Test	Combination	R^2 of All Data
				Group of Three Variables	
	Group of One Variable		16	T + S + V	0.9999
1	T	0.5123	17	T + S + D	0.6656
2	S	0.5206	18	T + S + F	0.6593
3	V	0.8766	19	T + V + D	0.9953
4	D	0.2875	20	T + V + F	0.9164
5	F	0.3904	21	S + V + D	0.9895
			22	S + V + F	0.9998
	Group of Two Variables		23	V + D + F	0.9947
6	T + S	0.6368	24	S + D + F	0.6087
7	T + V	0.9036	25	T + D + F	0.6302
8	T + D	0.5926			
9	T + F	0.5312		Group of Four Variables	
10	S + V	0.9895	26	T + S + V + D	0.9999
11	S + D	0.6848	27	T + S + V + F	0.9999
12	S + F	0.6111	28	S + V + D + F	0.9998
13	V + D	0.9782	29	T + V + D + F	0.9963
14	V + F	0.8958	30	T + S + D + F	0.6606
15	D + F	0.4883			
				Group of Five Variables	
			31	T + S + V + D + F	1

Source: Heidari et al. [2016].
Note: T: temperature, S: particle size, V: particle volume, D: particle density, F: fluid viscosity.

Heidari et al. [2016] reported the AARD for water-based nanofluid from the ANN model to be equal to 0.47, which is much lower than 11.2 obtained from correlation by Meybodi et al. [2015]. This observation indicates the superior performance of ANN in the prediction of nanofluid properties.

A sensitivity analysis was also carried out to determine the relative importance of different input variables (T, S, Φ, D, F) used by the ANN model. The performance evaluations of various possible interactions of input variables i.e. one, two, three, four and five variables were investigated. Table 7.2 compares the network performance for different input types i.e. group of input variables. The volume fraction of nanoparticles is observed to be the most effective variable in the group of one variable models due to its higher R^2 value (0.87662) among all parameters. The highest

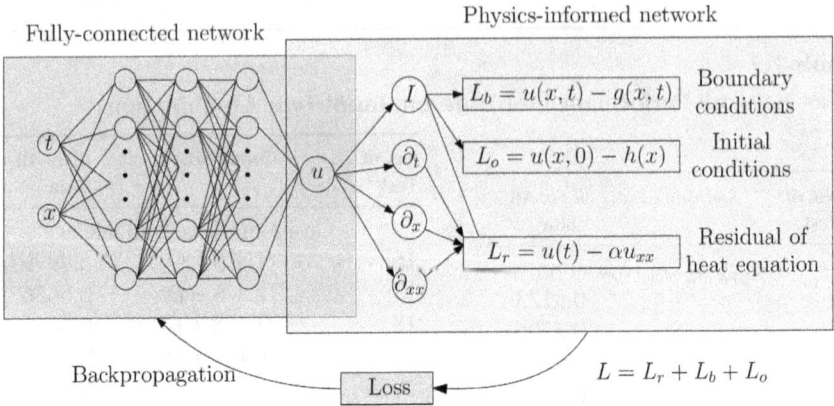

Figure 7.9 Schematic of physics-informed neural network (PINN). A fully connected neural network is used to approximate the solution u(x, t), which is then applied to construct the residual loss L_r, boundary conditions loss L_b and initial conditions loss L_o. (Cai et al. [2021].)

R^2 value observed for 2, 3, 4 and 5 input parameters are equal to 0.98946, 0.99989, 0.99994 and 0.99998, respectively. Overall, this study demonstrated the capability of the ANN technique for accurate prediction of nanofluid viscosity over wide ranges of operating parameters.

7.6 PHYSICS INFORMED ANNs

Both hard computing and soft computing techniques are useful for the design of thermal systems. The shortcomings of the hard computing approach are absence of accurate numerical model for complex problems, non-specific boundary conditions and large computational requirements. The shortcomings of the soft computing technique are the exclusion of the underlying physics of the process and the large resource requirement for the generation of the training data for ANN. The physics informed neural network (PINN) technique can accommodate these shortcomings and will be discussed in this section.

Cai et al. [2021] reported the application of PINNs to various prototype heat transfer problems with realistic conditions not readily tackled by traditional computational methods. A schematic of the PINN framework is demonstrated in Figure 7.9, in which a simple heat equation ($u_t = u_{xx}$) is used to solve a heat transfer problem. The fully connected neural network is used to predict the solution $u(x,t)$ inside the domain, where x and t denotes space and time variable, respectively. The solution $u(x,t)$ is used to calculate different partial derivative terms of the governing equation. Subsequently, the residual loss terms of governing equation are calculated as

$$L_r = \frac{1}{N_r} \sum_{i=1}^{N_r} \left| \frac{\partial u}{\partial t}(x^i t^i) - \alpha \frac{\partial^2 u}{\partial x^2}(x^i t^i) \right|^2 \tag{7.24}$$

where, N_b is the number of data points inside the domain. Residual loss term of the boundary conditions is calculated as

$$L_b = \frac{1}{N_b} \sum_{i=1}^{N_b} \left| u(x^i, t^i) - g^i \right|^2 \tag{7.25}$$

where, N_r is the number of data points in boundary condition and g^i is the boundary condition. Residual loss terms of initial conditions are given as

$$L_o = \frac{1}{N_o} \sum_{i=1}^{N_o} \left| u(x^i t^i) - h^i \right|^2 \tag{7.26}$$

where, N_o is the number of data points representing the initial condition and h^i is the initial condition.

The backpropagation training of the ANN is carried out for minimizing the total loss. The sensor locations, where the data sets are collected, also affect the inferred results. The best sensor locations are generally decided by trial-and-error, which is a costly process. Cai et al. [2021] also proposed an iterative method for optimizing the sensor location. The residual of the heat transfer equation is considered the criterion for selecting the sensor location. They compared the solution from the ANN with an independent reference CFD solution and reported the difference using $L2$ error defined as

$$\varepsilon_v = |V^p - V^*||_2 / ||V^*||_2 \tag{7.27}$$

where, V^p represents one of the predicted quantities and V^* is the corresponding reference solution. Several different types of problems were solved using PINN i.e. (1) forced and mixed convection with unknown thermal boundary conditions on the heated surfaces, (2) Stefan problem for two-phase flow and (3) industrial applications related to power electronics. The results presented for different sample problems demonstrated that PINN can solve ill-posed problems, which are beyond the reach of traditional computational methods. PINN also helps to bridge the gap between computational and experimental heat transfer.

7.7 ANNs FOR DYNAMIC THERMAL SYSTEMS

Most of the thermal systems operate in dynamic conditions responding to changes in the operating parameters and boundary conditions. The performance of a device in a thermal system is also affected by other devices and components that are directly or indirectly connected to it. Therefore, there is a need for dynamic models for prediction of performance under dynamic conditions. There is also a more critical role for such dynamic models in adaptive control systems. Many known control schemes that are accurate require dynamic plant models for their implementation. The traditional available techniques cannot develop accurate dynamic models even for simple thermal devices. However, ANN-based techniques offer a viable alternative. The central scheme in the dynamic modeling by ANN is the addition of

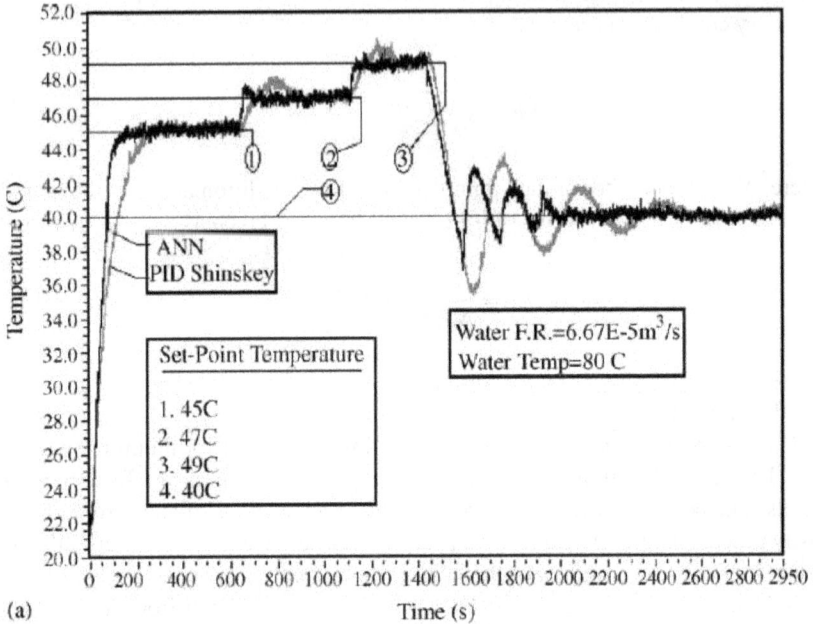

(a)

Figure 7.10 The behavior of the test cell temperature for ANN- and PID-based control with air flow actuation for four changes at the set point. (Varshney and Panigrahi [2005].)

time as a variable for both training and prediction. Another scheme is to provide variables at time t as inputs, and the values of the same variables at later times, $t + \Delta t$ as the outputs.

Varshney and Panigrahi [2005] demonstrated the dynamic modeling and control capability of ANN for a heat exchanger. The air flow rate over a fin tube heat exchanger was controlled by varying the voltage supply to the motor driving the blower. The water flow rate inside the heat exchanger tube was controlled by varying the supply voltage to the motor driving the pump. A thermocouple rack was used to monitor the air flow temperature downstream of the heat exchanger. The water temperature to the heat exchanger was maintained constant using a solid-state relay-based temperature controller. Figure 7.11 shows the general control structure of the neural network-based control. The neural network model and the inverse neural network model are the two important components of the control methodology. The neural network model uses the future process variable i.e. outlet air temperature at later time as output. The process variable i.e. air temperature at previous time step and the corresponding actuator output i.e. supply voltage are used as input. The future actuator output i.e. voltage supply to the motor is the output of the inverse neural network model. The process variables i.e. air temperature and actuator outputs i.e.

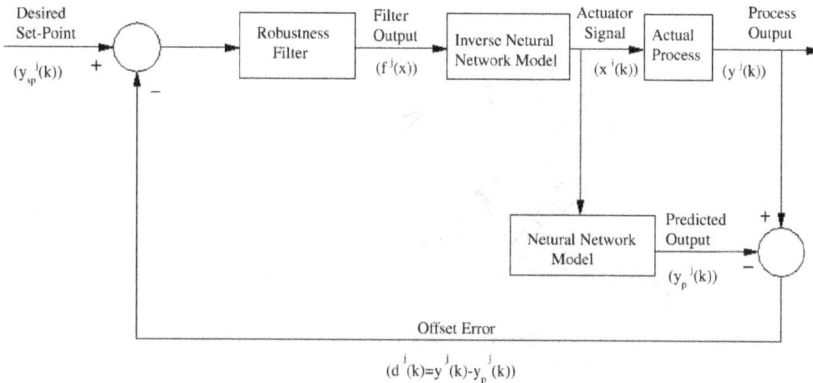

Figure 7.11 Schematic of the control structure of the ANN-based control of a heat exchanger. (Varshney and Panigrahi [2005].)

supply voltage of motor are the input to the inverse neural network model. Both the neural network model and the inverse neural model control use five inputs and one output for single actuation control. The five inputs for the inverse neural network model are the present temperature, the control voltage and temperature at two previous time steps, while the present actuation voltage is the output of the inverse neural network model. The output of the neural network model is the present temperature, while the five inputs are the temperature and voltage at two previous time steps and the present voltage. The robustness filter with a single tuning parameter is implemented to eliminate the steady-state offset due to the mismatch between the plant and the neural network model.

Figure 7.12 shows a typical feed-forward multilayer perceptron neural network used for modeling and control of the heat exchanger. The training data for the ANN were collected using the Labview program supplied by the National Instruments. The voltage to the motor controller from the DAQ card was varied in the range of 0–10 V in a step of 0.2 V and brought back from 10 V to 0 V in the same step size. The time interval was set equal to 60 s. This procedure was repeated with lower step sizes i.e. 0.1 V, 0.05 V and 0.02 V. The data file for the training of single actuation ANN control model was arranged in six columns; the first and second columns show current voltage and temperature data, the third and fourth columns show voltage and temperature data after 60 s and the last two columns show voltage and temperature data after 120 s. The training was carried out with the first five columns as input to the ANN and the last column as the output of the ANN model. The testing data file was generated using the same procedure at a different time i.e. one or two days after the acquisition of the training data. The bias and weight obtained from the training of ANN are used during the testing of the ANN model. The maximum and the rms value of the difference between the output from the ANN model and the testing data are considered as a performance indicator. When using the dual control, both air

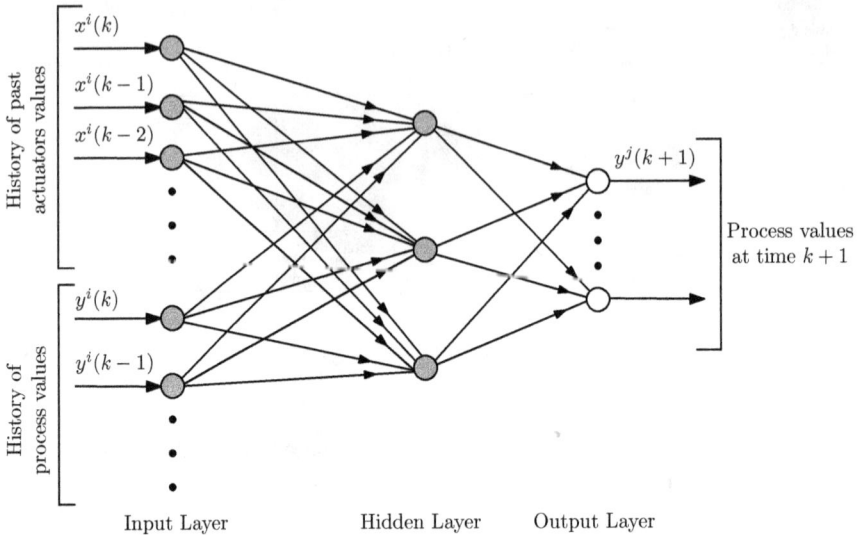

Figure 7.12   Schematic of the multilayer perceptron neural network. (Varshney and Panigrahi [2005].)

Table 7.3

Rise Time Comparison between the ANN- and PID-Based Controls at Different Set Points of Temperature

Method Control	Set Point Temperature (°C)	Rise Time (s)		
		Air Flow Actuation	Water Flow Actuation	Dual Actuation
ANN	45	150	140	75
PID	45	240	170	110
ANN	40	125	130	70
PID	40	210	150	105

Source: Varshney and Panigrahi [2005].

flow and water flow actuator voltages were varied simultaneously and arranged in nine columns of a data file for the training of the ANN, out of which the last two columns are the output of the ANN. The learning rate was set equal to 0.4. The momentum factor was set equal to 0.9. Table 7.3 compares the rise time between proportional-integral-derivative (PID) and ANN controls for different set points of temperature and actuation conditions. The rise time is observed to be significantly lower for ANN control compared to PID control at all actuation conditions. Similarly, the steady-state rms error was also observed to be lower for ANN than PID control.

The ANN- and PID-based control with air flow actuation for multiple changes in set points (40 °C, 45 °C, 47 °C and 49 °C) are presented in Figure 7.10 for both increase and decrease in the set point temperature. Both the ANN- and PID-based control are successful in maintaining multiple changes in set points. The ANN-based control shows a faster response with less overshoot in comparison to the PID-based control for all changes in set points. Overall, this result demonstrates the fast capability of ANN for modeling and control of a dynamic complex thermal system.

7.8   SUMMARY

This chapter introduces the concepts of AI for the design of thermal systems. The concept of an ES was introduced. Sample examples of an ES are discussed to illustrate its role in the design of thermal systems. Subsequently, the soft computing technique was introduced for the design of thermal systems with an emphasis on the ANN technique. The implementation procedure i.e. structure and training of ANN was presented. The capability of ANN for the sample examples i.e. modeling of heat exchanger, representation of physical property database of fluid, physics informed ANN and dynamics of thermal system was also presented.

The other soft computing techniques i.e. FL and GA are also useful for thermal systems depending on the nature of applications. The logic of fuzzy sets is useful in approximate reasoning in ESs. The FL deals with fuzzy sets and logical statements for modeling human-like reasoning problems of the real world. A fuzzy set includes all elements of the universal set of the domain with varying membership values in the interval [0,1]. GA is a stochastic algorithm that mimics the natural process of biological evolution. It is inspired by the way living organisms are adapted to the harsh realities of a hostile world i.e. by evolution and inheritance. The algorithm imitates the evolution of the population by selecting only fit individuals for reproduction. It is an optimum search technique based on the concepts of natural selection and survival of the fittest. It works with a fixed size population of possible solutions to a problem, called individuals, which evolves in time. GAs are denoted by chromosomes, which are usually represented by binary strings. A GA utilizes three principal genetic operators: selection, crossover and mutation.

Hybrid systems can combine different soft computing techniques to enhance overall effectiveness. The hybrid systems can overcome the limitations of single methods, enhance the prediction performance and have a faster response with lower error. Yang [2008] discussed some of the hybrid ANN technologies. Bahiraei et al. [2019] presented several hybrid AI techniques for different applications related to nanofluids. Bakhtiyari et al. [2021] reported the implementation of a hybrid technique for modeling and optimization of laser beam machining. These references and other resources can be reviewed for more details of improved design of thermal systems.

REFERENCES

M. Bahiraei, S. Heshmatian, and H. Moayedi. Artificial intelligence in the field of nanofluids: A review on applications and potential future directions. *Powder Technology*, 353:276–301, Jul 2019. doi: 10.1016/j.powtec.2019.05.034.

A. N. Bakhtiyari, Z. Wang, L. Wang, and H. Zheng. A review on applications of artificial intelligence in modeling and optimization of laser beam machining. *Optics & Laser Technology*, 135:106721, 2021. ISSN 0030-3992. doi: 10.1016/j.optlastec.2020.106721.

S. Cai, Z. Wang, S. Wang, P. Perdikaris, and G. E. Karniadakis. Physics-informed neural networks for heat transfer problems. *Journal of Heat Transfer*, 143(6), Apr 2021. ISSN 0022-1481. doi: 10.1115/1.4050542. 060801.

T. Gonciarz. An expert system for supporting the design and selection of mechanical equipment for recreational crafts. *TransNav, the International Journal on Marine Navigation and Safety of Sea Transportation*, 8(2):275–280, 2014. ISSN 2083-6473. doi: 10.12716/1001.08.02.13.

D. L. Gray and R. L. Webb. Heat transfer and friction correlations for plate finned-tube heat exchangers having plain fins. In *Proceeding of International Heat Transfer Conference 8*. Begellhouse, 1986. doi: 10.1615/ihtc8.1200.

E. Heidari, M. A. Sobati, and S. Movahedirad. Accurate prediction of nanofluid viscosity using a multilayer perceptron artificial neural network (MLP-ANN). *Chemometrics and Intelligent Laboratory Systems*, 155:73–85, 2016. ISSN 0169-7439. doi: 10.1016/j.chemolab.2016.03.031. https://www.sciencedirect.com/science/article/pii/S0169743916300740.

F. C. Mcquiston. Correlation of heat, mass, and momentum trans-port coefficients for plate-fin-tube heat transfer surfaces with staggered tubes. *ASHRAE Trans.*, 84:294–309, 1978a.

F. C. Mcquiston. Heat, mass and momentum transfer data for five plate-fin-tube heat transfer surfaces. *ASHRAE Trans.*, 84:266–293, 1978b.

M. K. Meybodi, S. Naseri, A. Shokrollahi, and A. Daryasafar. Prediction of viscosity of water-based Al2O3, TiO2, SiO2, and CuO nanofluids using a reliable approach. *Chemometrics and Intelligent Laboratory Systems*, 149(Part A):60–69, 2015. doi: 10.1016/j.chemolab.2015.10.001.

M. Mohanraj, S. Jayaraj, and C. Muraleedharan. Applications of artificial neural networks for thermal analysis of heat exchangers – A review. *International Journal of Thermal Sciences*, 90:150–172, 2015. ISSN 1290-0729. doi: 10.1016/j.ijthermalsci.2014.11.030.

A. Pacheco-Vega, G. Diaz, M. Sen, K. T. Yang, and R. L. McClain. Heat rate predictions in humid air-water heat exchangers using correlations and neural networks. *Journal of Heat Transfer*, 123(2):348–354, Oct 2001. ISSN 0022-1481. doi: 10.1115/1.1351167.

B. Paoletti and E. Sciubba. *Artificial Intelligence in Thermal Systems Design: Concepts and Applications*, page 234–278. Cambridge University Press, 1997. doi: 10.1017/CBO9780511529528.011.

K. Varshney and P. Panigrahi. Artificial neural network control of a heat exchanger in a closed flow air circuit. *Applied Soft Computing*, 5(4):441–465, 2005. ISSN 1568-4946. doi: 10.1016/j.asoc.2004.10.004. https://www.sciencedirect.com/science/article/pii/S156849460400105X.

K.-T. Yang. Artificial neural networks (ANNs): A new paradigm for thermal science and engineering. *Journal of Heat Transfer*, 130(9), Jul 2008. ISSN 0022-1481. doi: 10.1115/1.2944238. 093001.

8 Numerical Linear Algebra

In this chapter, we focus on the numerical solution of equation involving one unknown. The general form of such equation is written as

$$f(x) = 0 \tag{8.1}$$

This equation may have many real and complex roots. For brevity, however, here we focus only on finding a single real root using two methods: bisection method and Newton–Raphson method. Being iterative in nature, both methods start from an arbitrary initial guess and are likely to converge to a solution that is close to the initial guess. Let us assume $x = x_1$ to be a solution of equation 8.1. We can test our assumption very easily by evaluating $f(x = x_1)$. If we find $f(x = x_1) \approx 0$, we have just solved the problem. Unfortunately, one has to be very lucky to find the solution just from the first guess. In general, we are likely to find $f(x = x_1) \neq 0$.

8.1 BISECTION METHOD

In the bisection method, our initial goal is to find x_1 and x_2 such that

$$f(x = x_1) f(x = x_2) < 0 \tag{8.2}$$

Let us assume $f(x = x_1) < 0$ and $f(x = x_2) > 0$. If $f(x)$ is continuous in $[x_1, x_2]$, there must be a root $x = x_0$, such that $x_1 < x_0 < x_2$. The goal now is to reduce the interval $[x_1, x_2]$ while maintaining $f(x = x_1) f(x = x_2) < 0$, terminating the procedure when $f(x = x_1)$ or $f(x = x_2)$ is sufficiently close to zero. The procedure replaces either x_1 or x_2 with $x_0 = (x_1 + x_2)/2$. If $f(x = x_0) < 0$, we set $x_1 = x_0$, while $f(x = x_0) > 0$, we set $x_2 = x_0$. The algorithm of the bisection method may be written as follows.

Example: Find one root of the following equation:

$$f(x) = x^2 - 7x + 10 = 0 \tag{8.3}$$

We note that

$$f(x = 0) = 10 \text{ and } f(x = 3) = -2 \Rightarrow f(x = 0) f(x = 3) < 0 \tag{8.4}$$

Clearly equation 8.3 has a root between $x = 0$ and $x = 3$. If we now follow the bisection method, we iterate to generate Table 8.1. It is evident from the table that we are approaching toward $x = 2$, which is a solution of equation 8.3.

The computer program for the above problem is added in the course website.

DOI: 10.1201/9781003049272-8

Algorithm 8.1: Bisection Method

Data: $f(x), x_1, x_2$ where $f(x = x_1) f(x = x_2) < 0$, *maximumError*
Result: x_0 where $f(x = x_0) < maximumError$
$y_1 = f(x = x_1)$
$y_2 = f(x = x_2)$
if $(y_1 * y_2) < 0$ **then**
$\quad$ $error = \min(|y_1|, |y_2|)$
$\quad$ **while** $error > maximumError$ **do**
$\quad\quad$ $x0 = (x1 + x2)/2; y1, y0 = y(x1), y(x0)$
$\quad\quad$ $error = \min(|y_1|, |y_0|)$
$\quad\quad$ **if** $error < maximumError$ **then**
$\quad\quad\quad$ **if** $|y1| < maximumError$ **then**
$\quad\quad\quad$ | $x0 = x1$
$\quad\quad\quad$ **end**
$\quad\quad$ **else if** $(y1 * y0) > 0$ **then**
$\quad\quad\quad$ | $x1 = x0$
$\quad\quad$ **else**
$\quad\quad\quad$ $x2 = x0$
$\quad\quad\quad$ $print(error, x0, x1, x2)$
$\quad\quad$ **end**
$\quad$ **end**
else
$\quad$ | $print(y1, y2); print("invalid input")$
end

8.1.1 CONVERGENCE OF BISECTION METHOD

Consider a function $f(x)$ is continuous in $[x_1, x_2]$, where $f(x_1) f(x_2) < 0$. In the bisection method, described above, in every iteration, we update x_1 and x_2. We denote $x_1^{(n)}$ as the value of x_1 after the nth iteration. Thus, $x_1^{(0)}$ and $x_2^{(0)}$ are the initial values of x_1 and x_2. Following the iterative procedure described above, we know that after the first iteration

$$\text{either } x_1^{(1)} = x_1^{(0)}, x_2^{(1)} = \frac{x_1^{(0)} + x_2^{(0)}}{2} \tag{8.5}$$

$$\text{or } x_1^{(1)} = \frac{x_1^{(0)} + x_2^{(0)}}{2}, x_2^{(1)} = x_2^{(0)} \tag{8.6}$$

From these equations, we can write

$$x_1^{(1)} - x_2^{(1)} = \frac{x_1^{(0)} - x_2^{(0)}}{2} \tag{8.7}$$

Let us proceed to another iteration to find that

$$x_1^{(2)} - x_2^{(2)} = \frac{x_1^{(1)} - x_2^{(1)}}{2} = \frac{x_1^{(0)} - x_2^{(0)}}{2^2} \tag{8.8}$$

Table 8.1

Bisection Method Iterations

Iteration	x_1	x_2	$x_0 = (x_1 + x_2)/2$	$f(x_0)$
0	0	3	1.5	1.75
1	1.5	3	2.25	0.6875
2	1.5	2.25	1.875	0.3906
3	1.875	2.25	2.0625	0.1836
4	1.875	2.0625	1.9688	0.0947
5	1.9688	2.0625	2.0156	0.0466
6	1.9688	2.0156	1.9921	0.0235
7	1.9921	2.0156	2.0039	0.0117
8	1.9921	2.0039	1.9980	0.0059
9	1.9980	2.0039	2.0009	0.0029
10	1.9980	2.0009	1.9995	0.0015

Generalizing this procedure, we find that after the n-th iteration

$$x_1^{(n)} - x_2^{(n)} = \frac{x_1^{(0)} - x_2^{(0)}}{2^n} \tag{8.9}$$

We therefore find that

$$\lim_{n \to \infty} \left[x_1^{(n)} - x_2^{(n)} \right] = 0 \tag{8.10}$$

Since $f(x_1) f(x_1) < 0$, equation 8.10 indicates that the bisection algorithm always ensures convergence to the acceptable solution.

8.2 NEWTON–RAPHSON METHOD

The Newton–Raphson method presents a powerful alternative to the bisection method. The method is generally much faster than the bisection method. Further, as opposed to the *closed* bisection method that requires two initial guesses, the *open* Newton–Raphson method starts with one initial guess. While solving the equation $f(x) = 0$, the Newton–Raphson method relies on linearization of $f(x)$. The linearization is usually achieved by expanding the function via the Taylor series expansion of $f(x)$. Neglecting higher-order terms, a function $f(x)$ may be expanded in the Taylor series as

$$f(x + \Delta x) = f(x) + \Delta x f'(x) + \cdots \tag{8.11}$$

For $f(x + \Delta x) = 0$, we have

$$\Delta x = -\frac{f(x)}{f'(x)} \tag{8.12}$$

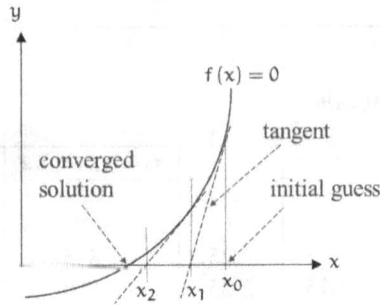

Figure 8.1   Geometric interpretation of the Newton–Raphson method.

The procedure for the Newton–Raphson method, therefore, proceeds as follows:

1. Start iteration from an initial guess $x = x_0$
2. Set $\Delta x = -\dfrac{f(x)}{f'(x)}$
3. Calculate $x = x_0 + \Delta x$
4. Go to step 2 and continue till $f(x) \approx 0$

An algorithm (pseudocode) of the Newton–Raphson method is shown below.

Algorithm 8.2: Newton–Raphson Method

Data: $f(x), f'(x), x_0, maximumError > 0$
Result: x_0 where $|f(x_0)| < maximumError$
$y_1 = f(x_0)$
while $|y_1| \geq maximumError$ **do**
  $\quad y_2 = f'(x_0)$
  $\quad \Delta x = -\dfrac{y_1}{y_2}$
  $\quad x_0 = x_0 + \Delta x$
  $\quad y_1 = f(x_0)$
end
print (x_0); print("solution converged")

The geometric interpretation of the Newton–Raphson method is described in Figure 8.1. Here, we intend to find the solution of the equation $f(x) = 0$. The initial guess, of the solution, is x_0. If we draw a tangent at the point $(x_0, f(x_0))$, the tangent cuts the x-axis at the point x_1. x_1 is the value of the solution after one iteration. In the same manner, we have the value of the solution after two iterations is x_2. If we proceed this way, we will eventually reach the converged solution with acceptable accuracy. It should be noted here that, unlike the bisection method, the convergence of the Newton–Raphson method is not guaranteed. A detailed discussion of the Newton–Raphson method is beyond the scope of this book.

Example: For an example, let us try to evaluate $\sqrt{2}$ using the Newton–Raphson method. As seen before, one way to find the value of $\sqrt{2}$ would be to solve the following equation and pick the positive root

$$f(x) = x^2 - 2 = 0 \qquad (8.13)$$

Table 8.2

Newton–Raphson Method Iterations

Iteration	x_0	Δx	x_1
0	2	−0.5	1.5
1	1.5	−0.0833	1.4167
2	1.4167	−0.00245	1.414215
3	1.414215	-2.1239×10^{-6}	1.414213

SOLUTION

We note that

$$f'(x) = 2x \tag{8.14}$$

and thus

$$\Delta x = -\frac{f(x)}{f'(x)} = -\frac{x^2 - 2}{2x} = \frac{1}{x} - \frac{x}{2} \tag{8.15}$$

For an initial guess of x_0, the solution after the first iteration x_1 would be

$$x_1 = x_0 + \Delta x(x = x_0) = x_0 + \frac{1}{x_0} - \frac{x_0}{2} = \frac{1}{x_0} + \frac{x_0}{2} \tag{8.16}$$

If we now follow the procedure suggested by the Newton–Raphson algorithm, we iterate to generate Table 8.2. It is evident from the table that we are approaching toward $x = \sqrt{2}$ which again is a solution of equation 8.13

The computer program for the above problem is added in the course website.

PROBLEMS

1. Modify the PYTHON program, shown in the book, to find the value of $\sqrt{2}$ using the bisection method. The value should be correct up to four decimal places.
2. Modify the PYTHON program, shown in the book, to find a root of $x^2 - 7x + 10$ using the Newton–Raphson method. The value should be correct up to four decimal places.

8.3   EIGENVALUES AND EIGENVECTORS

Any matrix $\mathbf{A}$, with eigenvalue λ and the corresponding eigenvector $\mathbf{x}$, must satisfy

$$\mathbf{A}\mathbf{v} = \lambda \mathbf{v} \tag{8.17}$$

Thus, for a matrix $\mathbf{A}$, eigenvalues are the roots λ of the polynomial $p(\lambda)$, where

$$p(\lambda) = det(\mathbf{A} - \lambda \mathbf{I}) \tag{8.18}$$

where, $\mathbf{I}$ is the identity matrix. The polynomial $p(\lambda)$ is called the characteristic polynomial of matrix $\mathbf{A}$.

For small matrices, we may solve the equation $p(\lambda) = 0$ to calculate the eigenvalues λ. Once we know λ, we find the eigenvector $\mathbf{x}$ by solving the equation $\mathbf{Ax} = \lambda\mathbf{x}$. Unfortunately, the same methods of finding eigenvalues and eigenvectors cannot be used for large matrices. For large matrices, the calculation of the determinant is a daunting task and so is finding the solution of $p(\lambda) = 0$.

8.4 POWER ITERATIONS

The power iteration, also called the power method, is designed to find dominant eigenvalue of a diagonlizable matrix. Consider an $n \times n$ diagonalizable matrix $\mathbf{A}$. Since the matrix $\mathbf{A}$ is diagonalizable, it has linearly independent eigenvectors $\mathbf{v}_1, \mathbf{v}_2, \ldots, \mathbf{v}_n$, where the corresponding eignevalues are $\lambda_1, \lambda_2, \ldots, \lambda_n$. The existence of a dominant eigenvale λ_1 suggests that $|\lambda_1| > |\lambda_i|$ where $i = 2, 3, \ldots, n$. Power iteration numerically finds λ_1 and the corresponding eigenvector $\mathbf{v}_1$ using the following algorithm.

Algorithm 8.3: Power Iteration

Data: Diagonalizable $n \times n$ matrix $\mathbf{A}$; a randomly generated n-dimensional vector $\mathbf{v}$ such that, $\|\mathbf{v}\| \neq 0$; $maximumError > 0$

Result: Dominant eigenvalue λ, corrsponding eigenvector $\mathbf{v}$

$\mathbf{v} = \frac{\mathbf{v}}{\|\mathbf{v}\|}$

$error = 10$

```
/* value of error is assigned arbitrarily, such that error > maximumError
   */
```

while $error > maximumError$ **do**

$\qquad \mathbf{u} = \frac{\mathbf{Av}}{\|\mathbf{Av}\|}$

$\qquad error = \|\mathbf{u} - \mathbf{v}\|$

$\qquad \mathbf{v} = \mathbf{u}$

end

$\lambda = \|\mathbf{Av}\|$

Example: Find the dominant eigenvalue and the corresponding normalized eigenvector for the following matrix:

$$\mathbf{A} = \begin{bmatrix} 7 & 3 \\ 6 & 4 \end{bmatrix} \tag{8.19}$$

SOLUTION

Analytically, we may find the eigenvalues as follows:

$$\det(\mathbf{A} - \lambda\mathbf{I}) = 0 \Rightarrow \lambda = 10, 1 \tag{8.20}$$

The normalized eigenvector corresponding to the eigenvalue $\lambda = 10$ is

$$\mathbf{v} = \frac{1}{\sqrt{2}} \begin{bmatrix} 1 \\ 1 \end{bmatrix} = \begin{bmatrix} 0.7071 \\ 0.7071 \end{bmatrix} \tag{8.21}$$

Table 8.3

Power Iteration

Iteration	v^T	Error
0	1,0	–
1	0.7592566, 0.65079137	0.693892
2	0.71240324, 0.70177035	0.0692394
3	0.70763704, 0.70657612	0.00676845
4	0.70715981, 0.70705374	0.0006751855
5	0.70711208, 0.70710148	6.7502×10^{-5}
6	0.70710731, 0.70710625	6.75×10^{-6}
6	0.70710683, 0.70710673	6.75×10^{-7}

For the initial guess of the eigenvector, we assume

$$\mathbf{v}^{(0)} = \begin{bmatrix} 1 \\ 0 \end{bmatrix} \tag{8.22}$$

Thus,

$$\mathbf{A}\mathbf{v}^{(0)} = \begin{bmatrix} 7 & 3 \\ 6 & 4 \end{bmatrix} \begin{bmatrix} 1 \\ 0 \end{bmatrix} = \begin{bmatrix} 7 \\ 6 \end{bmatrix} \Rightarrow \left\| \mathbf{A}\mathbf{v}^{(0)} \right\| = \sqrt{(7^2 + 6^2)} = 9.22 \tag{8.23}$$

Therefore,

$$\mathbf{v}^{(1)} = \frac{\mathbf{A}\mathbf{v}^{(0)}}{\left\| \mathbf{A}\mathbf{v}^{(0)} \right\|} = \begin{bmatrix} 0.76 \\ 0.65 \end{bmatrix} \tag{8.24}$$

We can calculate the error as follows:

$$error = \left\| \mathbf{v}^{(1)} - \mathbf{v}^{(0)} \right\| = 0.694 \tag{8.25}$$

The results of the iteration are given in Table 8.3. The computer program of the above problem is added in the course website.

Clearly, the results match the analytical solution, shown in equation 8.21.

8.5 CONVERGENCE

For a diagonalizable $n \times n$ matrix $\mathbf{A}$, the eigenvectors $\mathbf{v}_1, \mathbf{v}_2, \cdots, \mathbf{v}_n$ are linearly independent. Thus, any n-dimensional vector $\mathbf{x}$ may be represented as

$$\mathbf{x} = \sum_{j=1}^{n} (a_j \mathbf{v}_j) \text{ where } a_j \text{ are scalars} \tag{8.26}$$

Thus, we can write

$$\mathbf{A}\mathbf{x} = \sum_{j=1}^{n} (a_j \mathbf{A}\mathbf{v}_j) \tag{8.27}$$

We know

$$\mathbf{A}\mathbf{v}_j = \lambda_j \mathbf{v}_j \qquad (8.28)$$

Using equations 8.27 and 8.28, we have

$$\mathbf{A}\mathbf{x} = \sum_{j=1}^{n} \left(a_j \lambda_j \mathbf{v}_j \right) \qquad (8.29)$$

Similarly, we can show

$$\mathbf{A}^2\mathbf{x} = \sum_{j=1}^{n} \left(a_j \lambda_j^2 \mathbf{v}_j \right) \qquad (8.30)$$

Continuing in the same manner, we write for a positive integer k

$$\mathbf{A}^k\mathbf{x} = \sum_{j=1}^{n} \left(a_j \lambda_j^k \mathbf{v}_j \right) = \lambda_1^k \sum_{j=1}^{n} \left(\frac{\lambda_j^k}{\lambda_1^k} a_j \mathbf{v}_j \right) \qquad (8.31)$$

We now assume that

$$|\lambda_1| > |\lambda_i| \text{ where } i = 2,3,\ldots,n \qquad (8.32)$$

The above assumption suggests that

$$\lim_{k \to \infty} \frac{\lambda_j^k}{\lambda_1^k} = 0 \text{ for } j = 2,3,\ldots,n \qquad (8.33)$$

Using equations 8.31 and 8.33,

$$\lim_{k \to \infty} \mathbf{A}^k\mathbf{x} = \sum_{j=1}^{n} \left(a_j \lambda_j^k \mathbf{v}_j \right) = \lambda_1^k a_1 \mathbf{v}_1 \qquad (8.34)$$

Furthermore,

$$\lim_{k \to \infty} \left\| \mathbf{A}^k\mathbf{x} \right\| = \lambda_1^k a_1 \|\mathbf{v}_1\| \qquad (8.35)$$

Using equations 8.34 and 8.35, we have the unit eigenvector $\hat{\mathbf{v}}_1$

$$\lim_{k \to \infty} \frac{\mathbf{A}^k\mathbf{x}}{\left\| \mathbf{A}^k\mathbf{x} \right\|} = \frac{\mathbf{v}_1}{\|\mathbf{v}_1\|} = \hat{\mathbf{v}}_1 \qquad (8.36)$$

We also know

$$\mathbf{A}\hat{\mathbf{v}}_1 = \lambda_1 \hat{\mathbf{v}}_1 \Rightarrow \|\mathbf{A}\hat{\mathbf{v}}_1\| = \lambda_1 \|\hat{\mathbf{v}}_1\| = \lambda_1 \qquad (8.37)$$

Power iteration thus converges for all diagonalizable matrices that contain a dominant eigenvalue λ_1. The rate of converge, howvever, depends on the ratios λ_i/λ_1 for $i > 1$. High values of the ratios λ_i/λ_1 will lead to faster convergence. While the convergence of power method does not depend on the initial guess, the above discussion indicates that if the initial guess of eigenvector exactly matches with an eigenvector $\hat{\mathbf{v}}_i$, then the power method converges to λ_i and $\hat{\mathbf{v}}_i$ even if the eigenvalue λ_i is not the dominant one.

8.6 INVERSE POWER ITERATIONS

Recall the classical eigenvalue problem for an $n \times n$ matrix $\mathbf{A}$ with eigenvalues λ_i and the corresponding eigenvectors $\mathbf{v}_i$ for $i = 1, 2, \ldots, n$

We know

$$\mathbf{A}\mathbf{v} = \lambda \mathbf{v} \tag{8.38}$$

Consider a real number q such that

$$q \neq \lambda_i \text{ for } i = 1, 2, \ldots, n \tag{8.39}$$

From equations 8.38 and 8.39,

$$(\mathbf{A} - q\mathbf{I})\mathbf{v} = (\lambda - q)\mathbf{v} \Rightarrow (\mathbf{A} - q\mathbf{I})^{-1}\mathbf{v} = \frac{1}{\lambda - q}\mathbf{v} \tag{8.40}$$

From equations 8.38 and 8.40, we find that the eigenvectors of the matrices $\mathbf{A}$ and $\mathbf{A} - q\mathbf{I}$ are the same. We also see that for the same eigenvector $\mathbf{v}_i$, if the eigenvalue of the matrix $\mathbf{A}$ is λ_i, the eigenvalue of $\mathbf{A} - q\mathbf{I}$ is $\mu = 1/(\lambda - q)$. Now with careful selection of q, we may be able to find all the eigenvalues of λ.

For instance, if we apply the power method to find the eigenvalue of the matrix $(\mathbf{A} - q\mathbf{I})^{-1}$ for $q = 0$, we obtain maximum μ and thus the minimum λ. Similarly, if we select q in such a way that $|q - \lambda_i| < |q - \lambda_j|$ where $j \neq i$ and $j = 1, 2, \cdots, n$, applying the power method on the matrix $(\mathbf{A} - q\mathbf{I})^{-1}$, we may find the eigenvalue λ_i. Overall, the inverse power method provides all the eigenvalues of matrix $\mathbf{A}$.

PROBLEMS

1. Manually calculate the dominant eigenvalue of the following matrix:

$$\mathbf{A} = \begin{bmatrix} 2 & 1 & 1 \\ 1 & 2 & 1 \\ 1 & 1 & 2 \end{bmatrix}$$

2. Write a PYTHON program to find the dominant eigenvalue of the following matrix:

$$\mathbf{A} = \begin{bmatrix} 2 & 1 & 1 \\ 1 & 2 & 1 \\ 1 & 1 & 2 \end{bmatrix}$$

3. Write a PYTHON program for the inverse power method and thus find the minimum eigenvalue of the following matrix:

$$\mathbf{A} = \begin{bmatrix} 7 & 3 \\ 6 & 4 \end{bmatrix}$$

4. Write a PYTHON program for the inverse power method and thus find the all the eigenvalues of the following matrix:

$$A = \begin{bmatrix} 2 & 1 & 1 \\ 1 & 2 & 1 \\ 1 & 1 & 2 \end{bmatrix}$$

5. Write a PYTHON program for the inverse power method and thus find the all the eigenvalues of the following matrix:

$$A = \begin{bmatrix} 1 & 1 & 1 \\ 1 & 1 & 0 \\ 1 & 0 & 1 \end{bmatrix}$$

8.7 CURVE FITTING

Consider an experiment where we are varying a certain variable **x** and measuring another variable **y**. At the end of the experiment, we have generated a discrete data set (x_i, y_i) for $i = 1, 2, \ldots, n$ where n is a positive integer. No matter how carefully we conduct the experiments, all experimental data contain some noise. Since the data are noisy, fitting all the data exactly through a smooth curve may not provide a physically meaningful trend. In curve fitting, we do not intend to fit all the data exactly, rather we try to fit the noisy data, as closely as possible, through a curve $Y = f(x, a_0, a_1, \ldots, a_m)$ that is physically meaningful. Here, $a_i, i = 1, 2, \ldots, m$ are adjustable parameters. Adjusting the values of a_i provides the *best fit*. While there are many options to define and obtain the best fit, the most common technique is *least square fitting*, where the best fit is obtained by minimizing the following objective function:

$$q(a_0, a_1, \ldots, a_m) = \sum_{i=1}^{n} (r_i)^2 = \sum_{i=1}^{n} (y_i - Y_i)^2 \tag{8.41}$$

where, r_i are the residuals, quantify the differences between fit and the measured data, defined as

$$r_i = y_i - Y_i \text{ for } i = 1, 2, \ldots, n \tag{8.42}$$

To minimize the objective function, we enforce

$$\frac{\partial q}{\partial a_i} = 0 \text{ for } i = 1, 2, \ldots, m \tag{8.43}$$

The above $m+1$ equations, in principle, may be solved to obtain $m+1$ unknowns $a_0, a_1, \ldots, a_m$. However, to deduce a mathematical expression of $\frac{\partial q}{\partial a_i}$, we must first specify a physically meaningful form of $Y = f(x)$. Finally, equation 8.43 may be linear or nonlinear depending on the choice of the function $Y = f(x)$. In this chapter, we will discuss linear and polynomial curve fittings, both of which are in the domain of linear least square problems.

8.8 FITTING OF A STRAIGHT LINE

Let us assume a discrete data set (x_i, y_i) where x_i are the independent variables. We wish to fit the above data in a straight line

$$Y = a_0 + a_1 x \tag{8.44}$$

The residuals r_i quantify the differences between fit and the measured data

$$r_i = y_i - a_0 - a_1 x_i \text{ for } i = 1, 2, \dots, n \tag{8.45}$$

Thus, the objective function q is defined as

$$q(a_0, a_1) = \sum_{i=1}^{n} (r_i)^2 = \sum_{i=1}^{n} (y_i - a_0 - a_1 x_i)^2 \tag{8.46}$$

Therefore,

$$\frac{\partial q}{\partial a_0} = -2 \sum_{i=1}^{n} (y_i - a_0 - a_1 x_i) = 0 \tag{8.47}$$

$$\frac{\partial q}{\partial a_1} = -2 \sum_{i=1}^{n} \left[x_i (y_i - a_0 - a_1 x_i) \right] = 0 \tag{8.48}$$

Using these equations, we have

$$\sum_{i=1}^{n} (y_i - a_0 - a_1 x_i) = 0 \tag{8.49}$$

$$\sum_{i=1}^{n} \left[x_i (y_i - a_0 - a_1 x_i) \right] = 0 \tag{8.50}$$

Rearranging

$$n a_0 + \left(\sum x_i \right) a_1 = \sum y_i \tag{8.51}$$

$$\left(\sum x_i \right) a_0 + \left(\sum x_i^2 \right) a_1 = \sum (x_i y_i) \tag{8.52}$$

Solving the above two equations, we obtain

$$a_0 = \frac{\sum x_i^2 \sum y_i - \sum x_i \sum (x_i y_i)}{n \sum X_i^2 - (\sum x_i)^2} \tag{8.53}$$

$$a_1 = \frac{n \sum (x_i y_i) - \sum x_i \sum y_i}{n \sum X_i^2 - (\sum x_i)^2} \tag{8.54}$$

8.9   FITTING OF A POLYNOMIAL

Let us once again use the same discrete data set (x_i, y_i) for $i = 1, 2, \ldots, n$ where x_i are the independent variables. We wish to fit the above data with a polynomial of degree m, where $m < n$.

$$Y = \sum_{j=0}^{m} \left(a_j x^j \right) \tag{8.55}$$

The residuals r_i quantify the differences between fit and the measured data

$$r_i = y_i - \sum_{j=0}^{m} \left(a_j x_i^j \right) \text{ for } i = 1, 2, \ldots, n \tag{8.56}$$

Thus, the objective function q is defined as

$$q(a_0, a_1, \ldots, a_m) = \sum_{i=1}^{n} (r_i)^2 \tag{8.57}$$

Therefore,

$$\frac{\partial q}{\partial a_0} = \sum_{i=1}^{n} \left(2r_i \frac{\partial r_i}{\partial a_0} \right) = 0 \tag{8.58}$$

$$\frac{\partial q}{\partial a_1} = \sum_{i=1}^{n} \left(2r_i \frac{\partial r_i}{\partial a_1} \right) = 0$$

$$\cdots\cdots\cdots\cdots\cdots$$

$$\frac{\partial q}{\partial a_m} = \sum_{i=1}^{n} \left(2r_i \frac{\partial r_i}{\partial a_m} \right) = 0$$

Simplifying this equation, we obtain the following set of $m+1$ linear equations:

$$\sum_{j=0}^{m} \left(a_j \sum_{i=1}^{n} x_i^{j+k} \right) = \sum_{i=1}^{n} \left(x_i^k y_i \right) \text{ for } k = 0, 1, 2, \ldots, m \tag{8.59}$$

We may expand the above equations to have the following set of equations:

$$a_0 \sum_{i=1}^{n} x_i^0 + a_1 \sum_{i=1}^{n} x_i^1 + a_2 \sum_{i=1}^{n} x_i^2 + \cdots + a_m \sum_{i=1}^{n} x_i^m = \sum_{i=1}^{n} \left(x_i^0 y_i \right) \tag{8.60}$$

$$a_0 \sum_{i=1}^{n} x_i^1 + a_1 \sum_{i=1}^{n} x_i^2 + a_2 \sum_{i=1}^{n} x_i^3 + \cdots + a_m \sum_{i=1}^{n} x_i^{m+1} = \sum_{i=1}^{n} \left(x_i^1 y_i \right)$$

$$\vdots$$

$$a_0 \sum_{i=1}^{n} x_i^m + a_1 \sum_{i=1}^{n} x_i^{m+1} + a_2 \sum_{i=1}^{n} x_i^{m+2} + \cdots + a_m \sum_{i=1}^{n} x_i^{2m} = \sum_{i=1}^{n} \left(x_i^m y_i \right)$$

Solving the above $m+1$ equations, we evaluate $a_0, a_1, \cdots, a_m$.

Example: Fit a straight line and a second-degree polynomial through the data points shown in Table 8.4.

Table 8.4

Data Points in the Example in Section 8.4

x_i	0	1	2	3	4	5	6	7	8
y_i	5	4	4	7	12	8	13	16	18

SOLUTION

To fit a straight line, we use equation 8.53 to find the parameters a_0 and a_1. For the second-order polynomial fitting, we use equation 8.59 to obtain a_0, a_1 and a_2. The computer program of the above problem is given in the course website. The result is shown in Figure 8.2.

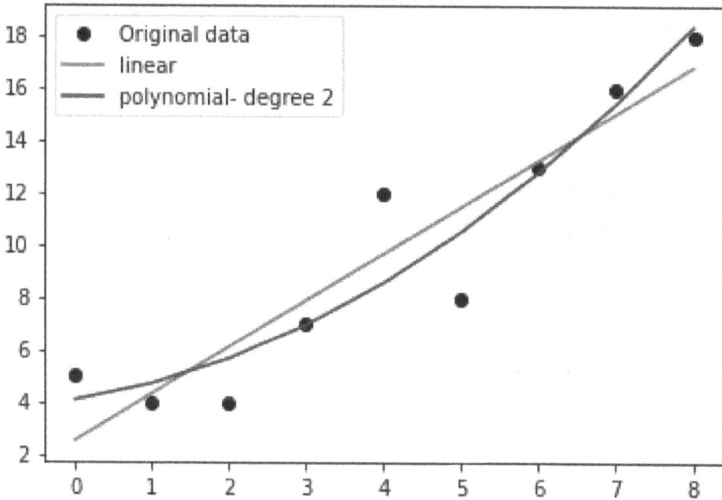

Figure 8.2 Fitting of arbitrary data using a straight line and a second degree polynomial.

8.10   ERROR ESTIMATION

Assuming that all the fitting parameters have been computed with high accuracy, there are multiple ways to look at the errors in curve fitting. First we would like to quantify *goodness of fit* that can be defined simply as the objective function q itself. A small value of q indicates the fitted curve closely follows the discrete points that are used for the curve fitting. In reality, the discrete points, used in the curve fitting, are coming from experiments and thus the discrete data set (x_i, y_i) for $i = 1, 2, \ldots, n$ contains noise. A small q, therefore, may lead to creeping of noise in our fitting. So, while a small value of q indicates the closeness of the fitted curve with the discrete points, a small value of q does not necessarily indicate physically meaningful results.

Another important discussion point in curve fitting is the extent of dependence of the variable **y** on the independent variable **x**. Most promising way to quantify such dependence is to evaluate the *correlation coefficient*, as defined below

$$r = \frac{covariance\,(\mathbf{x}, \mathbf{y})}{\sqrt{variance\,(\mathbf{x}) \times variance\,(\mathbf{y})}} = \frac{\sum_{i=1}^{n}\left(d_{x,i}d_{y,i}\right)}{\sqrt{\left(\sum_{i=1}^{n}d_{x,i}^2\right)\left(\sum_{i=1}^{n}d_{y,i}^2\right)}} \qquad (8.61)$$

where,

$$d_{x,i} = x_i - \frac{\sum_{i=1}^{n}x_i}{n}\,;\, d_{y,i} = y_i - \frac{\sum_{i=1}^{n}y_i}{n} \qquad (8.62)$$

The value of the correlation coefficient is limited to $-1 \le r \le 1$. Small values of $|r|$ indicate little dependence of **y** on **x**. On the other hand, the higher value of $|r|$, closes to 1, indicates significant dependence of **x** on **y**. For Example 8.9, the value of r is found to be 0.927, indicating a good choice of variables for the curve fitting.

PROBLEMS

1. Write a PYTHON program to fit a straight line and a second-degree polynomial through the data points shown in Table 8.4.

x_i	1	2	3	4	5	6	7	8	9	10
y_i	1.3	3.5	4.2	5	7	8.8	10.1	12.5	13	15.6

2. For the data set shown in the previous problem, fit the following curve:

$$Y = a_0 \exp(a_1 x)$$

where a_0 and a_1 are the fitting parameter, to be evaluated.

3. Comparing the above three curve fittings, from the previous two problems, compare their *goodness of fit*.

8.11   SOLUTION OF ALGEBRAIC EQUATIONS

In this chapter, we focus on the numerical solution of a set of linear equation. The general form of such equation is written as

$$\mathbf{Ax} = \mathbf{b} \qquad (8.63)$$

Here, in equation 8.63, $\mathbf{A}$ is an $n \times n$ square matrix while $\mathbf{x}$ and $\mathbf{b}$ are column vectors of size n. To begin with, we assume that the elements of the matrix $\mathbf{A}$ and the vector $\mathbf{b}$ are real constants. Such assumptions create a system of linear equations where the components of the vector $\mathbf{x}$ are unknowns. We will first see direct and iterative methods for the solution of such systems of linear equations. Later in this chapter, we will extend our understanding to the system of nonlinear equations.

For an $\mathbf{Ax} = \mathbf{b}$ system, we may have unique solution, no solution and many solutions. For brevity, in this chapter, we will only discuss the cases of unique solution.

8.12   GAUSSIAN ELIMINATION

Gaussian elimination intends to solve the system of linear equations. Let us first see the idea behind Gaussian elimination by taking a system with two equations and two unknowns.

$$3x + 2y = 5 \tag{8.64}$$

$$4x + 5y = 9 \tag{8.65}$$

The goal now is to eliminate the coefficients below the diagonal elements, a step known as *forward elimination*. Therefore, in this case, the first step is

$$\text{no change in first row} \quad 3x + 2y = 5 \tag{8.66}$$

$$\text{row2 (new)=row2 (old)-row1} \times \frac{4}{3} \qquad \frac{7}{3}y = \frac{7}{3} \tag{8.67}$$

Now the second step; from equation 8.67,

$$y = 1 \tag{8.68}$$

and now the third and final step, known as *back substitution*; using equations 8.66 and 8.68

$$x = \frac{5 - 2y}{3} = 1 \tag{8.69}$$

Let's extend our understanding to a 3×3 system

$$x_1 + x_2 - x_3 = 1 \tag{8.70}$$
$$3x_1 + x_2 + x_3 = 9$$
$$x_1 - x_2 + 4x_3 = 8$$

Now after the first step of forward elimination, we have

$$x_1 + x_2 - x_3 = 1 \tag{8.71}$$
$$-2x_2 + 4x_3 = 6$$
$$-2x_2 + 5x_3 = 7$$

Elimination one more time and we have

$$x_1 + x_2 - x_3 = 1 \tag{8.72}$$
$$-2x_2 + 4x_3 = 6$$
$$x_3 = 1$$

Now the back substitution provides

$$x_1 = 3 \tag{8.73}$$
$$x_2 = -1$$
$$x_3 = 1$$

We argue that the systems 8.70–8.73 are *equivalent* to each other. Two linear systems are equivalent to each other if one can be deduced from the other through *elementary operations*. Elementary operations involve: multiplication of an equation by a constant, exchange of two equations of the same system and replacing an equation with a *linear combination* of the equation with other equations of the system.

We will revisit the simple linear systems once again. But for now, let's see how the Gaussian elimination works for the most general system of equations

$$a_{11}x_1 + a_{12}x_2 + \cdots + a_{1n}x_n = b_1 \tag{8.74}$$
$$a_{21}x_1 + a_{22}x_2 + \cdots + a_{2n}x_n = b_2$$
$$\vdots$$
$$a_{n1}x_1 + a_{n2}x_2 + \cdots + a_{nn}x_n = b_n$$

These equations may be written in a simpler way, known as an index notation or Einstein notation:

$$a_{ij}x_j = b_i \quad \text{where} \quad a_{ij}x_j = \sum_{j=1}^{n} \left(a_{ij}x_j \right) \quad \text{for} \quad i = 1, 2, \ldots, n \tag{8.75}$$

8.12.1　FORWARD ELIMINATION

In the forward elimination step, our goal is to do something to make $a_{k+1,k} = 0$, where $k = 1, 2, \ldots, n$. To do so, the first step is to set all elements of column 1, except the a_{11}, to zero, as follows:

$$row2(new) = row2(old) - row1 \times \frac{a_{21}}{a_{11}} \tag{8.76}$$
$$row3(new) = row3(old) - row1 \times \frac{a_{31}}{a_{11}}$$
$$\vdots$$
$$rown(new) = rown(old) - row1 \times \frac{a_{n1}}{a_{11}}$$

Repeating the same procedure, we can make all the elements below diagonal to be zero. The above step will change the **A** matrix and **b** vector as follows:

$$a_{11}x_1 + a_{12}x_2 + \cdots + a_{1n}x_n = b_1 \tag{8.77}$$
$$a'_{22}x_2 + \cdots + a'_{2n}x_n = b'_2$$
$$\vdots$$
$$a'_{nn}x_n = b'_n$$

8.12.2   BACK SUBSTITUTION

Back substitution starts from the last row:

$$x_n = \frac{b'_n}{a'_{nn}} \tag{8.78}$$

The $(n-1)$th row now provides the value of x_{n-1}:

$$x_{n-1} = \frac{b'_{n-1} - a'_{n-1,n}x_n}{a'_{n-1,n-1}} \tag{8.79}$$

In general, x_i for $i = n-1, n-2, \ldots, 1$ may be obtained from

$$x_i = \frac{b'_i - \sum_{j=i+1}^{n}\left(a'_{ij}x_j\right)}{a'_{ii}} \tag{8.80}$$

The solution procedure, described above, may be summarized in the algorithm below. The procedure is called *naive* since it is a very basic form of Gaussian elimination and suffers from several shortcomings.

8.12.3   HOW TO IMPROVE THE SOLUTION

The errors associated with Gaussian elimination stem from what we call *round-off error*. Any realistic calculation is *finite precision* in nature. Finite-precision calculations, while dealing with real numbers, can take only few digits after the decimal leading to round-off errors in the calculation. Round-off error is common in all numerical calculations. Solution algorithm must be carefully crafted to control such errors. In Gaussian elimination, we can reduce round-off errors using the following methods.

Techniques for improving Gaussian elimination

1. We can simply increase the number of significant digits used in the calculation. As the number of significant digits increases, the solution calls for more computation time and storage space. Resource limitations, therefore, prohibits the increase in the number of significant digits beyond some optimum value.

Algorithm 8.4: Naive Gaussian Elimination

Data: $a_{ij} \in \mathscr{R}$ where $a_{ii} \neq 0$ and $b_i \in \mathscr{R}$ for $i = 1, 2, \ldots, n$ and $j = 1, 2, \ldots, n$
Result: x_i for $i = 1, 2, \ldots, n$ such that $\sum_{j=1}^{n} a_{ij}x_j = b_i$

```
/* Forward elimination starts                                                    */
```
for $k = 1, n-1$ **do**
    for $i = k+1, n$ **do**
        $factor = a_{ik}/a_{kk}$
        for $j - k+1, n$ **do**
            $a_{ij} = a_{ij} - factor * a_{kj}$
        end
        $b_i = b_i - factor * b_k$
    end
end
```
/* Forward elimination ends, back substitution starts                            */
```
$x_n = b_n/a_{nn}$
for $i = n-1, 1, -1$ **do**
    $sum = b_i$
    for $j = i+1, n$ **do**
        $sum = sum - a_{ij} * x_j$
    end
    $x_i = sum/a_{ii}$
end
```
/* Back substitution ends, program ends                                          */
```

---

2. *Partial Pivoting*: The goal of partial pivoting is to rearrange the equations in a way such that $|a_{ii}| \geq |a_{ij}|$ where $i, j = 1, 2, \ldots n$, and $j > i$.
3. *Scaling*: Partial pivoting requires a comparison of $|a_{ii}|$ with $|a_{ij}|$ for $j > i$. Such a comparison may break down if we multiply an equation with a large number. Therefore, before we compare $|a_{ii}|$ with $|a_{ij}|$, we make $\max|a_{ij}| = 1$ before partial pivoting. Such an operation is called *scaling*.

8.13   JACOBI AND GAUSS–SEIDEL ITERATIONS

We wish to solve for **x** in the following system of linear equations, where **A** is an $n \times n$ matrix with elements a_{ij} and **b** is an n-dimensional column vector of elements b_i

$$\mathbf{Ax} = \mathbf{b} \Rightarrow \sum_{j=1}^{n} \left(a_{ij}x_j \right) = b_i \text{ for } i = 1, 2, \ldots, n \tag{8.81}$$

The Jacobi iteration uses the following algorithm to solve the above problem:

$$x_i^{(k)} = \frac{b_i - \sum_{j=1, j \neq i}^{n} \left[a_{ij}x_j^{(k-1)} \right]}{a_{ii}} \tag{8.82}$$

where, k is the iteration number and $k = 0$ indicates initial guess of the unknown vector x. In matrix notation, the above algorithm may be written as

$$\mathbf{x}^{(k)} = \mathbf{M}\mathbf{x}^{(k-1)} + \mathbf{C} \tag{8.83}$$

where $\mathbf{A} = \mathbf{D} + \mathbf{L} + \mathbf{U}$ and $\mathbf{M} = -\mathbf{D}^{-1}(\mathbf{L} + \mathbf{U}), \mathbf{C} = \mathbf{D}^{-1}\mathbf{b}$

$\mathbf{D}$: diagonal matrix

$\mathbf{L}$: strictly (zero-diagonal) lower-triangular matrix,

$\mathbf{U}$: strictly (zero-diagonal) upper-triangular matrix

The exact solution of equation 8.81 is given by $\mathbf{x} = \mathbf{x}^{(*)}$, where

$$\mathbf{x}^{(*)} = \mathbf{M}\mathbf{x}^{(*)} + \mathbf{C} \tag{8.84}$$

The error vector after the kth iteration is given by

$$\mathbf{e}^{(k)} = \mathbf{x}^{(k)} - \mathbf{x}^{(*)} = \mathbf{M}\left(\mathbf{x}^{(k-1)} - \mathbf{x}^{(*)}\right) = \mathbf{M}\mathbf{e}^{(k-1)} = \mathbf{M}^2\mathbf{e}^{(k-2)} = \cdots = \mathbf{M}^k\mathbf{e}^{(0)} \tag{8.85}$$

The problems of the Jacobi iteration is to understand where to stop iteration and how to confirm that the solution converges. To inquire these questions, we need to revisit some basic concepts of vector and matrix norms.

8.13.1   VECTOR AND MATRIX NORMS

Definition 8.13.1 *An inner product space is a vector space where the inner product between two vectors* $\boldsymbol{u}$ *and* $\boldsymbol{v}$, *expressed as* $\boldsymbol{u} \cdot \boldsymbol{v}$, *is defined as follows:*

$$\boldsymbol{u} \cdot \boldsymbol{v} = \boldsymbol{v} \cdot \boldsymbol{u}$$
$$\boldsymbol{u} \cdot \boldsymbol{u} = 0 \quad \textit{for } \boldsymbol{u} = \boldsymbol{0}$$
$$> 0 \quad \textit{otherwise}$$
$$(a\boldsymbol{u} + b\boldsymbol{v}) \cdot \boldsymbol{w} = a(\boldsymbol{u} \cdot \boldsymbol{w}) + b(\boldsymbol{v} \cdot \boldsymbol{w}) \quad \textit{where } a, b \textit{ are scalars}$$

Definition 8.13.2 *Inner product space also includes the norm of a vector* $\boldsymbol{u}$, *expressed as* $\|\boldsymbol{u}\|$ *as follows:*

$$\|a\boldsymbol{u}\| = a\|\boldsymbol{u}\|$$
$$\|\boldsymbol{u}\| = 0 \quad \textit{for } \boldsymbol{u} = \boldsymbol{0}$$
$$> 0 \quad \textit{otherwise}$$
$$\|\boldsymbol{u} + \boldsymbol{v}\| \leq \|\boldsymbol{u}\| + \|\boldsymbol{v}\|$$

For an n-dimensional vector $\mathbf{u}$ with components $u_1, u_2, \ldots, u_n$, the l_p norm, for p being a positive integer, is defined as

$$\|\mathbf{u}\|_p = \left[\sum_{i=1}^{n} |u_i|^p\right]^{\frac{1}{p}} \quad \text{where } p = 1, 2, \ldots, \infty \tag{8.86}$$

The commonly used norms are as follows:

$$\|\mathbf{u}\|_1 = |u_1| + |u_1| + \cdots + |u_n|$$
$$\|\mathbf{u}\|_2 = \sqrt{u_1^2 + u_1^2 + \cdots + u_n^2}$$
$$\|\mathbf{u}\|_\infty = \max_{1 \le i \le n} |u_i| \tag{8.87}$$

Definition 8.13.3 *The l_p norm of a matrix A, denoted by $\|A\|_p$, is defined as*

$$\|A\|_p = \max_{\|\mathbf{x}\|_p = 1} \|A\mathbf{x}\|_p = \max_{\|\mathbf{x}\|_p \ne 0} \frac{\|A\mathbf{x}\|_p}{\|\mathbf{x}\|_p} \tag{8.88}$$

The above definition of matrix norm suggests

$$\|\mathbf{Ax}\|_p \le \|\mathbf{A}\|_p \|\mathbf{x}\|_p \text{ for all } \|\mathbf{x}\|_p \ne 0 \tag{8.89}$$

Thus, for all $\|\mathbf{x}\|_p \ne 0$, for two matrices $\mathbf{A}$ and $\mathbf{B}$, we have

$$\|\mathbf{ABx}\|_p \le \|\mathbf{A}\|_p \|\mathbf{Bx}\|_p \le \|\mathbf{A}\|_p \|\mathbf{B}\|_p \|\mathbf{x}\|_p \tag{8.90}$$

For all $\|\mathbf{x}\|_p = 1$, we

$$\|\mathbf{ABx}\|_p \le \|\mathbf{A}\|_p \|\mathbf{B}\|_p \tag{8.91}$$

Using equations 8.88 and 8.91, we now find

$$\|\mathbf{AB}\|_p \le \|\mathbf{A}\|_p \|\mathbf{B}\|_p \tag{8.92}$$

Setting $\mathbf{A} = \mathbf{B}$, we have

$$\left\| \mathbf{A}^2 \right\|_p \le \|\mathbf{A}\|_p^2 \tag{8.93}$$

In general, using equation 8.93 successively, we can show that for k being a positive integer

$$\left\| \mathbf{A}^k \right\|_p \le \|\mathbf{A}\|_p^k \tag{8.94}$$

Further, we can prove (left as an exercise) that

$$\|\mathbf{A}\|_\infty = \max_{1 \le i \le n} \left(\max_{|x_j| \le 1} \sum_{j=1}^n |u_{ij} x_j| \right) = \max_{1 \le i \le n} \sum_{j=1}^n |u_{ij}| \tag{8.95}$$

8.13.2 CONVERGENCE OF THE JACOBI ITERATION

The Jacobi method generates a sequence of vectors $\mathbf{x}^{(k)}$. The sequence converges to the exact solution $\mathbf{x}^{(*)}$ if for any positive real number ε, we find an integer $N(\varepsilon)$, such that

$$\left\| \mathbf{x}^{(k)} - \mathbf{x}^{(*)} \right\|_p < \varepsilon \text{ for } k > N \tag{8.96}$$

Since, in general, we don't know the exact solution, we take resort to equation 8.84 or 8.81 to use any one of the following *stopping criteria*:

$$\frac{\left\|\mathbf{x}^{(k)} - \mathbf{x}^{(k-1)}\right\|_p}{\left\|\mathbf{x}^{(k-1)}\right\|_p} < \varepsilon \quad \text{or} \quad \left\|\mathbf{A}\mathbf{x}^{(k)} - \mathbf{b}\right\| < \varepsilon, \text{ where } \varepsilon \text{ is a positive real number}$$

(8.97)

Now we focus on the condition that will enable the Jacobi iteration to provide a convergent solution. Looking back at equation 8.85, we suggest that the condition for convergence is given by

$$\lim_{k \to \infty} \left\|\mathbf{e}^{(k)}\right\|_\infty = \lim_{k \to \infty} \left\|\mathbf{M}^k \mathbf{e}^{(0)}\right\|_\infty = 0 \tag{8.98}$$

Using equations 8.89 and 8.94, we can write the *sufficient* condition for convergence as

$$\lim_{k \to \infty} \left\|\mathbf{M}\right\|_\infty^k = 0 \Rightarrow \left\|\mathbf{M}\right\|_\infty < 1 \tag{8.99}$$

From equations 8.83, and 8.95, we can, therefore, write the convergence condition as

$$\sum_{\substack{j=1 \\ j \neq i}}^{n} \left|\frac{a_{ij}}{a_{ii}}\right| < 1 \text{ for all } i \tag{8.100}$$

The Jacobi iteration, therefore, converges if the matrix $\mathbf{A}$ has the property of *diagonal dominance*, as shown below:

$$|a_{ii}| > \sum_{\substack{j=1 \\ j \neq i}}^{n} |a_{ij}| \text{ for all } i \tag{8.101}$$

Please note that the above condition is *sufficient*, but may not be necessary.

8.13.3   GAUSS–SEIDEL ITERATION

The Gauss–Seidel iteration is almost similar to the Jaobi method with minor modifications to improve the rate of convergence. Here again, we wish to solve for $\mathbf{x}$ in the following system of linear equations $\mathbf{A}\mathbf{x} = \mathbf{b}$, where $\mathbf{A}$ is an $n \times n$ matrix with elements a_{ij} while $\mathbf{b}$ is an n-dimensional column vector of elements b_i

$$\mathbf{A}\mathbf{x} = \mathbf{b} \Rightarrow \sum_{j=1}^{n} (a_{ij}x_j) = b_i \text{ for } i = 1, 2, \ldots, n \tag{8.102}$$

The Jacobi iteration uses the following algorithm to solve the above problem:

$$x_i^{(k)} = \frac{b_i - \sum_{j=1}^{i-1} \left[a_{ij}x_j^{(k)}\right] - \sum_{j=i+1}^{n} \left[a_{ij}x_j^{(k-1)}\right]}{a_{ii}} \tag{8.103}$$

where, k is the iteration number and $k = 0$ indicates the initial guess of the unknown vector x. The distinction of the Gauss–Seidel from the Jacobi iteration is now quite clear. While the Jacobi iteration uses previous iteration values of the vector $\mathbf{x}$ to find the current $\mathbf{x}$, in the Gauss–Seidel iteration, we always use the latest values of $\mathbf{x}$. The advantages of the Gauss–Seidel algorithm over the Jacobi method are that Gauss–Seidel requires fewer iterations and fewer storage space than the latter.

In matrix notation, the Gauss–Seidel algorithm may be written as

$$\mathbf{x}^{(k)} = \mathbf{M}\mathbf{x}^{(k-1)} + \mathbf{C} \tag{8.104}$$

$$\tag{8.105}$$

where $\mathbf{A} = \mathbf{D} + \mathbf{L} + \mathbf{U}$ and $\mathbf{M} = -(\mathbf{D}+\mathbf{L})^{-1}\mathbf{U}, \mathbf{C} = (\mathbf{D}+\mathbf{L})^{-1}\mathbf{b}$

Here, $\mathbf{D}$ is diagonal matrix, $\mathbf{L}$ is strictly (zero-diagonal) lower-triangular matrix and $\mathbf{U}$ is strictly (zero-diagonal) upper-triangular matrix.

The exact solution of equation 8.102 is given by $\mathbf{x} = \mathbf{x}^{(*)}$ where

$$\mathbf{x}^{(*)} = \mathbf{M}\mathbf{x}^{(*)} + \mathbf{C} \tag{8.106}$$

The error vector after the kth iteration is given by

$$\mathbf{e}^{(k)} = \mathbf{x}^{(k)} - \mathbf{x}^{(*)} = \mathbf{M}\left(\mathbf{x}^{(k-1)} - \mathbf{x}^{(*)}\right) = \mathbf{M}\mathbf{e}^{(k-1)} = \mathbf{M}^2\mathbf{e}^{(k-2)} = \cdots = \mathbf{M}^k\mathbf{e}^{(0)} \tag{8.107}$$

The Gauss–Seidel algorithm may be written as follows.

Algorithm 8.5: Gauss–Seidel Method

Data: $a_{ij} \in \mathscr{R}$ where $a_{ii} \neq 0$ and
$\quad\quad x_i, b_i, maximumError \in \mathscr{R}, error > maximumError$ for $i, j = 1, 2 \ldots n$
Result: x_i for $i = 1, 2 \ldots n$ such that $\sum_{j=1}^{n} a_{ij}x_j = b_i$

while $error > maximumError$ **do**
    $error = 0$
    for $i = 1, n$ **do**
        $res = b_i$
        for $j = 1, n$ **do**
            $res = res - a_{ij}/x_j$
        end
        $error = error + res * res$
        $x_i = x_i + res/a_{ii}$
    end
end

8.14 EXTENSION TO NONLINEAR SYSTEMS

So far, we have learned to solve single nonlinear equations using the Newton–Raphson method and a system of linear equations using Gaussian elimination. If we combine these two techniques, we can solve the system of nonlinear equations. Consider the following system of nonlinear equations:

$$f_1(x_1, x_2, \ldots, x_n) = 0 \qquad (8.108)$$
$$f_2(x_1, x_2, \ldots, x_n) = 0$$
$$\vdots$$
$$f_n(x_1, x_2, \ldots, x_n) = 0$$

Consider the initial guess of the solution is given by $x_i^{(0)}$, where $i = 1, 2, \ldots, n$ and the superscript indicates the iteration counter. Thus, after the kth iteration, the solution is $x_i^{(k)}$. We now assume

$$x_i^{(k+1)} = x_i^{(k)} + \Delta x_i \qquad (8.109)$$

$$f_1(x_1 + \Delta x_1, x_2 + \Delta x_2, \cdots, x_n + \Delta x_n) \qquad (8.110)$$
$$= f_1(x_1, x_2, \ldots, x_n) + f_{1,1}\Delta x_1 + f_{1,2}\Delta x_2 + \cdots + f_{1,n}\Delta x_n + \cdots$$

where,

$$f_{1,1} = \frac{\partial f_1}{\partial x_1}; f_{1,2} = \frac{\partial f_1}{\partial x_2} \text{ etc.} \qquad (8.111)$$

Assuming $f_1(x_1 + \Delta x_1, x_2 + \Delta x_2, \cdots, x_n + \Delta x_n) = 0$, we have

$$f_1(x_1, x_2, \ldots, x_n) + f_{1,1}\Delta x_1 + f_{1,2}\Delta x_2 + \cdots + f_{1,n}\Delta x_n = 0 \qquad (8.112)$$

Leading to

$$f_{1,1}\Delta x_1 + f_{1,2}\Delta x_2 + \cdots + f_{1,n}\Delta x_n = -f_1 \qquad (8.113)$$

In general,

$$f_{i,1}\Delta x_1 + f_{i,2}\Delta x_2 + \cdots + f_{i,n}\Delta x_n = -f_i \text{ where } i = 1, 2, \ldots, n \qquad (8.114)$$

The system of equations is thus given by

$$f_{1,1}\Delta x_1 + f_{1,2}\Delta x_2 + \cdots + f_{1,n}\Delta x_n = f_1 \qquad (8.115)$$
$$f_{2,1}\Delta x_1 + f_{2,2}\Delta x_2 + \cdots + f_{2,n}\Delta x_n = f_2$$
$$\vdots$$
$$f_{n,1}\Delta x_1 + f_{n,2}\Delta x_2 + \cdots + f_{n,n}\Delta x_n = f_n$$

In this system of equations, we evaluate $f_{i,j}$ and f_i using the previous iteration values $x_i^{(k)}$. The above system of equation is, therefore, a system of linear equations that can be solved to evaluate Δx_i. Once we have the Δx_i, we can evaluate the $x_i^{(k+1)}$ using equation 8.109. The procedure proceeds till convergence.

PROBLEMS

1. Write a **PYTHON** program to solve the following system of equations:

$$3x + 2y + z = 6$$
$$4x + 6y + 5z = 15$$
$$7x + 8y + 9z = 24$$

   The computer program is provided with the companion, please check your program and the solution.

2. Write a **PYTHON** program to solve the above system of equations using the Gauss–Seidel iteration.

9 Ordinary Differential Equations

9.1 INTRODUCTION

In general, an n-th order ordinary differential equation (ODE) is expressed as

$$F\left(x, y(x), \frac{dy}{dx}, \frac{d^2y}{dx^2}, \cdots, \frac{d^ny}{dx^n}\right) = 0 \tag{9.1}$$

Let us begin with the following first-order ODE:

$$y' = \frac{dy}{dx} = f\left(x, y(x)\right); y\left(x = x_0\right) = y_0 \tag{9.2}$$

This ODE solves for the unknown $y(x)$ for known values of $x > x_0$, where the value of $(x = x_0) = y_0$ is known as the initial condition (IC). The above ODE is thus called an initial value problem (IVP). In many physical problems, x represents time. For simplicity, we assume both x and y to be real-valued scalar variables. For brevity, in this chapter, we only discuss the ODEs where the existance and uniqueness of solutions are guaranteed.

9.2 EULER METHOD

Numerical differentiation is the primary tool for numerical solution of an ODE. Let us now try to solve the ODE shown in equation 9.2. The most primitive method for solving this equation is known as the Euler method. In the Euler method, we first discretize x, preferably with uniform grid size (also called stepsize) of $\Delta x = h$. The Euler method is then applied, with the first-order forward difference approximation of y', as

$$\frac{y_{i+1} - y_i}{h} + \mathcal{O}(h) = f(x_i, y_i) \Rightarrow y_{i+1} = y_i + hf(x_i, y_i) \tag{9.3}$$

for $i = 0, 2, \ldots, n-1$ where $y_0 = y(x = x_0)$ is known from the initial condition (IC).

Equation 9.3, derived using the finite difference method, is the *difference equation* or the *discretized form* of equation 9.2. While in equation 9.2, unknown y appears more than once, in the discretized equation 9.3, only y' uses y_{i+1}, all other terms involving y are discretized with y_i. Such a discretization method is called *explicit discretization*. It is now quite clear that the solution of y_{i+1} for $i = 0, 1, \cdots, n-1$ requires sequential calculations as shown in the following algorithm:

DOI: 10.1201/9781003049272-9

Algorithm 9.1: Solution of ODE Using the Euler Method

Data: $f(x,y), x_0, y_0, h, n$
Result: $y1, y2, \cdots, y_n$
$i = 0$
while $i < n$ **do**
   | $y_{i+1} = y_i + h * f(x_i, y_i)$
   | $x_{i+1} = x_i + h$
   | $i = i + 1$
end

While the Euler method is very simple to implement, the truncation error $\mathcal{O}(h)$, in the Euler method, may grow rapidly. We will learn a group of methods, known as the Runge–Kutta (RK) methods. RK methods fall under the umbrella of *predictor-corrector* method, where we first *predict* a reasonable value of y_{i+1} and then *correct* the value.

9.3 RUNGE–KUTTA METHOD

Consider the second-order RK method, also known as the Huen method. Now, we once again try to solve equation 9.2 i.e. $y' = f(x,y)$. Here, the initial prediction from the Euler method is corrected. The second-order RK method suggests the following difference equation:

$$y_{i+1} = y_i + \frac{h}{2}(k_1 + k_2) + \mathcal{O}\left(h^2\right) \tag{9.4}$$

$$\text{where, } k_1 = f(x_i, y_i)$$
$$k_2 = f(x_i + h, y_i + k_1 h)$$

Let us now see how we reach the above difference equation. Consider the following general form of difference equation:

$$y_{i+1} = y_i + h(ak_1 + bk_2) + \mathcal{O}(h^m) \tag{9.5}$$
$$\text{where, } k_1 = f(x_i, y_i)$$
$$k_2 = f(x_i + ph, y_i + qk_1 h)$$

Our goal is to select the constants a, b, p, and q, such that $m = 2$.

We can clearly see that for $a = 1$ and $b = 0$, the above algorithm reverts back to the first-order accurate Euler method. This is the predictor part of the algorithm. The corrector part, then adds the k_2, to improve the order of accuracy of the technique.

$$y_{i+1} = y_i + h(ak_1 + bk_2) \tag{9.6}$$
$$\text{predictor: } k_1 = f(x_i, y_i) \text{ predicts from the Euler method}$$
$$\text{corrector: } k_2 = f(x_i + ph, y_i + qk_1 h)$$

Using the Taylor series expansion of the corrector part and neglecting the higher-order terms,

$$k_2 = f(x_i + ph, y_i + qk_1 h) = k_1 + ph\frac{\partial f}{\partial x} + qk_1 h\frac{\partial f}{\partial y} \tag{9.7}$$

Using equations 9.6 and 9.7,

$$y_{i+1} = y_i + (a+b)hf + bph^2\frac{\partial f}{\partial x} + bqk_1h^2\frac{\partial f}{\partial x} \tag{9.8}$$

The Taylor series expansion of y provides

$$y_{i+1} = y_i + hy_i' + \frac{h^2}{2}y_i'' + \mathcal{O}\left(h^3\right) \tag{9.9}$$

We know

$$y' = f(x,y) \Rightarrow y'' = \frac{df}{dx} = \frac{\partial f}{\partial x} + y'\frac{\partial f}{\partial y} = \frac{\partial f}{\partial x} + f\frac{\partial f}{\partial y} \tag{9.10}$$

Combining equations 9.9 and 9.10, we have

$$y_{i+1} = y_i + hf + \frac{h^2}{2}\frac{\partial f}{\partial x} + \frac{h^2 f}{2}\frac{\partial f}{\partial y} \tag{9.11}$$

Comparing equations 9.8 and 9.11, we have

$$a+b=1, bph^2 = \frac{h^2}{2}, bqk_1h^2 = \frac{h^2 f}{2} \Rightarrow p=q=\frac{1}{2b}, a=1-b \tag{9.12}$$

Setting an arbitrary value of b, we get the values of a,p,q. The family of method thus obtained is known as the second-order RK method. For instance $b=1/2$, we have a second-order accurate RK method, popularly known as the Huen method, which is shown in equation 9.4.

Extending the above procedure, we can device RK method of various orders. The generalized RK method is given by

$$y_{i+1} = y_i + h\sum_{j=1}^{n}(a_jk_j) \tag{9.13}$$

$$k_1 = f(x_i, y_i)$$
$$k_j = f\left(x_i + p_{j-1}h, y_i + q_{j-1,1}k_1h, \dots, y_i + q_{j-1,j-1}k_{j-1}h\right)$$
$$\text{for } j = 2,3,\dots,n$$

The most common of them is the fourth-order RK method. The difference equation of fourth-order RK method is described as follows:

$$y_{i+1} = y_i + \frac{h}{6}(k_1 + 2k_2 + 2k_3 + k_4) \tag{9.14}$$

$$k_1 = f(x_i, y_i)$$
$$k_2 = f\left(x_i + \frac{h}{2}, y_i + \frac{k_1 h}{2}\right)$$
$$k_3 = f\left(x_i + \frac{h}{2}, y_i + \frac{k_2 h}{2}\right)$$
$$k_4 = f(x_i + h, y_i + k_3 h)$$

Example: *Solve the following IVP using the fourth-order RK method in the interval [0,2]. Use ten (10) uniform grids*

$$y' = x + y; y(x = 0) = 0 \qquad (9.15)$$

SOLUTION

The exact solution of the problem may be shown as $y(x) = \exp(x) - x - 1$. Results are plotted in Figure 9.1. The PYTHON program is given in the course website.

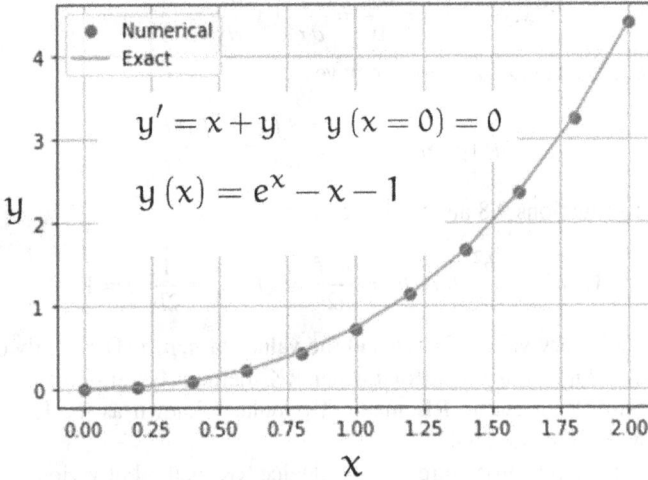

Figure 9.1   Numerical versus exact solutions of $y' = x + y$.

9.4   HIGHER-ORDER IVP

So far we have discussed the solution techniques for the IVP $y' = f(x, y)$ assuming y to be a real scalar. The solution procedures, described so far, are also applicable if y is a vector. As such, a higher-order IVP may be soled as a first-order IVP, if we allow y to be a vector. Consider the following second-order IVP:

$$u'' + au' + bu = f(x, u), u(x = 0) = p, u'(x = 0) = q \qquad (9.16)$$

This IVP may be written as

$$u' = v \quad \text{with the initial condition} \quad u(x = 0) = p$$
$$v' = f - av - bu \quad \text{with the initial condition} \quad v(x = 0) = q \qquad (9.17)$$

In vector form,

$$y' = F \text{ where } y = \begin{bmatrix} u \\ v \end{bmatrix} \Rightarrow y' = \frac{d}{dx}\begin{bmatrix} u \\ v \end{bmatrix} \text{ and } y(x = 0) = \begin{bmatrix} p \\ q \end{bmatrix} \qquad (9.18)$$

and

$$\mathbf{F} = \begin{bmatrix} v \\ g \end{bmatrix} \text{ where } g = g(x, u, v) = f - av - bu \qquad (9.19)$$

The difference equation of the fourth-order RK method, for equation 9.16, is described as follows:

$$u_{i+1} = u_i + \frac{h}{6}(k_1 + 2k_2 + 2k_3 + k_4); v_{i+1} = v_i + \frac{h}{6}(s_1 + 2s_2 + 2s_3 + s_4) \qquad (9.20)$$

$$k_1 = v_i; s_1 = g(x_i, u_i, v_i)$$

$$k_2 = v_i + \frac{s_1 h}{2}; s_2 = g\left(x_i + \frac{h}{2}, u_i + \frac{k_1 h}{2}, v_i + \frac{s_1 h}{2}\right)$$

$$k_3 = v_i + \frac{s_2 h}{2}; s_3 = g\left(x_i + \frac{h}{2}, u_i + \frac{k_2 h}{2}, v_i + \frac{s_2 h}{2}\right)$$

$$k_4 = f(v_i + s_3 h); s_4 = g(x_i + h, u_i + k_3 h, v_i + s_3 h)$$

In general, any higher-order IVP may be converted into a set of first-order IVPs. The basic concept of the solution techniques, shown for the above second-order IVP, may be extended to any higher-order IVP.

9.5 BOUNDARY VALUE PROBLEMS: SHOOTING METHOD

So far, we have solved IVPs to find the unkown variable $y(x)$ in the interval $x_0 < x \le x_n$, where the initial condition are given at $x = x_0$. In case of boundary value problems (BVPs), the boundary conditions are available at both ends of the independent variable i.e. at $x = x_0$ and $x = x_n$. The solution methodologies, applicable for the IVPs, cannot be readily used for the BVPs. One option is to pose the BVP as an IVP with a reasonable assumption for the missing initial condition. The assumed initial condition is then iteratively updated to match the given boundary condition. Such a solution technique is known as the shooting method. Let us consider the following example.

Example: *Solve the following ODE, famously known as the Blasius equation, using the shooting method.*

$$y''' + \frac{1}{2}yy'' = 0 \qquad (9.21)$$

$$y(x = 0) = 0$$

$$y'(x = 0) = 0$$

$$y'(x \to \infty) = 1$$

SOLUTION

This problem is clearly a BVP since all the conditions are not available at $x = 0$. One way to solve the above BVP is to convert the problem in to an IVP by assuming a reasonable value of $y''(x = 0)$. The goal here is to iteratively update the value of $y''(x = 0)$, so that the boundary condition (BC) $y'(x \to \infty) = 1$ is satisfied. The first step is to pose the above problem as a set of first-order IVPs as follows:

$$y' = y_1, \quad y(x = 0) = 0 \tag{9.22}$$

$$y_1' = y_2, \quad y_1(x = 0) = 0 \tag{9.23}$$

$$y_2' = -\frac{1}{2}yy_2, \quad y_2(x = 0) = y_{21} \tag{9.24}$$

Here, y_{21} is the value assumed for $y_2(x = 0)$ that has to be iteratively updated to match the boundary condition $y'(x \to \infty) = 1$. The procedure for updating the value of $y_2(x = 0)$ is as follows:

1. We assume $y_2(x = 0) = y_{21}$ and solve the above system of IVPs to have $y'(x \to \infty) = y_{21n}$. In general, $y_{21n} \neq 1$.
2. We assume another value of the initial condition $y_2(x = 0) = y_{22}$ and solve the above system of IVPs to have $y'(x \to \infty) = y_{22n}$. In general, $y_{22n} \neq 1$.
3. The values of the initial condition y_{21} and y_{21} are chosen such that $(1 - y_{21n})(1 - y_{21n}) < 0$. This condition ensures that if $y_{21n} < 1$, then $y_{22n} > 1$ or vice versa.
4. We now solve the above system of IVPs using $y_2(x = 0) = y_{23} = (y_{21} + y_{22})/2$ to have $y'(x \to \infty) = y_{23n}$. If $(1 - y_{21n})(1 - y_{23n}) > 0$, we set $y_{21} = y_{23}$. Conversely, if $(1 - y_{22n})(1 - y_{23n}) > 0$, we set $y_{22} = y_{23}$.
5. We follow the above bisection-type procedure until $y'(x \to \infty) \approx 1$.

The plots of the numerical solution of the Blasius equation are shown in Figure 9.2. The PYTHON program of the solution is given in course website.

9.6 BOUNDARY VALUE PROBLEMS: FINITE DIFFERENCE METHOD

As an alternative to the shooting method, we can discretize the BVP using finite differences to generate a system of linear equations. The system of equation may then be solved using a standard linear equations solver, described before. Let us solve the following equation using the finite difference method.

$$y'' + ay' + by = f(x), y(x = 0) = p, y'(x = 1) = q \tag{9.25}$$

Discretization of this equation leads to

$$\frac{y_{i-1} - 2y_i + y_{i+1}}{h^2} + a\frac{y_{i+1} - y_{i-1}}{2h} + by_i = f_i$$

$$\left(\frac{1}{h^2} - \frac{a}{2h}\right)y_{i-1} + \left(-\frac{2}{h^2}\right)y_i + \left(\frac{1}{h^2} + \frac{a}{2h}\right)y_{i+1} = f_i \tag{9.26}$$

For $i = 2, 3, \ldots, n - 1$, we have now $n - 2$ equations. The boundary conditions provide two more equations. Thus, we have an $\mathbf{Ax} = \mathbf{b}$ system containing n simultaneous

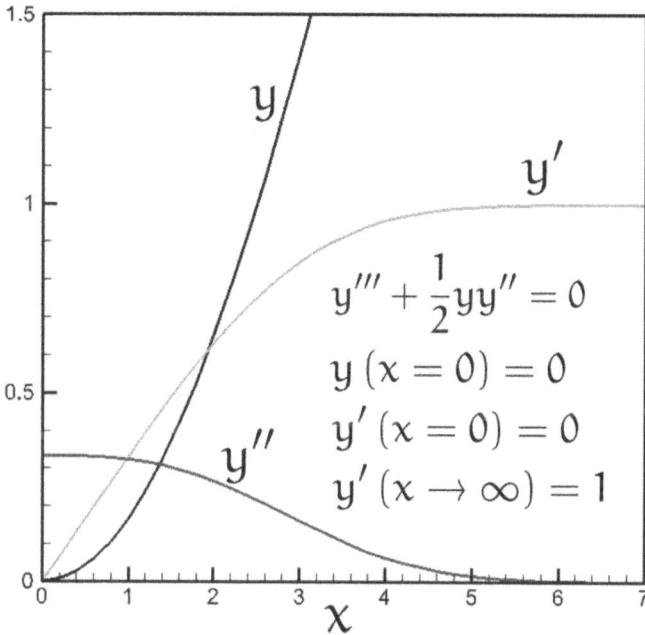

Figure 9.2 Numerical solution of the BVP using the shooting method.

linear equations with n unknowns. The system may then be solved using standard linear solvers, developed before. Depending on the value of n, matrix $\mathbf{A}$, could be of large size. Matrix $\mathbf{A}$, however, contains mostly zeroes and thus is called a sparse matrix. The matrix inversion is conducted in a way that the sparsity of matrix $\mathbf{A}$ is preserved.

PROBLEMS

1. Write a computer program to solve the following ODE using the fourth-order RK method:

$$y' = xy^2, \qquad y(x=0) = 1 \tag{9.27}$$

2. Write a computer program to solve the following ODE using the fourth-order RK and shooting method:

$$y'' = xy^2, \qquad y(x=0) = 1, \qquad y'(x=0) = 1 \tag{9.28}$$

3. Write a computer program to solve the following ODE using the fourth-order RK method:

$$y' = x - y, \qquad y(x=0) = 1 \tag{9.29}$$

4. Write a computer program to solve the following ODE using the finite difference method:

$$y'' = x - y, \qquad y(x = 0) = 1, \qquad y'(x = 0) = 1 \qquad (9.30)$$

5. Write a computer program to solve the following ODE using the fourth-order RK method:

$$y' + y \tan x = \sin(2x), \qquad y(x = 0) = 1 \qquad (9.31)$$

10 Numerical Differentiation and Integration

10.1 INTRODUCTION

The idea of continuity, as we learned in calculus, does not readily apply in numerical analysis. Numerical methods deal with discrete numbers and thus a function $y = f(x)$ provides certain discrete values of y for certain discrete inputs of x. In other sense, the independent variable x can attain only some discrete values. Once we get familiar with this idea, the numerical differentiation, of any order, becomes very easy.

10.2 NUMERICAL DIFFERENTIATION

In this section, we study the procedure for numerical differentiation of $y = f(x)$ in some interval $[a, b]$ using the *finite difference method*. When we analytically evaluate f', we get f' as a function of x. For any given x within the interval $[a, b]$, we can then evaluate f'. When we evaluate f' numerically, we can calculate f' only at few discrete points. The procedure of selecting these discrete x is called *discretization*. To begin with, we assume x is dicretized in n points $x_i, i = 1, 2, \ldots, n$, where $x_{i+1} - x_i = \Delta x$ for $i = 1, 2, \ldots, n-1$. The discrete divisions in x form a *grid* structure, and Δx is called *grid size*. When Δx is independent of x, we call the grid to be *uniform*, while in case of a *nonuniform* grid, Δx varies with x. To find the derivative in the discretized domain, we start by expanding $f(x + \Delta x)$ and $f(x - \Delta x)$ in the Taylor series as follows:

$$f_{i+1} = f(x + \Delta x) = f(x) + \Delta x f'(x) + \frac{(\Delta x)^2}{2!} f''(x) + \cdots = f_i + \Delta x f'_i + \ldots$$

$$f_{i-1} = f(x - \Delta x) = f(x) - \Delta x f'(x) + \frac{(\Delta x)^2}{2!} f''(x) - \cdots = f_i - \Delta x f'_i + \ldots \quad (10.1)$$

From equation 10.1, we can evaluate f' as follows:

$$\text{Forward difference: } f'_i = \frac{f_{i+1} - f_i}{\Delta x} + \mathcal{O}(\Delta x)$$

$$\text{Backward difference: } f'_i = \frac{f_i - f_{i-1}}{\Delta x} + \mathcal{O}(\Delta x)$$

$$\text{Central difference: } f'_i = \frac{f_{i+1} - f_{i-1}}{2\Delta x} + \mathcal{O}(\Delta x^2) \quad (10.2)$$

The errors that appear in the evaluation of derivatives due to the truncation of the Taylor series are known as *truncation errors*. Equation 10.2 indicates that truncation errors in forward and backward differences are of the *orders* of Δx, while the central

DOI: 10.1201/9781003049272-10

difference leads to a truncation error of the order of Δx^2. The forward and backward differences are thus called *first-order accurate* numerical schemes, while the central difference is *second-order accurate*. For a first-order accurate scheme, the truncation error reduces linearly as we reduce Δx, while for second-order accurate scheme, the same reduces quadratically. Please note that we can specify only the orders of truncation error but not the absolute value of the truncation error. The following example illustrates how we can create higher-order numerical schemes.

Example: Express f_i' in terms of f_i, f_{i+1}, f_{i+2}.

Solution:

$$f_{i+1} = f_i + \Delta x f_i' + \frac{\Delta x^2}{2} f_i'' + \frac{\Delta x^3}{6} f_i''' + \frac{\Delta x^4}{24} f_i'''' + \cdots$$

$$f_{i+2} = f_i + 2\Delta x f_i' + \frac{(2\Delta x)^2}{2} f_i'' + \frac{(2\Delta x)^3}{6} f_i''' + \frac{(2\Delta x)^4}{24} f_i'''' + \cdots \qquad (10.3)$$

Assuming $f_i' = a f_i + b f_{i+1} + c f_{i+2}$, where a, b, c are constants, we have

$$a + b + c = 0$$
$$b\Delta x + 2c\Delta x = 1$$
$$b\frac{\Delta x^2}{2} + 4c\frac{\Delta x^2}{2} = 0 \qquad (10.4)$$

Solving the above set of equations, we have

$$a = -\frac{3}{2\Delta x} \qquad b = \frac{2}{\Delta x} \qquad c = -\frac{1}{2\Delta x} \qquad (10.5)$$

Thus, the f_i' may be written as

$$f_i' = a f_i + b f_{i+1} + c f_{i+2} - b\frac{\Delta x^3}{6} f_i''' - c\frac{(2\Delta x)^3}{6} f_i'''$$

$$= \frac{-3 f_i + 4 f_{i+1} - f_{i+2}}{2\Delta x} + \mathcal{O}\left(\Delta x^2\right) \qquad (10.6)$$

10.3 NONUNIFORM GRID

For a nonuniform grid, we assume x is dicretized in n points $x_i, i = 1, 2, \ldots, n$, where $x_{i+1} - x_i = \Delta x_i$ for $i = 1, 2, \ldots, n-1$, and in general, $\Delta x_i \neq \Delta x_j$ for $i \neq j$. While the usual expressions of forward differences, as shown in equation 10.1, do not change, complication appears for higher order accurate schemes as shown in the example in Section 10.2.

SOLUTION

Example: Express f_i', in a nonuniform grid, in terms of f_i, f_{i+1}, f_{i+2}.

$$f_{i+1} = f_i + \Delta x_i f_i' + \frac{\Delta x_i^2}{2} f_i'' + \frac{\Delta x_i^3}{6} f_i''' + \cdots$$

$$f_{i+2} = f_i + (\Delta x_i + \Delta x_{i+1}) f_i' + \frac{(\Delta x_i + \Delta x_{i+1})^2}{2} f_i'' + \frac{(\Delta x_i + \Delta x_{i+1})^3}{6} f_i''' + \cdots \qquad (10.7)$$

Assuming $f_i' = af_i + bf_{i+1} + cf_{i+2}$, where a, b, c are constants, we have

$$a + b + c = 0$$
$$b\Delta x_i + c(\Delta x_i + \Delta x_{i+1}) = 1$$
$$b\frac{\Delta x_i^2}{2} + c\frac{(\Delta x_i + \Delta x_{i+1})^2}{2} = 0 \tag{10.8}$$

Solving this set of equations, we have

$$a = -\frac{2\Delta x_i + \Delta x_{i+1}}{\Delta x_i(\Delta x_i + \Delta x_{i+1})}$$

$$b = \frac{\Delta x_i + \Delta x_{i+1}}{\Delta x_i \Delta x_{i+1}}$$

$$c = -\frac{\Delta x_i}{\Delta x_{i+1}(\Delta x_i + \Delta x_{i+1})} \tag{10.9}$$

To obtain the order of accuracy, we evaluate

$$-\frac{b\Delta x_i^3}{6} - \frac{c(\Delta x_i + \Delta x_{i+1})^3}{6} = \frac{1}{6}\left(\Delta x_i^2 + \Delta x_i \Delta x_{i+1}\right) \tag{10.10}$$

Thus, f_i' may be written as

$$f_i' = -\frac{2\Delta x_i + \Delta x_{i+1}}{\Delta x_i(\Delta x_i + \Delta x_{i+1})}f_i + \frac{\Delta x_i + \Delta x_{i+1}}{\Delta x_i \Delta x_{i+1}}f_{i+1} - \frac{\Delta x_i}{\Delta x_{i+1}(\Delta x_i + \Delta x_{i+1})}f_{i+2}$$
$$+ \mathcal{O}\left(\Delta x_i^2 + \Delta x_i \Delta x_{i+1}\right) \tag{10.11}$$

This example indicates that three-point derivative in a nonuniform grid may not be second-order accurate. In fact, if $\Delta x_i \ll \Delta x_{i+1}$, equation 10.11 shows that scheme would be almost first-order accurate. Therefore, in a nonuniform grid, grids should be stretched very conservatively to maintain a higher order of accuracy.

10.4 DOUBLE DERIVATIVE

To find f'', we follow a similar procedure and start from expanding f_{i+1} and f_{i-1} in the Taylor series as follows:

$$f_{i+1} = f_i + \Delta x f_i' + \frac{(\Delta x)^2}{2!}f_i'' + \frac{(\Delta x)^3}{3!}f_i''' + \frac{(\Delta x)^4}{4!}f_i'''' + \cdots$$
$$f_{i-1} = f_i - \Delta x f_i' + \frac{(\Delta x)^2}{2!}f_i'' - \frac{(\Delta x)^3}{3!}f_i''' + \frac{(\Delta x)^4}{4!}f_i'''' + \cdots \tag{10.12}$$

Adding this set of equations, we find

$$f_{i+1} + f_{i-1} = 2f_i + (\Delta x)^2 f_i'' + 2\frac{(\Delta x)^4}{4!}f_i'''' + \cdots$$
$$\Rightarrow f_i'' = \frac{f_{i-1} - 2f_i + f_{i+1}}{(\Delta x)^2} + \mathcal{O}\left(\Delta x^2\right) \tag{10.13}$$

While equation 10.13 shows the central difference discretization of double derivative in a uniform grid, extending the procedure for forward or backward differences as well as in nonuniform grids are left as your exercise.

10.5 NUMERICAL INTEGRATION: NEWTON–COTES FORMULAS

In numerical integration, we intend to evaluate I given below:

$$I = \int_{x=a}^{x=b} f(x)\,dx \qquad (10.14)$$

We further assume that function $f(x)$ is known to us probably as a bunch of discrete points (x_i) for $i = 1, 2, \cdots, n$. Even if we know function $f(x)$, for the purpose of integration, we would use only discrete points $f(x_i)$.

The fundamental idea behind numerical integration is to replace function $f(x)$ with an n-th order polynomial such that

$$f(x) \approx P_n(x) = \sum_{i=0}^{n} a_i x^i \qquad (10.15)$$

Depending on the value of n, we may have a variety of expressions of the integral, known as the *Newton–Cotes Formulas*. These formulas are, in genral, of two types: *open* Newton–Cotes formula and *closed* Newton–Cotes formula. To evaluate the integral in equation 10.14 using n points in the closed Newton–Cotes formula, the first point $x_1 = a$ and the last point $x_n = b$ are used. For evaluating the same integral, in the open Newton–Cotes formula, however, excludes the end points such that $a < x_i < b$. In the subsequent sections, we will see two instances of the closed Newton–Cotes approach, trapezoidal rule and Simpson's one-third rule.

10.5.1 TRAPEZOIDAL RULE

The trapezoidal rule uses a first-order polynomial i.e. a straight line to approximate a curve, as shown in Figure 10.1.

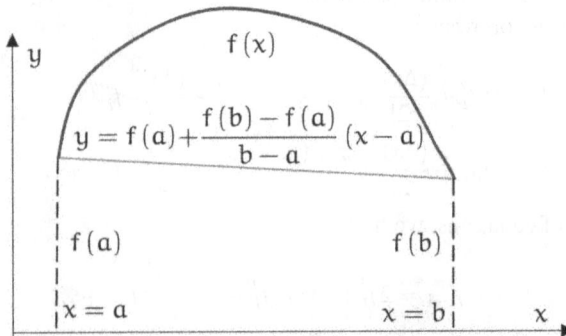

Figure 10.1 Trapezoidal rule: approximating a curve with linear approximation.

Thus, we evaluate the integral, shown in equation 10.14, as

$$I = \int_{x=a}^{x=b} f(x)\,dx \approx \int_{x=a}^{x=b} \left[f(a) + \frac{f(b)-f(a)}{b-a}(x-a) \right] dx$$

$$= \left[xf(a) + \frac{1}{2}\frac{f(b)-f(a)}{b-a}(x-a)^2 \right]_a^b$$

$$= \frac{b-a}{2}\left[f(a) + f(b) \right] \tag{10.16}$$

Assuming $b - a = h$, we can rewrite equation 10.16 as

$$I = \int_{x=a}^{a+h} f(x)\,dx = \frac{h}{2}\left[f(a) + f(a+h) \right] \tag{10.17}$$

This idea may be generalized by assuming the curve shown in Figure 10.1 as a portion of another curve. Thus, instead of a linear approximation, described earlier, we now have piecewise linear approximation. This idea is illustrated in Figure 10.2.

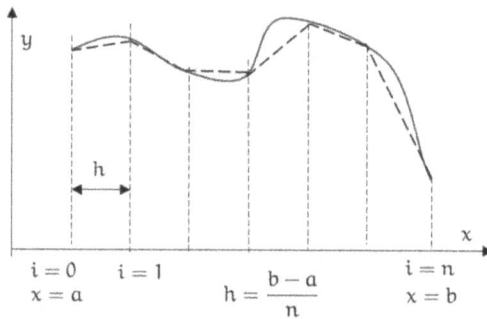

Figure 10.2   Trapezoidal rule: approximating a curve with piecewise linear approximation.

Following the idea expressed in equation 10.16 and Figure 10.2, we can now write the following general expression for the trapezoidal rule:

$$I = \int_{x=a}^{b} f(x)\,dx$$

$$= \frac{h}{2}\left[f(a) + f(a+h) \right] + \frac{h}{2}\left[f(a+h) + f(a+2h) \right] + \cdots$$

$$+ \frac{h}{2}\left[f(a+nh-h) + f(a+nh) \right]$$

$$= \frac{h}{2}\left[f(a) + 2f(a+h) + \cdots + 2f(a+nh-h) + f(a+nh) \right]$$

$$= \frac{h}{2}\left[f(x_0) + 2\sum_{i=1}^{n-1} f(x_i) + f(x_n) \right] \tag{10.18}$$

Example: Evaluate the following integral using the trapezoidal rule:

$$I = \int_0^{\pi} \sin x \, dx \tag{10.19}$$

Solution:

To evaluate the above integral numerically, we first divide the interaval $[0, \pi]$ in n equal divisons and then apply equation 10.18. A PYTHON program of this problem in given in the course website.

10.5.2 SIMPSON'S ONE-THIRD RULE

While the trapezoidal rule uses a first-order polynoial i.e. a straight line to approximate a curve, the Simpson rule, as shown in Figure 10.3, uses a second-order (quadratic) polynomial i.e. a parabola for the same approximation.

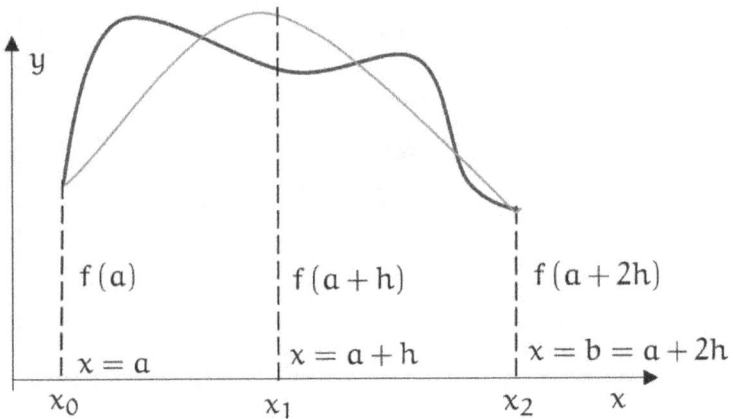

Figure 10.3 Simpson's rule: approximating a curve with quadratic approximation.

Thus, the equation of a quadratic polynomial passing through (x_0, f_0), (x_1, f_1) and (x_2, f_2) is

$$y = \frac{(x-x_1)(x-x_2)}{(x_0-x_1)(x_0-x_2)} f_0 + \frac{(x-x_0)(x-x_2)}{(x_1-x_0)(x_1-x_2)} f_1 + \frac{(x-x_0)(x-x_1)}{(x_2-x_0)(x_2-x_1)} f_2 \tag{10.20}$$

In the present case,

$$x_1 - x_0 = h; x_2 - x_0 = 2h; x_2 - x_1 = h \tag{10.21}$$

Using equations 10.20 and 10.21, we have

$$y = \frac{1}{2h^2} \left[(x-x_1)(x-x_2) f_0 - 2(x-x_0)(x-x_2) f_1 + (x-x_0)(x-x_1) f_2 \right] \tag{10.22}$$

Further assuming $x - x_1 = hz$, we obtain

$$x - x_0 = x_1 + hz - x_0 = hz + h = h(z+1); x = x_0 \Rightarrow z = -1 \quad (10.23)$$
$$x - x_2 = x_1 + hz - x_2 = hz - h = h(z-1); x = x_2 \Rightarrow z = 1 \quad (10.24)$$

equations 10.22–10.24, we have

$$
\begin{aligned}
y &= \frac{1}{2h^2} \left[h^2 z(z-1) f_0 - 2h^2(z+1)(z-1) f_1 + h^2 z(z+1) f_2 \right] \\
&= \frac{1}{2} \left(z^2 - z \right) f_0 - \left(z^2 - 1 \right) f_1 + \frac{1}{2} \left(z^2 + z \right) f_2 \quad (10.25)
\end{aligned}
$$

Now, using $dx = hdz$, we can evaluate the integral

$$
\begin{aligned}
I &= \int_{x=a}^{x=b} f(x)\,dx \approx \int_{x=x_0}^{x=x_2} y\,dx \\
&= \int_{z=-1}^{z=1} \left[\frac{1}{2} \left(z^2 - z \right) f_0 - \left(z^2 - 1 \right) f_1 + \frac{1}{2} \left(z^2 + z \right) f_2 \right] hdz \\
&= \frac{h}{3} (f_0 + 4f_1 + f_2) \\
&= \frac{h}{3} \left[f(a) + 4f(a+h) + f(a+2h) \right] \quad (10.26)
\end{aligned}
$$

This idea may be generalized assuming the curve shown in Figure 10.1 as a portion of another curve. Thus, instead of a quadratic approximation, described earlier, we now have piecewise quadratic approximation. We can now write the following general expression for the Simpson rule:

$$
\begin{aligned}
I &= \int_{x=a}^{b} f(x)\,dx \\
&= \frac{h}{3} \left[f(a) + 4f(a+h) + 2f(a+2h) + \cdots + 2f(a+nh-h) + f(a+nh) \right] \\
&= \frac{h}{3} \left[f(x_0) + 4 \sum_{i=1,3,5}^{n-1} f(x_i) + 2 \sum_{i=2,4,6}^{n-2} f(x_i) + f(x_n) \right] \quad (10.27)
\end{aligned}
$$

Clearly, the Simpson rule of computation of the integral will be applicable only when n is even i.e. when we have odd number of data points.

Example: Evaluate the following integral using trapezoidal rule:

$$I = \int_0^\pi \sin x\,dx \quad (10.28)$$

SOLUTION

To evaluate this integral numerically, we first divide the interaval $[0, \pi]$ in n equal divisons, where n is an even number, and then apply equation 10.27. A PYTHON program of this problem in given in the course website.

PROBLEMS

1. Express f_i'' in terms of f_i, f_{i+1}, f_{i+2} in a uniform grid.
2. Express f_i'' in terms of f_i, f_{i-1}, f_{i-2} in a uniform grid.
3. Express f_i'' in terms of f_i, f_{i+1}, f_{i+2} in a nonuniform grid.
4. Express f_i'' in terms of f_i, f_{i-1}, f_{i-2} in a nonuniform grid.
5. Find the order of accuracy in the trapezoidal rule and Simpson's rule.
6. Evaluate the following integral using the trapezoidal rule, and calculate the difference between analytical and numerical solutions:

$$I = \int_0^{10} \left(x^2 + 2x + 5 \right) dx \tag{10.29}$$

7. Evaluate the following integral using the Simpson rule, and calculate the difference between analytical and numerical solutions:

$$I = \int_0^{10} (2x + 5) \, dx \tag{10.30}$$

11 Partial Differential Equations

11.1 INTRODUCTION

For a differentiable function $u = u(x,y)$ and n being a nonzero positive integer, the n-th order partial differential equation (PDE) involving u is given by

$$f\left(u, x, y, \frac{\partial u}{\partial x}, \frac{\partial u}{\partial y}, \frac{\partial^2 u}{\partial x^2}, \frac{\partial^2 u}{\partial x \partial y}, \frac{\partial^2 u}{\partial y^2}, \cdots, \frac{\partial^n u}{\partial y^n}\right) = 0 \tag{11.1}$$

Similar PDEs may also be written if u is a function of more than two independent variables. The PDE shown in equation 11.1 is linear if the unknown u is not multiplied by any function of u or derivative of u. When the above condition is violated, the PDE becomes nonlinear.

Partial derivatives are often symbolized as u_x, u_{xx} etc., creating a shorter form of equation 11.1. Similarly, PDEs are also written in operator form. For example, the following three equations mean the same:

$$\frac{\partial u}{\partial x} + \frac{\partial v}{\partial y} + \frac{\partial w}{\partial z} = 0 \tag{11.2}$$

$$u_x + v_y + w_z = 0 \tag{11.3}$$

$$\nabla . \mathbf{u} = 0, \text{ where } \mathbf{u} = (u, v, w) \tag{11.4}$$

In engineering practice, PDEs often appear in modeling physical phenomena. For instance, modeling of fluid flow produces the famous Navier–Stokes equation. We routinely encounter PDEs in continuum mechanics, electrodynamics, quantum mechanics and various other branches of science and engineering. Unlike ordinary differential equations (ODEs), very few PDEs can be solved analytically. Furthermore, PDE solutions are generally very sensitive to the imposed initial and boundary conditions. As you proceed, you will be able to appreciate the solution complexities of PDEs as compared to that of the ODEs. You will also see that unlike the ODEs, PDE solution often requires knowledge of the physical problem that has generated the PDE.

In this course, we will pick up problems related to continuum mechanics where a variable can be a function of time and three spatial directions. We will thus limit our discussion to second-order PDEs with a maximum of four independent variables, and three spatial and temporal coordinates. While we will primarily discuss linear PDEs, the methods, to be discussed, are general enough to solve nonlinear PDEs.

DOI: 10.1201/9781003049272-11

11.2   CLASSIFICATION

Based on the physical problem being modeled, PDEs can be classified into

1. Marching (propagation) problem
2. Equilibrium problem
3. Eigenvalue problem

11.2.1   MARCHING PROBLEM

A marching or propagation problem involves evolution of the unknown variable u as a function of time and space. A common example is the diffusion problem that models diffusive heat or mass transfer as follows:

$$\frac{\partial u}{\partial t} = \alpha \nabla^2 u + S = \alpha \left(\frac{\partial^2 u}{\partial x^2} + \frac{\partial^2 u}{\partial y^2} + \frac{\partial^2 u}{\partial z^2} \right) + S \qquad (11.5)$$

Here, the variable u evolves in time t and propagates in the 3-D space x, y, z. The extent of propagation is governed by a parameter $\alpha > 0$, known as diffusivity. A source term S, which in general could be a function of t, x, y, z, indicates the internal generation of $u(t, x, y, z)$. In marching problems, the initial condition $u(t = 0)$ is known and subsequently $u(t > 0)$ is calculated. The boundary conditions of u in the domain x, y must be known for all t. Please note that $u(t = t_0)$ depends on all $u(t < t_0)$ but not on $u(t > t_0)$. The marching problem mimics the initial value problems as we learned in ODE.

11.2.2   EQUILIBRIUM PROBLEM

Given sufficient time to evolve, as well as conducive source term and boundary conditions, the above marching problem may lead to a time-independent or *equilibrium* state as follows:

$$\nabla^2 u = \left(\frac{\partial^2 u}{\partial x^2} + \frac{\partial^2 u}{\partial y^2} + \frac{\partial^2 u}{\partial z^2} \right) = S(x, y, z) \qquad (11.6)$$

This problem is known as the equilibrium or *steady-state* problem, where variable $u(x, y, z)$ at any location depends on the value of $u(x, y, z)$ at all other locations. Equation 11.6 is known as the Poisson equation. If the source term in equation 11.6 is $S(x, y, z) = 0$, then that equation is called the Laplace equation. The operator ∇^2 is known as the Laplacian operator. The equilibrium problem, shown above, is somewhat similar to the boundary-value problems of ODE.

11.2.3   EIGENVALUE PROBLEM

An eigenvalue problem is a special type of equilibrium problem where the solution exists for certain values of the eigenvalue λ. Such problem appears usually for stability of a system. An example is the following Helmholtz equation:

$$\nabla^2 u = \left(\frac{\partial^2 u}{\partial x^2} + \frac{\partial^2 u}{\partial y^2} + \frac{\partial^2 u}{\partial z^2} \right) = \lambda u \tag{11.7}$$

where, $u = 0$ at the boundaries.

11.3 SECOND-ORDER LINEAR PDE

A second-order PDE is written as

$$A u_{xx} + B u_{xy} + C u_{yy} + D u_x + E u_y + F u + G = 0 \tag{11.8}$$

The boundary condition of this PDE is usually of the form

$$a u_x + b u_y + c u + d = 0 \tag{11.9}$$

This PDE is linear if A, B, C, D, E, F, G are constants or functions of x and y only. The PDE is homogeneous if $G = 0$. A boundary condition is linear if a, b, c, d are constants or function of x and y in that boundary. Similarly, a boundary condition is homogeneous if $d = 0$ in that boundary. Second-order linear PDEs are classified into the following three groups:

1. Parabolic problem
2. Hyperbolic problem
3. Elliptic problem

11.3.1 PARABOLIC PROBLEM

A parabolic problem is governed by $B^2 - 4AC = 0$. A typical parabolic problem is exemplified by equation 11.5 that describes evolution of heat or species concentration over a domain.

11.3.2 HYPERBOLIC PROBLEM

A hyperbolic problem is governed by $B^2 - 4AC > 0$. A typical hyperbolic problem is exemplified by wave equation as follows:

$$\frac{\partial^2 u}{\partial t^2} = \alpha \nabla^2 u = \alpha \left(\frac{\partial^2 u}{\partial x^2} + \frac{\partial^2 u}{\partial y^2} + \frac{\partial^2 u}{\partial z^2} \right) \tag{11.10}$$

This equation describes a finite speed propagation and may show sharp discontinuity between different values of u.

11.3.3 ELLIPTIC PROBLEM

An elliptic problem is governed by $B^2 - 4AC < 0$. A typical elliptic problem is exemplified by equation 11.6 that describes steady-state solution of heat or species concentration over a domain.

Since, in general, $B^2 - 4AC$ could be a function of x and y, the nature of a PDE may vary over the domain.

11.4 ONE-DIMENSIONAL TRANSIENT DIFFUSION

Consider a metallic rod of length L, having constant thermal conductivity, is subjected to heat transfer from two ends. There is no heat loss from its surface other than the two ends and the radius is much smaller compared to the length of the rod. If the temperature of the rod is u_1 at $x = 0$ and u_2 at $x = L$, we know that the steady-state temperature (u) distribution of the rod is given by

$$\frac{d^2u}{dx^2} = 0 \Rightarrow u(x) = u_1 + \frac{x}{L}(u_2 - u_1) \tag{11.11}$$

Note that the steady-state temperature distribution does not depend on the material properties. Therefore, the steady-state temperature distribution of a glass rod and an iron rod will be the same if we maintain same length and same end temperatures of both the rods. However, if we now suddenly (time $t = 0$) change temperature at one end of the rod, let's say we change u_2 to u_3, we know that the steady-state thermal field will be disturbed and both the rods will now show temporal variation in temperature. In long time $(t \rightarrow \infty)$, both the rods will again reach a new steady state very similar to that shown in equation 11.11. We, of course, know that both glass and iron rods will take their own sweet times to reach the steady state. Intuitively, we know that glass rod will respond much slower than the iron rod. The material property that determines which reaches the steady state first is called thermal diffusivity. Therefore, during intermediate time $(0 < t < \infty)$, the temperature in both the rods cannot be same due to the difference in thermal diffusivities of glass and iron. The equation that governs the temporal temperature variation is given by

$$\frac{\partial u}{\partial t} = \alpha \frac{\partial^2 u}{\partial x^2} \text{ for } 0 < t < \infty, 0 < x < L \tag{11.12}$$

where α is the thermal diffusivity.

Needless to say this PDE, known as a 1-D diffusion equation, requires one initial and two boundary conditions. Let's say

Initial condition: $u(t = 0, x) = u_0(x)$ \hfill (11.13)

Boundary conditions: $u(t > 0, x = 0) = u_1$, $u(t > 0, x = L) = u_2$ \hfill (11.14)

The analytical solution of the above problem, though tedious, is possible. You may try to find the following analytical solution:

$$u(t,x) = v(x) + w(t,x) \qquad (11.15)$$

$$\text{where, } v(x) = u_1 + \frac{x}{L}(u_2 - u_1)$$

$$w(t,x) = \sum_{n=1}^{\infty} a_n \exp\left(-n^2 \pi^2 \frac{\alpha t}{L^2}\right) \sin\left(\frac{n\pi x}{L}\right)$$

$$a_n = \frac{2}{L} \int_0^L q(z) \sin\left(\frac{n\pi z}{L}\right) dz$$

$$q(z) = u_0(z) - u(z)$$

Please note that the solution comprises a time independent part $v(x)$ and a transient part $w(t,x)$. In large time $(t \to \infty)$, the transient part $w(t,x) \to 0$ and the solution reaches the steady state $u(x) = v(x)$. Note that we start with a *marching* problem and end up with an *equilibrium* problem. In other words, we start with a *parabolic* problem and reach an *elliptic* problem at the end. In subsequent sections, we will learn to solve 1-D diffusion equations numerically.

11.5 NUMERICAL SCHEMES

There are various numerical schemes to solve 1-D diffusion equations. Let us first discretize domain x in $m-1$ equal divisions creating m points x_i, where $i = 1, 2, \cdots, m$ and a constant grid size $x_{i+1} - x_i = \Delta x$. Note that $(m-1)\Delta x = L$. We now discretize time t via constant time step Δt. Recalling our discussion of numerical differentiation, we discretize the LHS of equation 11.12 as follows:

$$\frac{\partial u}{\partial t} = \frac{u_i^{n+1} - u_i^n}{\Delta t} \qquad (11.16)$$

Here, the superscript indicates temporal and subscript indicates spatial locations. Suppose u_i^0 is the initial condition and u_1^n, u_m^n are the boundary conditions at the n-th time step. In the present case, u_1^n, u_m^n are assumed to be independent of n for $n > 0$. While the above discretization is only first-order accurate, the accuracy is sufficient for a wide variety of engineering problem solving.

Before we discretize the spatial derivative, given by the RHS of equation 11.12, we have to decide at which time level we will take the variables. Based on this decision, we will now see there are three major approaches: explicit, implicit and Crank–Nicolson.

11.5.1 EXPLICIT SCHEME

The simplest idea to discretize the spatial derivative u_{xx} is to take all terms at the n-th time step. Such a discretization is known as *explicit* and is shown as follows:

$$\frac{\partial^2 u}{\partial x^2} = \frac{u_{i-1}^n - 2u_i^n + u_{i+1}^n}{\Delta x^2} \qquad (11.17)$$

Overall, the discretized form of equation 11.12 can be written as

$$\frac{u_i^{n+1} - u_i^n}{\alpha \Delta t} = \frac{u_{i-1}^n - 2u_i^n + u_{i+1}^n}{\Delta x^2} \tag{11.18}$$

This equation shows explicit discretization of 1-D diffusion equation. A Taylor series analysis shows that the explicit scheme has an order of accuracy of $\mathcal{O}\left(\Delta t, \Delta x^2\right)$.

Now to solve equation 11.12, we rewrite the discretized equation 11.18 as

$$u_i^{n+1} = u_i^n + \frac{\alpha \Delta t}{\Delta x^2}\left(u_{i-1}^n - 2u_i^n + u_{i+1}^n\right) \tag{11.19}$$

This difference equation is valid for $n > 0$ and $1 < i < m$. T_i^0 will come from the initial condition while u_1^n and u_m^n will be taken from the boundary conditions. Overall, the solution methodology is given by

1. Initailization: $u_i^n = u_0$ for $n = 0$
2. Boundary: $x = 0$: $u_i^n = u_1$ for $i = 1, n > 0$
3. Interior: $u_i^{n+1} = u_i^n + \frac{\alpha \Delta t}{\Delta x^2}\left(u_{i-1}^n - 2u_i^n + u_{i+1}^n\right)$ for $1 < i < m, n > 0$
4. Boundary: $x = L$: $u_i^n = u_2$ for $i = m, n > 0$
5. Continue until desired time or steady state is reached

The above procedure clearly shows that given the initial and boundary conditions, the value of T may be calculated at any subsequent time at any location. The calculation may be stopped when we reach a specified time or steady state. To define steady state, we know that at steady state, the time derivative would vanish. Thus, if we check the norm of T at two subsequent time, we declare steady state when the norm reaches less than a specified small number, such that

$$\sum_{i=1}^{i=m}\left(u_i^{n+1} - u_i^n\right)^2 < \varepsilon \tag{11.20}$$

where, ε is a small number specified by the programmer.

While the explicit method is quite easy to implement and most importantly the method does not require solution of equations, there are several limitations of this method. In the subsequent exercises, we will discuss other methods and compare all of them. Before we do so, I encourage you to complete the exercise below to have a feeling about the nature of solution.

Exercise: Solve equation 11.12 with initial and boundary conditions 11.13, 11.14 using explicit scheme. Use $L = 1, \alpha = 1, u_0 = u_1 = 0, u_2 = 1$.

11.5.2 IMPLICIT SCHEME

The *implicit* method, unlike the explicit approach, discretizes the spatial derivative at the $(n+1)$-th time level as

$$\frac{\partial^2 u}{\partial x^2} = \frac{u_{i-1}^{n+1} - 2u_i^{n+1} + u_{i+1}^{n+1}}{\Delta x^2} \tag{11.21}$$

Overall, the discretized form of equation 11.12 can be written as

$$\frac{u_i^{n+1} - u_i^n}{\alpha \Delta t} = \frac{u_{i-1}^{n+1} - 2u_i^{n+1} + u_{i+1}^{n+1}}{\Delta x^2} \tag{11.22}$$

This equation shows implicit discretization of 1-D diffusion equation with an order of accuracy of $\mathcal{O}\left(\Delta t, \Delta x^2\right)$, similar to that of the explicit method.

Now to solve equation 11.12, we rewrite the above discretized equation 11.22 as

$$au_{i-1}^{n+1} + bu_i^{n+1} + cu_{i+1}^{n+1} = u_i^n \tag{11.23}$$

$$a = c = -\frac{\alpha \Delta t}{\Delta x^2}, b = 1 + 2\frac{\alpha \Delta t}{\Delta x^2} \tag{11.24}$$

Equation 11.24 is valid for $n > 0$ and $1 < i < m - 1$. The boundary conditions provide

$$u_{i=1}^{n+1} = u_1, u_{i=m}^{n+1} = u_2 \tag{11.25}$$

Now combining equations 11.24 and 11.25, we find that u_i^{n+1} for $1 \leq i \leq m$ generates a tridiagonal system, which may be solved the using the Thomas algorithm. Overall the solution procedure may be written as follows:

1. Initailization: $u_i^n = u_0$ for $n = 0$
2. Solve equations 11.24 and 11.25 for $(n+1)$-th time step
3. Continue until desired time or steady state is reached

This procedure shows how the PDE in converted into a system of linear equations that has to be solved in every time step. While the implicit method has the same order of accuracy as that of the explicit method, there are considerable advantages of the implicit method that we will discuus shortly.

11.5.3 CRANK–NICOLSON SCHEME

In the Crank–Nicolson method, we discretize the governing equation as follows:

$$\frac{u_i^{n+1} - u_i^n}{\alpha \Delta t} = \frac{1}{2}\left(\frac{u_{i-1}^n - 2u_i^n + u_{i+1}^n}{\Delta x^2}\right) + \frac{1}{2}\left(\frac{u_{i-1}^{n+1} - 2u_i^{n+1} + u_{i+1}^{n+1}}{\Delta x^2}\right) \tag{11.26}$$

You can see that the spatial derivative is discretized as an average of explicit and implicit discretizations. To understand the solution methodology, rearrange equation 11.26 taking all u^{n+1} on LHS and remaining terms on the RHS. Now considering the bounday conditions, please note that like implicit procedure, the PDE is once again transformed into a tridiagonal system of linear equations. The solution algorithm of such a system is now well known to us.

The question, however, remains why we are creating a complicated scheme when we already have explicit and implicit methods in our hand. One reason is that the order of accuracy of the Crank–Nicolson scheme is $\mathcal{O}\left(\Delta t^2, \Delta x^2\right)$, which is higher than that of the explicit or implicit methods. The scheme is also a bit more *stable* compared to explicit method, a feature that we will discuss later.

Example: *Solve the following 1-D transient diffusion equation using the implicit method:*

$$\frac{\partial u}{\partial t} = \frac{\partial^2 u}{\partial x^2} + S(x, u) \tag{11.27}$$

$$\text{The source term } S(x, u) = \frac{1}{x-1}\frac{\partial u}{\partial x} + \frac{10u}{x-1}$$

The initial and boundary conditions are

$$u(t = 0) = 0; u(x = 0) = 1; \frac{\partial u}{\partial x}(x = 1) = 0 \tag{11.28}$$

SOLUTION

The PYTHON program is included in the course website. The results are shown in Figure 11.1.

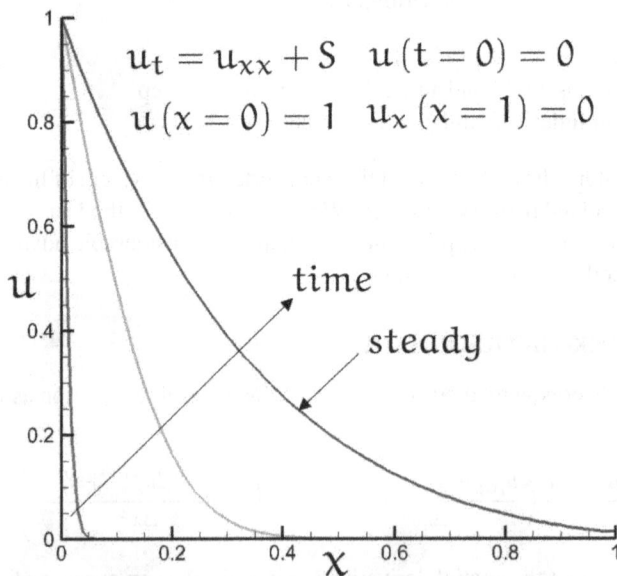

$$u_t = u_{xx} + S \quad u(t = 0) = 0$$
$$u(x = 0) = 1 \quad u_x(x = 1) = 0$$

Figure 11.1 Transient variation of u.

11.6 STABILITY AND CONSISTENCY

While discussing the 1-D diffusion equation solution methodologies, we have seen a variety of solution schemes. Every scheme converts the PDE into a finite difference equation (FDE). We then solve the FDE and hope that our solution is correct. The question, however, comes what makes a solution scheme right. If we wish to make our own solution schemes, what are the properties we should look into. The characteristic of a solution scheme that makes it acceptable is its *convergent* nature. A solution scheme is convergent if

$$\lim_{\Delta t, \Delta x \to 0} (\text{FDE Solution}) = \text{PDE Solution} \tag{11.29}$$

Lax's equivalence theorem suggests that for a scheme to be convergent it must be *consistent* and *stable*. Here, we will not get into the equivalence theorem, rather we concentrate on these two elegant and useful concepts: consistency and stability. However, before we proceed, let us recall our discussion on computational errors.

We already know numerical solutions are susceptible to errors from two sources, namely, round-off error (RE) and truncation error (TE).

11.6.1 ROUND-OFF ERROR

A real number often contains infinte number of digits after the decimal point. However, for practical computation, we can take only finite number of digits. A typical computer is a *finite-precision* machine that truncates a real number. Such truncation of real number saves memory space in a computer and speeds up computation. At the same time, such truncation brings inaccuracies in our computed results. We can imagine a computer-like machine that does not truncate a real number after few digits, rather use all the digits after the decimal point. We name our imaginary machine an *infinite-precision* machine. If we now solve the same equation in these two machines, we expect a difference in result. This difference is known as RE.

Recall the following FDE generated from the Crank–Nicolson discretization of the diffusion equation:

$$\frac{u_i^{n+1} - u T_i^n}{\alpha \Delta t} = \frac{1}{2} \left(\frac{u_{i-1}^n - 2u_i^n + u_{i+1}^n}{\Delta x^2} \right) + \frac{1}{2} \left(\frac{u_{i-1}^{n+1} - 2u_i^{n+1} + u_{i+1}^{n+1}}{\Delta x^2} \right) \tag{11.30}$$

Let us say we have a solution $u = D$ if we compute in an infinite-precision machine. Then, we can write

$$\frac{D_i^{n+1} - D_i^n}{\alpha \Delta t} = \frac{1}{2} \left(\frac{D_{i-1}^n - 2D_i^n + D_{i+1}^n}{\Delta x^2} \right) + \frac{1}{2} \left(\frac{D_{i-1}^{n+1} - 2D_i^{n+1} + D_{i+1}^{n+1}}{\Delta x^2} \right) \tag{11.31}$$

Let us say we have a solution $u = N$ if we compute in a finite-precision machine. Then, we can write

$$\frac{N_i^{n+1} - N_i^n}{\alpha \Delta t} = \frac{1}{2} \left(\frac{N_{i-1}^n - 2N_i^n + N_{i+1}^n}{\Delta x^2} \right) + \frac{1}{2} \left(\frac{N_{i-1}^{n+1} - 2N_i^{n+1} + N_{i+1}^{n+1}}{\Delta x^2} \right) \tag{11.32}$$

We can now define RE as RE $= \varepsilon = N - D$. The exact calculation of RE is neither possible nor necessary. The behavior of RE, especially the accumulation of RE with iterations/time-stepping, would be of serious interest and we will discuss that shortly.

11.6.2 TRUNCATION ERROR

Truncation error (TE), on the other hand, is defined as the difference between the PDE and the FDE. For instance, for the earlier example, the TE in the Crank–Nicolson scheme is given by

$$\frac{1}{\alpha}\frac{\partial u}{\partial t} - \frac{\partial^2 u}{\partial x^2} = \frac{u_i^{n+1} - u_i^n}{\alpha \Delta t} - \frac{1}{2}\left(\frac{u_{i-1}^n - 2u_i^n + u_{i+1}^n}{\Delta x^2}\right) - \frac{1}{2}\left(\frac{u_{i-1}^{n+1} - 2u_i^{n+1} + u_{i+1}^{n+1}}{\Delta x^2}\right) + TE$$

(11.33)

As we already know, the exact calculation of TE is neither possible nor necessary, rather we focus on the *order* of the TE. We have also discussed before that the order of TE indicates only the behavior of TE with Δt and Δx but not the absolute value of TE.

Based on this discussion, we can now define consistency. A numerical scheme is consistent if

$$\lim_{\Delta t, \Delta x \to 0} (TE) = 0 \qquad (11.34)$$

11.6.3 CONSISTENCY

Example: *Find the TE of the explicit scheme for the 1-D transient diffusion equation and comment on the consistency of the scheme.*

SOLUTION

Recall the explicit discretization of the 1-D transient diffusion equation

$$\frac{1}{\alpha}\frac{\partial u}{\partial t} - \frac{\partial^2 u}{\partial x^2} = \frac{u_i^{n+1} - u_i^n}{\alpha \Delta t} - \frac{u_{i-1}^n - 2u_i^n + u_{i+1}^n}{\Delta x^2} + TE \qquad (11.35)$$

Using the Taylor series expansion, we have

$$u_{i+1} = u_i + \Delta x u_i' + \frac{(\Delta x)^2}{2!}u_i'' + \frac{(\Delta x)^3}{3!}u_i''' + \frac{(\Delta x)^4}{4!}u_i'''' + \dots$$

$$u_{i-1} = f_i - \Delta x u_i' + \frac{(\Delta x)^2}{2!}u_i'' - \frac{(\Delta x)^3}{3!}u_i''' + \frac{(\Delta x)^4}{4!}u_i'''' + \dots$$

(11.36)

Adding these equations, we find

$$u_{i+1} + u_{i-1} = 2u_i + (\Delta x)^2 u_i'' + 2\frac{(\Delta x)^4}{4!}u_i'''' + \dots$$

$$\Rightarrow u_i'' = \left(\frac{\partial^2 u}{\partial x^2}\right)_i^n = \frac{u_{i-1}^n - 2u_i^n + u_{i+1}^n}{(\Delta x)^2} - \frac{(\Delta x)^2}{12}\left(\frac{\partial^4 u}{\partial x^4}\right)_i^n + \dots \qquad (11.37)$$

Similarly, from the Taylor series expansion

$$\left(\frac{\partial u}{\alpha \partial t}\right)_i^n = \frac{u_i^{n+1} - u_i^n}{\alpha\,(\Delta t)} - \frac{\Delta t}{2}\left(\frac{\partial^2 u}{\partial t^2}\right)_i^n + \cdots \tag{11.38}$$

Combining equations 11.35,11.37,11.38, we now have

$$\frac{1}{\alpha}\frac{\partial u}{\partial t} - \frac{\partial^2 u}{\partial x^2} = \frac{u_i^{n+1} - u_i^n}{\alpha \Delta t} - \frac{u_{i-1}^n - 2u_i^n + u_{i+1}^n}{\Delta x^2} + \text{TE}$$

$$\text{where, TE} = -\frac{\Delta t}{2}\left(\frac{\partial^2 u}{\partial t^2}\right)_i^n + \frac{(\Delta x)^2}{12}\left(\frac{\partial^4 u}{\partial x^4}\right)_i^n + \cdots \tag{11.39}$$

From equation 11.39, it is now quite clear that

$$\lim_{\Delta t, \Delta x \to 0} (\text{TE}) = 0 \tag{11.40}$$

Thus, the explicit scheme for the 1-D diffusion equation is consistent.

Example: Find the TE of the Crank–Nicolson scheme for the 1-D transient diffusion equation and comment on the consistency of the scheme.

SOLUTION

Recall the Crank–Nicolson discretization of the 1-D transient diffusion equation

$$\frac{1}{\alpha}\frac{\partial u}{\partial t} - \frac{\partial^2 u}{\partial x^2} = \frac{u_i^{n+1} - u_i^n}{\alpha \Delta t} - \frac{u_{i-1}^n - 2u_i^n + u_{i+1}^n}{2\Delta x^2} - \frac{u_{i-1}^{n+1} - 2u_i^{n+1} + u_{i+1}^{n+1}}{2\Delta x^2} + \text{TE} \tag{11.41}$$

Using the Taylor series expansion, we have

$$u_{i+1}^n = u_i^n + \Delta x \left(\frac{\partial u}{\partial x}\right)_i^n + \frac{(\Delta x)^2}{2!}\left(\frac{\partial^2 u}{\partial x^2}\right)_i^n + \frac{(\Delta x)^3}{3!}\left(\frac{\partial^3 u}{\partial x^3}\right)_i^n + \frac{(\Delta x)^4}{4!}\left(\frac{\partial^4 u}{\partial x^4}\right)_i^n + \cdots \tag{11.42}$$

$$u_{i-1}^n = u_i^n - \Delta x \left(\frac{\partial u}{\partial x}\right)_i^n + \frac{(\Delta x)^2}{2!}\left(\frac{\partial^2 u}{\partial x^2}\right)_i^n - \frac{(\Delta x)^3}{3!}\left(\frac{\partial^3 u}{\partial x^3}\right)_i^n + \frac{(\Delta x)^4}{4!}\left(\frac{\partial^4 u}{\partial x^4}\right)_i^n + \cdots \tag{11.43}$$

Adding these equations, we find

$$u_{i+1}^n + u_{i-1}^n = 2u_i^n + 2\frac{(\Delta x)^2}{2!}\left(\frac{\partial^2 u}{\partial x^2}\right)_i^n + 2\frac{(\Delta x)^4}{4!}\left(\frac{\partial^4 u}{\partial x^4}\right)_i^n + \cdots$$

$$\Rightarrow \frac{u_{i-1}^n - 2u_i^n + u_{i+1}^n}{2(\Delta x)^2} = \frac{1}{2}\left(\frac{\partial^2 u}{\partial x^2}\right)_i^n + \frac{(\Delta x)^2}{4!}\left(\frac{\partial^4 u}{\partial x^4}\right)_i^n + \cdots \tag{11.44}$$

Further using the multivariable Taylor series expansion, we have

$$
u_{i+1}^{n+1} = u_i^n + \Delta t \left(\frac{\partial u}{\partial t} \right)_i^n + \Delta x \left(\frac{\partial u}{\partial x} \right)_i^n
$$

$$
+ \frac{(\Delta t)^2}{2!} \left(\frac{\partial^2 u}{\partial t^2} \right)_i^n + \frac{(\Delta t)(\Delta x)}{1!1!} \left(\frac{\partial^2 u}{\partial t \partial x} \right)_i^n + \frac{(\Delta x)^2}{2!} \left(\frac{\partial^2 u}{\partial x^2} \right)_i^n
$$

$$
+ \frac{(\Delta t)^3}{3!} \left(\frac{\partial^3 u}{\partial t^3} \right)_i^n + \frac{(\Delta t)^2(\Delta x)}{2!1!} \left(\frac{\partial^3 u}{\partial t^2 \partial x} \right)_i^n + \frac{(\Delta t)(\Delta x)^2}{1!2!} \left(\frac{\partial^3 u}{\partial t \partial x^2} \right)_i^n
$$

$$
+ \frac{(\Delta x)^3}{3!} \left(\frac{\partial^3 u}{\partial x^3} \right)_i^n + \frac{(\Delta t)^4}{4!} \left(\frac{\partial^4 u}{\partial t^4} \right)_i^n + \frac{(\Delta t)^3(\Delta x)}{3!1!} \left(\frac{\partial^4 u}{\partial t^3 \partial x} \right)_i^n
$$

$$
+ \frac{(\Delta t)^2(\Delta x)^2}{2!2!} \left(\frac{\partial^4 u}{\partial t^2 \partial x^2} \right)_i^n + \frac{(\Delta t)(\Delta x)^3}{1!3!} \left(\frac{\partial^4 u}{\partial t \partial x^3} \right)_i^n + \frac{(\Delta x)^4}{4!} \left(\frac{\partial^4 u}{\partial x^4} \right)_i^n + \dots
$$

$$
\tag{11.45}
$$

$$
u_{i-1}^{n+1} = u_i^n + \Delta t \left(\frac{\partial u}{\partial t} \right)_i^n - \Delta x \left(\frac{\partial u}{\partial x} \right)_i^n
$$

$$
+ \frac{(\Delta t)^2}{2!} \left(\frac{\partial^2 u}{\partial t^2} \right)_i^n - \frac{(\Delta t)(\Delta x)}{1!1!} \left(\frac{\partial^2 u}{\partial t \partial x} \right)_i^n + \frac{(\Delta x)^2}{2!} \left(\frac{\partial^2 u}{\partial x^2} \right)_i^n
$$

$$
+ \frac{(\Delta t)^3}{3!} \left(\frac{\partial^3 u}{\partial t^3} \right)_i^n - \frac{(\Delta t)^2(\Delta x)}{2!1!} \left(\frac{\partial^3 u}{\partial t^2 \partial x} \right)_i^n + \frac{(\Delta t)(\Delta x)^2}{1!2!} \left(\frac{\partial^3 u}{\partial t \partial x^2} \right)_i^n
$$

$$
- \frac{(\Delta x)^3}{3!} \left(\frac{\partial^3 u}{\partial x^3} \right)_i^n + \frac{(\Delta t)^4}{4!} \left(\frac{\partial^4 u}{\partial t^4} \right)_i^n - \frac{(\Delta t)^3(\Delta x)}{3!1!} \left(\frac{\partial^4 u}{\partial t^3 \partial x} \right)_i^n
$$

$$
+ \frac{(\Delta t)^2(\Delta x)^2}{2!2!} \left(\frac{\partial^4 u}{\partial t^2 \partial x^2} \right)_i^n - \frac{(\Delta t)(\Delta x)^3}{1!3!} \left(\frac{\partial^4 u}{\partial t \partial x^3} \right)_i^n + \frac{(\Delta x)^4}{4!} \left(\frac{\partial^4 u}{\partial x^4} \right)_i^n + \dots
$$

$$
\tag{11.46}
$$

and

$$
u_i^{n+1} = u_i^n + \Delta t \left(\frac{\partial u}{\partial t} \right)_i^n + \frac{(\Delta t)^2}{2!} \left(\frac{\partial^2 u}{\partial t^2} \right)_i^n + \frac{(\Delta t)^3}{3!} \left(\frac{\partial^3 u}{\partial t^3} \right)_i^n + \frac{(\Delta t)^4}{4!} \left(\frac{\partial^4 u}{\partial t^4} \right)_i^n
$$

$$
\tag{11.47}
$$

Using these equations, we find

$$u_{i+1}^{n+1} - 2u_i^{n+1} + u_{i-1}^{n+1} = 2\frac{(\Delta x)^2}{2!}\left(\frac{\partial^2 u}{\partial x^2}\right)_i^n + 2\frac{(\Delta t)(\Delta x)^2}{1!2!}\left(\frac{\partial^3 u}{\partial t \partial x^2}\right)_i^n$$

$$+ 2\frac{(\Delta t)^2(\Delta x)^2}{2!2!}\left(\frac{\partial^4 u}{\partial t^2 \partial x^2}\right)_i^n + 2\frac{(\Delta x)^4}{4!}\left(\frac{\partial^4 u}{\partial x^4}\right)_i^n + \cdots$$

$$\Rightarrow \frac{u_{i-1}^{n+1} - 2u_i^{n+1} + u_{i+1}^{n+1}}{2(\Delta x)^2} = \frac{1}{2}\left(\frac{\partial^2 u}{\partial x^2}\right)_i^n + \frac{\Delta t}{2}\left(\frac{\partial^3 u}{\partial t \partial x^2}\right)_i^n + \frac{(\Delta t)^2}{4}\left(\frac{\partial^4 u}{\partial t^2 \partial x^2}\right)_i^n$$

$$+ \frac{(\Delta x)^2}{4!}\left(\frac{\partial^4 u}{\partial x^4}\right)_i^n + \cdots \tag{11.48}$$

Thus,

$$\frac{u_{i-1}^n - 2u_i^n + u_{i+1}^n}{2(\Delta x)^2} + \frac{u_{i-1}^{n+1} - 2u_i^{n+1} + u_{i+1}^{n+1}}{2(\Delta x)^2}$$

$$= \left(\frac{\partial^2 u}{\partial x^2}\right)_i^n + \frac{\Delta t}{2}\left(\frac{\partial^3 u}{\partial t \partial x^2}\right)_i^n + \frac{(\Delta t)^2}{4}\left(\frac{\partial^4 u}{\partial t^2 \partial x^2}\right)_i^n + 2\frac{(\Delta x)^2}{4!}\left(\frac{\partial^4 u}{\partial x^4}\right)_i^n + \cdots \tag{11.49}$$

Similarly, recall the earlier Taylor series expansion

$$u_i^{n+1} = u_i^n + \Delta t\left(\frac{\partial u}{\partial t}\right)_i^n + \frac{(\Delta t)^2}{2!}\left(\frac{\partial^2 u}{\partial t^2}\right)_i^n + \frac{(\Delta t)^3}{3!}\left(\frac{\partial^3 u}{\partial t^3}\right)_i^n + \frac{(\Delta t)^4}{4!}\left(\frac{\partial^4 u}{\partial t^4}\right)_i^n \tag{11.50}$$

leading to

$$\frac{u_i^{n+1} - u_i^n}{\alpha \Delta t} = \frac{1}{\alpha}\left(\frac{\partial u}{\partial t}\right)_i^n + \frac{\Delta t}{2\alpha}\left(\frac{\partial^2 u}{\partial t^2}\right)_i^n + \frac{(\Delta t)^2}{6\alpha}\left(\frac{\partial^3 u}{\partial t^3}\right)_i^n + \frac{(\Delta t)^3}{24\alpha}\left(\frac{\partial^4 u}{\partial t^4}\right)_i^n \tag{11.51}$$

Combining equations 11.41 and 11.51, we now have

$$\frac{1}{\alpha}\frac{\partial u}{\partial t} - \frac{\partial^2 u}{\partial x^2} = \frac{u_i^{n+1} - u_i^n}{\alpha \Delta t} - \frac{u_{i-1}^n - 2u_i^n + u_{i+1}^n}{2(\Delta x)^2} - \frac{u_{i-1}^{n+1} - 2u_i^{n+1} + u_{i+1}^{n+1}}{2(\Delta x)^2} + \text{TE}$$

$$\text{where, TE} = \frac{\Delta t}{2\alpha}\left(\frac{\partial^2 u}{\partial t^2}\right)_i^n + \frac{(\Delta t)^2}{6\alpha}\left(\frac{\partial^3 u}{\partial t^3}\right)_i^n + \frac{(\Delta t)^3}{24\alpha}\left(\frac{\partial^4 u}{\partial t^4}\right)_i^n$$

$$- \frac{\Delta t}{2}\left(\frac{\partial^3 u}{\partial t \partial x^2}\right)_i^n - \frac{(\Delta t)^2}{4}\left(\frac{\partial^4 u}{\partial t^2 \partial x^2}\right)_i^n - \frac{(\Delta x)^2}{12}\left(\frac{\partial^4 u}{\partial x^4}\right)_i^n + \cdots$$

$$= \frac{\Delta t}{2}\left[\frac{\partial}{\partial t}\left(\frac{1}{\alpha}\frac{\partial u}{\partial t} - \frac{\partial^2 u}{\partial x^2}\right)\right]_i^n + \mathcal{O}\left[(\Delta t)^2, (\Delta x)^2\right]$$

$$= \mathcal{O}\left[(\Delta t)^2, (\Delta x)^2\right] \tag{11.52}$$

From equation 11.52, it is now quite clear that

$$\lim_{\Delta t, \Delta x \to 0} (\text{TE}) = 0 \tag{11.53}$$

Thus, the Crank–Nicolson scheme for the 1-D diffusion equation is consistent.

11.6.4 STABILITY

Recall our discussion of RE originating from finite-precision computation. As we keep on solving the FDE, at every time step, RE is added to our solution. A stable numerical scheme, we expect, will reduce the effects of RE as the time progresses. Conversely, an unstable scheme will pile up RE leading us toward an incorrect or often unphysical solution. To quantify the growth of RE, we define an *amplification factor* or *growth factor* G, such that

$$G = \frac{\varepsilon_i^{n+1}}{\varepsilon_i^n} \tag{11.54}$$

where, ε_i^n is the RE at location i at the n-th time step.

Note that for $|G| > 1$, the RE would increase in time. Therefore, we call a scheme stable for which $|G| \leq 1$ for all i and n.

Example: *Verify whether the explicit scheme for the 1-D transient diffusion equation is stable.*

SOLUTION

Recall the explicit discretization of 1-D transient diffusion equation

$$\frac{u_i^{n+1} - u_i^n}{\alpha \Delta t} = \frac{u_{i-1}^n - 2u_i^n + u_{i+1}^n}{\Delta x^2} \tag{11.55}$$

Let us say we have a solution $u = D$ if we compute in an infinite-precision machine. Then, we can write

$$\frac{D_i^{n+1} - D_i^n}{\alpha \Delta t} = \frac{D_{i-1}^n - 2D_i^n + D_{i+1}^n}{\Delta x^2} \tag{11.56}$$

Let us say we have a solution $u = N$ if we compute in a finite-precision machine (actual computer). Then, we can write

$$\frac{N_i^{n+1} - N_i^n}{\alpha \Delta t} = \frac{N_{i-1}^n - 2N_i^n + N_{i+1}^n}{\Delta x^2} \tag{11.57}$$

We now deduce the equation of RE noting that $\text{RE} = \varepsilon = N - D$. Using equations 11.56 and 11.57

$$\frac{\varepsilon_i^{n+1} - \varepsilon_i^n}{\alpha \Delta t} = \frac{\varepsilon_{i-1}^n - 2\varepsilon_i^n + \varepsilon_{i+1}^n}{\Delta x^2} \tag{11.58}$$

Looks like the equation of ε follows a similar PDE as does u

$$\frac{1}{\alpha} \frac{\partial \varepsilon}{\partial t} = \frac{\partial^2 \varepsilon}{\partial x^2} \tag{11.59}$$

We cannot solve this equation since we don't know the initial and boundary conditions. We can, however, be clever enough to understand that under linear initial/boundary conditions, this equation is amenable to separation of variables. You can consult an engineering mathematics book for the entire solution procedure. For this course, it is sufficient to appreciate that the solution of equation 11.59 would take the following form:

$$\varepsilon = \sum_m \exp(a_m t + jk_m x) \text{ where, } j = \sqrt{-1} \tag{11.60}$$

Here, m indicates mode and takes integer values up to infinity. Each m represents a wave, and each wave satisfies the governing equation. Therefore, linear superposition of all the waves provides the final solution. You can think of the whole solution as a vector, and each wave is like a component of that vector. Therefore, we can follow only one mode to track the nature of the solution. Dealing with the m-th mode only, we have

$$\varepsilon_{i,m}^n = \exp(a_m t + jk_m x) = \exp(a_m n \Delta t + jk_m x) \tag{11.61}$$

Thus, we can write

$$\varepsilon_{i-1,m}^n = \exp\left[a_m n \Delta t + jk_m (x - \Delta x)\right]$$
$$\varepsilon_{i+1,m}^n = \exp\left[a_m n \Delta t + jk_m (x + \Delta x)\right] \tag{11.62}$$

Combining equations 11.61 and 11.62, we have

$$\frac{\varepsilon_{i-1,m}^n}{\varepsilon_{i,m}^n} = \exp(-jk_m \Delta x)$$

$$\frac{\varepsilon_{i+1,m}^n}{\varepsilon_{i,m}^n} = \exp(jk_m \Delta x) \tag{11.63}$$

Dividing both sides of equation 11.63 with $\varepsilon_{i,m}^{n}$, we have

$$\frac{\varepsilon_i^{n+1}}{\varepsilon_i^n} - 1 = r\left[\exp\left(-jk_m\Delta x\right) - 2 + \exp\left(jk_m\Delta x\right)\right] \quad \text{where, } r = \frac{\alpha\Delta t}{\Delta x^2} \quad (11.64)$$

$$\Rightarrow G - 1 = 2r\left(\cos\beta - 1\right)$$

where,

$$G = \frac{\varepsilon_i^{n+1}}{\varepsilon_i^n} \text{ is the growth factor and } \beta = k_m\Delta x$$

$$G = 1 + 2r\left(\cos\beta - 1\right) = 1 - 4r\sin^2\frac{\beta}{2} \quad (11.65)$$

Applying the stability condition,

$$|G| \leq 1$$

$$\Rightarrow \left|1 - 4r\sin^2\frac{\beta}{2}\right| \leq 1$$

$$\Rightarrow -1 \leq 1 - 4r\sin^2\frac{\beta}{2} \leq 1$$

$$\Rightarrow 0 \leq 4r\sin^2\frac{\beta}{2} \leq 2$$

$$\Rightarrow 0 \leq r \leq \frac{1}{2\sin^2\frac{\beta}{2}} \quad (11.66)$$

Since the value of β could be anything between 0 to 2π, we try to identify the range of r that would satisfy the stability condition for any value of β. Please note that the expression of r includes user-defined variables Δt and Δx and thus we can vary the value of r to meet the stability criteria.

$$|G| \leq 1$$

$$\Rightarrow 0 \leq r \leq \left(\frac{1}{2\sin^2\frac{\beta}{2}}\right)_{\min}$$

$$\Rightarrow 0 \leq r \leq \frac{1}{2\left(\sin^2\frac{\beta}{2}\right)_{\max}}$$

$$\Rightarrow 0 \leq r \leq \frac{1}{2} \quad (11.67)$$

$$\text{since } \left(\sin^2\frac{\beta}{2}\right)_{\max} = 1$$

So for the 1-D transient diffusion equation, the explicit scheme is conditionally stable and the condition for stability is given by

$$\frac{\alpha\Delta t}{\Delta x^2} \leq \frac{1}{2} \quad (11.68)$$

Needless to mention is that r is always positive and thus $r \geq 0$ does not add to any additional condition. This procedure for stability analysis is known as von Neumann stability analysis to commemorate the famous mathematician and computer scienctist John von Neumann (1903–1957).

11.7 TWO-DIMENSIONAL TRANSIENT DIFFUSION

Consider a metallic slab of cross-section $L \times L$, which was initially at ambient temperature T_∞, is subjected to sudden heating from one side. The transient temperature of the slab $T(t,x,y)$ is given by the following governing equation:

$$\frac{\partial u}{\partial u} = \alpha \nabla^2 u = \alpha \left(\frac{\partial^2 u}{\partial x^2} + \frac{\partial^2 u}{\partial y^2} \right) \tag{11.69}$$

This 2-D diffusion equation requires one initial and four boundary conditions. Let's say

$$\text{Initial condition: } u(t=0,x,y) = u_\infty \tag{11.70}$$
$$\text{Boundary conditions: } u(t>0,x=0,y) = u_0, u(t>0,x=L,y) = u_\infty \tag{11.71}$$
$$u(t>0,x,y=0) = u_\infty, u(t>0,x,y=L) = u_\infty \tag{11.72}$$

The analytical solution of this problem, though tedious, is possible. Here, we will try to solve this problem numerically.

11.7.1 EXPLICIT SCHEME

Let us first discretize domain x in $m-1$ equal divisions creating m points x_i, where $i = 1,2,\cdots,m$ and a constant grid size $x_{i+1} - x_i = \Delta x$. Note that $(m-1)\Delta x = L$. Similarly, we discretize domain y in $p-1$ equal divisions creating p points y_j, where $j = 1,2,\cdots,p$ and a constant grid size $y_{j+1} - y_j = \Delta y$. Note that $(p-1)\Delta y = L$.

We now discretize time t via constant time step Δt. Recalling our discussion of numerical differentiation, we discretize the LHS of equation 11.69 as follows:

$$\frac{\partial u}{\partial t} = \frac{u_{i,j}^{n+1} - u_{i,j}^n}{\Delta t} \tag{11.73}$$

For the explicit method, we discretize the spatial derivative at the n-th time step.

$$\frac{\partial^2 u}{\partial x^2} + \frac{\partial^2 u}{\partial x^2} = \frac{u_{i-1,j}^n - 2u_{i,j}^n + u_{i+1,j}^n}{\Delta x^2} + \frac{u_{i,j-1}^n - 2u_{i,j}^n + u_{i,j+1}^n}{\Delta y^2} \tag{11.74}$$

Overall, the discretized form of equation 11.69 can be written as

$$\frac{u_{i,j}^{n+1} - u_{i,j}^n}{\alpha \Delta t} = \frac{u_{i-1,j}^n - 2u_{i,j}^n + u_{i+1,j}^n}{\Delta x^2} + \frac{u_{i,j-1}^n - 2u_{i,j}^n + u_{i,j+1}^n}{\Delta y^2} \tag{11.75}$$

This equation shows explicit discretization of 1-D diffusion equation with an order of accuracy of $\mathscr{O}\left(\Delta t, \Delta x^2\right)$.

Now, to solve equation 11.69, we rewrite the discretized equation 11.75 as

$$u_{i,j}^{n+1} = u_{i,j}^n + \alpha \Delta t \left(\frac{u_{i-1,j}^n - 2u_{i,j}^n + u_{i+1,j}^n}{\Delta x^2} + \frac{u_{i,j-1}^n - 2u_{i,j}^n + u_{i,j+1}^n}{\Delta y^2} \right) \qquad (11.76)$$

This difference equation is valid for $n > 0, 1 < i < m, 1 < j < p$. Overall, the solution methodology is given by

1. Initailization: $u_{i,j}^n = u_\infty$ for $n = 0$
2. Boundary: $x = 0$: $u_{i,j}^n = T_0$ for $i = 1, n > 0$
3. Boundary: $x = L$: $u_{i,j}^n = u_\infty$ for $i = m, n > 0$
4. Boundary: $y = 0$: $u_{i,j}^n = u_\infty$ for $j = 1, n > 0$
5. Boundary: $y = L$: $u_{i,j}^n = u_\infty$ for $j = p, n > 0$
6. Interior: $u_{i,j}^{n+1} = u_{i,j}^n + \alpha \Delta t \left(\frac{u_{i-1,j}^n - 2u_{i,j}^n + u_{i+1,j}^n}{\Delta x^2} + \frac{u_{i,j-1}^n - 2u_{i,j}^n + u_{i,j+1}^n}{\Delta y^2} \right)$ for $1 < i < m, n > 0$
7. Continue until desired time or steady state is reached

This procedure clearly shows that given the initial and boundary conditions, the value of u may be calculated at any subsequent time at any location. The calculation may be stopped when we reach a specified time or steady state. We know that at steady state, the time derivative would vanish. Thus, if we check the norm of u at two subsequent times, we declare steady state when the norm reaches less than a specified small number, such that

$$\sum_{i=1}^{i=m} \left(u_{ij}^{n+1} - u_{ij}^n \right)^2 < \varepsilon \qquad (11.77)$$

where, ε is a small number specified by the programmer.

As discussed earlier the explicit method is quite easy to implement and most importantly it does not require a solution of the equations.

11.7.2   IMPLICIT AND CRANK–NICOLSON SCHEMES

As we did in the explicit method, let us first discretize domain x in $m - 1$ equal divisions creating m points x_i, where $i = 1, 2, \cdots, m$ and a constant grid size $x_{i+1} - x_i = \Delta x$. Note that $(m-1)\Delta x = L$. Similarly, we discretize the domain y in $p - 1$ equal divisions creating p points y_j, where $j = 1, 2, \cdots, p$ and a constant grid size $y_{j+1} - y_j = \Delta y$. Note that $(p-1)\Delta y = L$.

We now discretize time t via constant time step Δt. Recalling our discussion of numerical differentiation, we discretize the LHS of equation 11.69 as follows:

$$\frac{\partial u}{\partial t} = \frac{u_{i,j}^{n+1} - u_{i,j}^n}{\Delta t} \qquad (11.78)$$

For the implicit method, we discretize the spatial derivative at the $(n+1)$-th time step.

$$\frac{\partial^2 u}{\partial x^2} + \frac{\partial^2 u}{\partial y^2} = \frac{u_{i-1,j}^{n+1} - 2u_{i,j}^{n+1} + u_{i+1,j}^{n+1}}{\Delta x^2} + \frac{u_{i,j-1}^{n+1} - 2u_{i,j}^{n+1} + u_{i,j+1}^{n+1}}{\Delta y^2} \tag{11.79}$$

Overall, the discretized form of equation 11.69 can be written as

$$\frac{u_{i,j}^{n+1} - u_{i,j}^{n}}{\alpha \Delta t} = \frac{u_{i-1,j}^{n+1} - 2u_{i,j}^{n+1} + u_{i+1,j}^{n+1}}{\Delta x^2} + \frac{u_{i,j-1}^{n+1} - 2u_{i,j}^{n+1} + u_{i,j+1}^{n+1}}{\Delta y^2} \tag{11.80}$$

Both the explicit and implicit discretizations of the 2-D diffusion equation show order of accuracy of $\mathcal{O}\left(\Delta t, \Delta x^2, \Delta y^2\right)$.

For the Crank–Nicolson method, we discretize the spatial derivative as the average of implicit and explicit as follows:

$$\frac{\partial^2 u}{\partial x^2} + \frac{\partial^2 u}{\partial y^2} = \frac{1}{2}\left(\frac{u_{i-1,j}^{n+1} - 2u_{i,j}^{n+1} + u_{i+1,j}^{n+1}}{\Delta x^2} + \frac{u_{i,j-1}^{n+1} - 2u_{i,j}^{n+1} + T_{i,j+1}^{n+1}}{\Delta y^2} \right)$$
$$+ \frac{1}{2}\left(\frac{u_{i-1,j}^{n} - 2u_{i,j}^{n} + u_{i+1,j}^{n}}{\Delta x^2} + \frac{u_{i,j-1}^{n} - 2u_{i,j}^{n} + u_{i,j+1}^{n}}{\Delta y^2} \right) \tag{11.81}$$

Overall, the discretized form of equation 11.69 can be written as

$$\frac{u_{i,j}^{n+1} - u_{i,j}^{n}}{\alpha \Delta t} = \frac{1}{2}\left(\frac{u_{i-1,j}^{n+1} - 2u_{i,j}^{n+1} + u_{i+1,j}^{n+1}}{\Delta x^2} + \frac{u_{i,j-1}^{n+1} - 2u_{i,j}^{n+1} + u_{i,j+1}^{n+1}}{\Delta y^2} \right)$$
$$+ \frac{1}{2}\left(\frac{u_{i-1,j}^{n} - 2u_{i,j}^{n} + u_{i+1,j}^{n}}{\Delta x^2} + \frac{u_{i,j-1}^{n} - 2u_{i,j}^{n} + u_{i,j+1}^{n}}{\Delta y^2} \right) \tag{11.82}$$

The Crank–Nicolson discretizations of the 2-D diffusion equation show order of accuracy of $\mathcal{O}\left(\Delta t^2, \Delta x^2, \Delta y^2\right)$.

Now, to solve equation 11.69, we rewrite the discretized equations 11.80 and 11.82 as

$$au_{i,j-1}^{n+1} + bu_{i-1,j}^{n+1} + cu_{i,j}^{n+1} + bu_{i+1,j}^{n+1} + au_{i,j+1}^{n+1} = d_{ij}^{n} \tag{11.83}$$

For implicit method,

$$a = -\frac{\alpha \Delta t}{\Delta y^2}, b = -\frac{\alpha \Delta t}{\Delta x^2}, c = 1 + 2\frac{\alpha \Delta t}{\Delta x^2} + 2\frac{\alpha \Delta t}{\Delta y^2}, d_{ij}^{n} = T_{ij}^{n} \tag{11.84}$$

Similarly, for the Crank–Nicolson method,

$$a = -\frac{\alpha \Delta t}{2\Delta y^2}, b = -\frac{\alpha \Delta t}{2\Delta x^2}, c = 1 + \frac{\alpha \Delta t}{\Delta x^2} + \frac{\alpha \Delta t}{\Delta y^2} \tag{11.85}$$

$$d_{ij}^{n} = u_{ij}^{n} + \frac{1}{2}\left(\frac{u_{i-1,j}^{n} - 2u_{i,j}^{n} + u_{i+1,j}^{n}}{\Delta x^2} + \frac{u_{i,j-1}^{n} - 2u_{i,j}^{n} + u_{i,j+1}^{n}}{\Delta y^2} \right) \tag{11.86}$$

We now see that the PDE is transformed into a system of linear equations. In case of 1-D diffusion, the linear equations formed a tridiagonal system of equations. But here we see a pentadiagonal system. We have $m \times p$ numbers of equations with the same number of unknowns, but each equation contains five unknowns only. In each equation, coefficients of $m \times p - 5$ unknowns are zero. Therefore, to solve the PDE, we have to solve an $\mathbf{Ax} = \mathbf{b}$ system, where $\mathbf{A}$ is a sparse matrix. As you know that such system can be efficiently solved by Gauss–Seidel or similar techniques. Please note that the system of equations must be solved for every time step. While matrix $\mathbf{A}$ will remain unchanged in time, $\mathbf{b}$ must be updated after every time step.

11.8 ELLIPTIC EQUATIONS

So far, in our discussion of the PDE solution, we have talked about 1-D or 2-D equations of the following nature:

$$\frac{\partial u}{\partial t} = \alpha \nabla^2 u \qquad (11.87)$$

$$\text{where, Laplacian } \nabla^2 = \frac{d^2}{dx^2} \text{ in 1-D} \qquad (11.88)$$

$$\text{and } \nabla^2 = \left(\frac{\partial^2}{\partial x^2} + \frac{\partial^2}{\partial y^2} \right) \text{ in 2-D} \qquad (11.89)$$

For reasons discussed earlier, this equation is *parabolic* in nature and the physical problem is known as the *marching* problem. In such problems, t is either time or a *time-like* variable. For solving this problem, an initial condition must be prescribed at $t = t_0$ and we solve the PDE to find $u(t > t_0)$. The initial condition *diffuses* at a *pace* determined by constant $\alpha > 0$. For any $\tau > t_0$, solution $u(t > \tau)$ depends on $u(t \leq \tau)$, while solution $u(t \leq \tau)$ does not depend on $u(t > \tau)$. In other words, future depends on past but the reverse is not true. The idea is consistent with our everyday perception of time as well as with the second law of thermodynamics.

Under compatible boundary conditions, this equation leads to an *equilibrium* or *steady-state* solution as $t \to \infty$. At this point, the solution $u(t \to \infty)$ does not undergo any more changes in time. The steady-state solution also seems to be independent of the initial condition. As such, at steady state, time does not have any meaning at all, the temporal derivative of equation 11.87 vanishes and the equilibrium distribution of u is given by the famous Laplace equation

$$\nabla^2 u = 0 \qquad (11.90)$$

If the original equation contains a source term $-f(x, y)$, the steady-state equation will be given by the following Poisson equation:

$$\nabla^2 u = f(x, y) \qquad (11.91)$$

Both these equations frequently appear in mathematical modeling of engineering applications and often call for highly accurate solution. For instance, Poisson's equation appears in solving the Navier–Stokes equation, where slight inaccuracy in

The Poisson solver leads to mass imbalance of the flowing fluid. Such equations are also quite common in computational electromagnetics for evaluating potentials and in potential flow for computing velocity potentials.

11.8.1 DISCRETIZATION

Consider the following Poisson equation:

$$\nabla^2 u = \left(\frac{\partial^2}{\partial x^2} + \frac{\partial^2}{\partial y^2} \right) u = f(x,y) \tag{11.92}$$

$$u = u_\infty \text{ in all boundaries } x = 0, L \text{ and } y = 0, L \tag{11.93}$$

It is obvious that for $f = 0$, this equation leads to trivial solution $u = u_\infty$. This is, however, not true for $f \neq 0$. To discretize the governing equation, let us first discretize domain x in $m - 1$ equal divisions creating m points x_i, where $i = 1, 2, \cdots, m$ and a constant grid size $x_{i+1} - x_i = \Delta x$. Note that $(m - 1)\Delta x = L$. Similarly, we discretize domain y in $p - 1$ equal divisions creating p points y_j, where $j = 1, 2, \cdots, p$ and a constant grid size $y_{j+1} - y_j = \Delta y$. Note that $(p - 1)\Delta y = L$. Now, we discretize equation 11.92 as follows:

$$\frac{u_{i-1,j} - 2u_{i,j} + u_{i+1,j}}{\Delta x^2} + \frac{u_{i,j-1} - 2u_{i,j} + u_{i,j+1}}{\Delta y^2} = f_{ij} \tag{11.94}$$

$$\text{for } i = 2, \cdots, m - 1 \text{ and } j = 2, \cdots, p - 1$$

Please note that time does not exist here and thus there is no superscript in this equation. The idea of explicit or implicit solvers does not arise here. For boundary, we write

$$u_{ij} = u_\infty \text{ for } i = 1, m \text{ and } j = 1, p \tag{11.95}$$

This scheme is second-order accurate. The von Neumann type stability analysis is not applicable here since the growth factor, in the von Neumann sense, cannot be defined for a steady case.

11.8.2 SOLUTION PROCEDURE

The discretized equation 11.94 may be written as

$$au_{i,j-1} + bu_{i-1,j} + cu_{i,j} + bu_{i+1,j} + au_{i,j+1} = f_{ij} \tag{11.96}$$

where,

$$a = \frac{1}{\Delta y^2}, b = \frac{1}{\Delta x^2}, c = -2 \left(\frac{1}{\Delta x^2} + \frac{1}{\Delta y^2} \right) \tag{11.97}$$

The discretized equations 11.95 and 11.96 provide $m \times p$ equations with the same numbers of unknowns leading to a linear system $\mathbf{A}\mathbf{x} = \mathbf{b}$. Every equation, however, contains only five unknowns. Thus, in the linear system $\mathbf{A}\mathbf{x} = \mathbf{b}$, matrix $\mathbf{A}$ is penta-diagonal. Thus, matrix $\mathbf{A}$ is usually large and sparse, and the linear system $\mathbf{A}\mathbf{x} = \mathbf{b}$

can be solved using a suitable algorithm such as the Gauss–Seidel method to find the value of u at all i, j. Note that unlike the transient problem, we need to solve $\mathbf{Ax} = \mathbf{b}$ system only once. You may, however, find the Poisson solver to be quite slow and matrix $\mathbf{A}$ may often fails to be diagonally dominant. Underrelaxation may be necessary in such cases. Further in a Poisson equation, large source terms (f_{ij}) often destabilize the solution. Underrelaxation, once again, may be necessary to arrest the unbounded growth of residuals arising in the linear solver.

11.8.3 PSEUDO-TRANSIENT APPROACH

As we now see that the numerical solution of the Poisson or Laplace equation always requires inversion of a large, sparse linear system. For a transient case, we know that we can avoid such complexities using an explicit scheme. We may also use alternate direction implicit (ADI) in a transient case, and we know that an ADI scheme uses a tridiagonal solver instead of a general matrix inverter. The question is, therefore, can we borrow such ideas from the transient case to solve the Laplace or Poisson equation? Recall, we are trying to solve the following equation:

$$\nabla^2 u = \left(\frac{\partial^2}{\partial x^2} + \frac{\partial^2}{\partial y^2} \right) u = f(x, y) \tag{11.98}$$

$$u = u_\infty \text{ in all boundaries } x = 0, L \text{ and } y = 0, L \tag{11.99}$$

We can rewrite the equation as

$$\frac{1}{\alpha} \frac{\partial u}{\partial t} = \left(\frac{\partial^2}{\partial x^2} + \frac{\partial^2}{\partial y^2} \right) u - f(x, y) \tag{11.100}$$

$$u = u_\infty \text{ in all boundaries } x = 0, L \text{ and } y = 0, L \tag{11.101}$$

$$u = u_0 \text{ at } t = 0 \tag{11.102}$$

If we now solve equation 11.100, we know that the solution $u(t \to \infty)$ must satisfy equation 11.98. For solving equation 11.100, we can use explicit, ADI or any other method we have discussed before. The problem here, of course, we do not know α and u_0. We, however, know that the equilibrium solution $u(t \to \infty)$ does not depend on the initial condition $u(t = 0)$. We can, therefore, select any value of u_0 to solve equation 11.100. The equilibrium solution also does not depend on the value of α and thus any reasonable value of α would be sufficient to reach the correct solution of the Poisson equation.

While the pseudo-transient approach may enable us to use a transient solver for the steady-state problem, for most problems, the pseudo-transient solver may consume more CPU time than the steady solver.

PROBLEMS

1. Consult any standard engineering mathematics book and solve the following PDE analytically:

$$\frac{\partial u}{\partial t} = \alpha \frac{\partial^2 u}{\partial x^2} \tag{11.103}$$

Initial and boundary conditions are given by

$$u(t=0) = u_0(x); u(x=0) = u_1, u(x=L) = u_2 \tag{11.104}$$

where, u_1 and u_2 are constants.

2. Consult any standard engineering mathematics book and solve the following PDE analytically:

$$\frac{\partial^2 u}{\partial x^2} + \frac{\partial^2 u}{\partial y^2} = 0 \tag{11.105}$$

The boundary conditions are given by

$$u(x=0) = u_1, u(x=L_x) = u_2, u(y=0) = u_3, u(y=L_y) = u_4 \tag{11.106}$$

where, u_1, u_2, u_3 and u_4 are constants.

3. Write a computer program, in any language, to solve 2-D transient diffusion equation using explicit method. Validate your result with an analytical solution. Clearly write the solution algorithm.

4. Solve the following PDE numerically using the pseudo-transient approach. Vary initial conditions to show that your final solution is independent of the initial condition.

$$\frac{\partial^2 u}{\partial x^2} + \frac{\partial^2 u}{\partial y^2} = 0 \tag{11.107}$$

The boundary conditions are given by

$$u(x=0) = 1, u(x=L_x) = 0, u(y=0) = 1, u(y=L_y) = 0 \tag{11.108}$$

    a. Solve this PDE numerically without using the pseudo-transient approach.
    b. Compare your results with the analytical solution.

5. Use von Neumann stability analysis to compare the stability criteria of explicit, implict and Crank–Nicolson methods for a 2-D transient diffusion equation. Use $\Delta x = \Delta y$.

12 Computational Fluid Dynamics

12.1 INTRODUCTION

The development of rapid computing technologies has created a new approach of numerically solving the basic equations of fluid mechanics and heat transfer. Computational fluid dynamics (CFD) deals with the numerical solution of the governing equations of fluid mechanics as well as heat and mass transfer. CFD being a vast subject, for brevity, in this chapter, we will confine our discussion to the incompressible mass and momentum equations shown as follows:

$$\nabla \cdot \mathbf{u} = 0 \tag{12.1}$$

$$\frac{\partial \mathbf{u}}{\partial t} + \nabla \cdot (\mathbf{uu}) = \mathbf{g} - \frac{1}{\rho}\nabla P + v\nabla^2 \mathbf{u} + \mathbf{S_u} \tag{12.2}$$

While equation 12.1 signifies the incompressible mass conservation also known as the continuity equation, equation 12.2 enforces the momentum conservation equation, also known as the Navier–Stokes (N-S) equation. Although the mass conservation equation is a scalar equation, the momentum conservation equation is a vector equation. Solving equations 12.1 and 12.2 we compute velocity field $\mathbf{u}$ and pressure field P. To solve these equations, therefore, we must know the density (ρ), kinematic viscosity (v), gravitational acceleration $(\mathbf{g})$, and the source term $(\mathbf{S_u})$. The source term, $\mathbf{S_u}$, indicates the external forcing that may appear in some cases, such as in buoyancy-driven flows.

Let us now consider a special case, where $\mathbf{u} = 0$ i.e. the fluid statics. In such a situation, the continuity equation is trivially satisfied. The momentum (N-S) equation is reduced to

$$\mathbf{g} - \frac{1}{\rho}\nabla P_h = 0 \Rightarrow \mathbf{g} = \frac{1}{\rho}\nabla P_h \tag{12.3}$$

where, P_h is the hydrostatic pressure. Combining equations 12.2 and 12.3, we have

$$\frac{\partial \mathbf{u}}{\partial t} + \nabla \cdot (\mathbf{uu}) = \frac{1}{\rho}\nabla P_h - \frac{1}{\rho}\nabla P + v\nabla^2 \mathbf{u} + \mathbf{S_u}$$

$$= -\frac{1}{\rho}\nabla p + v\nabla^2 \mathbf{u} + \mathbf{S_u} \tag{12.4}$$

Here, $p = P - P_h$ is the modified pressure.

DOI: 10.1201/9781003049272-12

12.1.1 NON-DIMENSIONALIZATION

To non-dimensionalize the governing equations, we consider the following dimensionless variables:

$$x^* = \frac{x}{L}, y^* = \frac{y}{L}, z^* = \frac{z}{L} \Rightarrow \nabla^* = L\nabla, \nabla^{2*} = L^2\nabla^2; \mathbf{u}^* = \frac{\mathbf{u}}{u_0}, t^* = \frac{tu_0}{L}, p^* = \frac{p}{\rho u_0^2}$$
$$(12.5)$$

Neglecting the source term and dropping the superscript * for simplicity, we have the following *dimensionless* governing equations:

$$\nabla \cdot \mathbf{u} = 0 \qquad\qquad (12.6)$$

$$\frac{\partial \mathbf{u}}{\partial t} + \nabla \cdot (\mathbf{uu}) = -\nabla p + \frac{1}{Re}\nabla^2 \mathbf{u} \qquad\qquad (12.7)$$

Here, $Re = u_0 L/\nu$ is the Reynolds number.

This non-dimensionalization reveals certain important features of fluid mechanics. For instance, high values of Re indicates the dominance of inertia (convection) $\nabla \cdot (\mathbf{uu})$ over viscous (diffusion) $\nabla^2 \mathbf{u}$ forces. In many cases, the dominance of inertia may eventually lead to the transition of laminar to turbulent flow. In this chapter, we will discuss the solution strategies of laminar flows only. While solving equations 12.6 and 12.7, we will follow the finite difference discretization as discussed in Chapters 10 and 11; we must deal with the nonlinearity of the N-S equation.

12.2 STREAM FUNCTION, VORTICITY ($\psi - \omega$) FORMULATION

Consider a 2-D flow field for which the scalar forms of the governing equations, in the Cartesian coordinate, are

$$\frac{\partial u}{\partial x} + \frac{\partial v}{\partial y} = 0 \qquad\qquad (12.8)$$

$$\frac{\partial u}{\partial t} + \frac{\partial \left(u^2\right)}{\partial x} + \frac{\partial (uv)}{\partial y} = -\frac{\partial p}{\partial x} + \frac{1}{Re}\left(\frac{\partial^2 u}{\partial x^2} + \frac{\partial^2 u}{\partial y^2}\right) \qquad (12.9)$$

$$\frac{\partial v}{\partial t} + \frac{\partial (uv)}{\partial x} + \frac{\partial \left(v^2\right)}{\partial y} = -\frac{\partial p}{\partial y} + \frac{1}{Re}\left(\frac{\partial^2 v}{\partial x^2} + \frac{\partial^2 v}{\partial y^2}\right) \qquad (12.10)$$

Here, $u = u(x,y)$ and $v = v(x,y)$ are the components of the velocity vector in x and y directions, respectively.

12.2.1 STREAM FUNCTION

The nature of the above two-dimensional, incompressible continuity equation suggests the existence of a scalar function $\psi(x,y)$, known as the stream function as follows:

$$u = \frac{\partial \psi}{\partial y} \text{ and } v = -\frac{\partial \psi}{\partial x} \tag{12.11}$$

With the above definition of stream function, the continuity equation is identically satisfied

$$\frac{\partial u}{\partial x} + \frac{\partial v}{\partial y} = \frac{\partial}{\partial x}\left(\frac{\partial \psi}{\partial y}\right) + \frac{\partial}{\partial y}\left(-\frac{\partial \psi}{\partial x}\right) = 0 \tag{12.12}$$

Also, we find

$$d\psi = \frac{\partial \psi}{\partial x}dx + \frac{\partial \psi}{\partial y}dy = -vdx + udy \tag{12.13}$$

which suggests

$$d\psi = 0 \Rightarrow -vdx + udy = 0 \Rightarrow \frac{dy}{dx} = \frac{v}{u} \tag{12.14}$$

Since $d\psi = 0 \Rightarrow$ indicates $\psi = $ constant, we find that

$$\psi = \text{constant} \Rightarrow \frac{dy}{dx} = \frac{v}{u} \tag{12.15}$$

The solution of this ODE leads to a function $y = y(x)$ that indicates a line on the 2-D Cartesian plane. Such a line is known as the *streamline*. Equation 12.15 and Figure 12.1 show the tangent on the streamline that indicates the direction of flow.

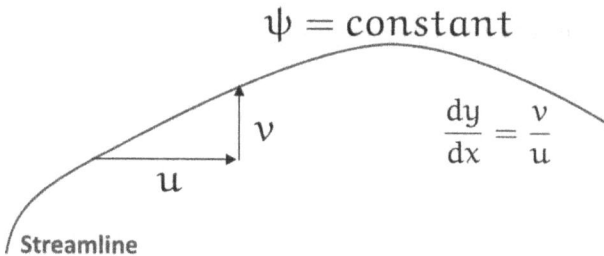

Figure 12.1 Velocity components over a typical streamline.

Furthermore, since streamline is always tangential to the velocity vectors, fluid particles cannot cross a streamline. In Figure 12.2, the volume flow rate dQ crossing the line AB is given by

$$dQ = udy - vdx = \frac{\partial \psi}{\partial y}dy + \frac{\partial \psi}{\partial x}dx = d\psi \tag{12.16}$$

Equation 12.16 is crucial for setting boundary conditions in the stream function, vorticity formulation.

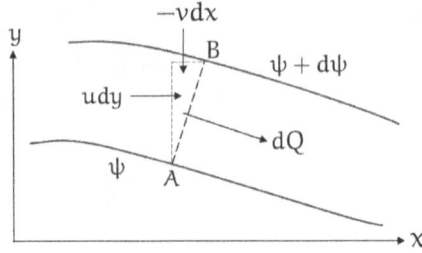

Figure 12.2 Flow between two streamlines.

12.2.2 VORTICITY

Vorticity vector *mathbf ω*, defined as the curl of the velocity vector, quantifies the rotation of a fluid particle with respect to its own axes.

$$\omega = \nabla \times \mathbf{u} \tag{12.17}$$

In a 3-D Cartesian system, the velocity vector has three components $\mathbf{u} = (u, v, w)$. The vorticity vector *mathbf ω* in such cases is given by

$$\omega = \nabla \times \mathbf{u} = \begin{vmatrix} \hat{\mathbf{i}} & \hat{\mathbf{j}} & \hat{\mathbf{k}} \\ \frac{\partial}{\partial x} & \frac{\partial}{\partial y} & \frac{\partial}{\partial z} \\ u & v & w \end{vmatrix}$$

$$= \left(\frac{\partial w}{\partial y} - \frac{\partial v}{\partial z}\right)\hat{\mathbf{i}} - \left(\frac{\partial w}{\partial x} - \frac{\partial u}{\partial z}\right)\hat{\mathbf{j}} + \left(\frac{\partial v}{\partial x} - \frac{\partial u}{\partial y}\right)\hat{\mathbf{k}} \tag{12.18}$$

where, $\hat{\mathbf{i}}, \hat{\mathbf{j}}, \hat{\mathbf{k}}$ are unit vectors in x, y and z directions, respectively. Further, in a 2-D velocity field, we know

$$\mathbf{u} = (u, v, 0) ; u = u(x, y) ; v = v(x, y) \tag{12.19}$$

Thus, in a 2-D velocity field,

$$\frac{\partial w}{\partial y} - \frac{\partial v}{\partial z} = 0; \frac{\partial w}{\partial x} - \frac{\partial u}{\partial z} = 0 \Rightarrow \omega = \omega\hat{\mathbf{k}} \text{ where, } \omega = \frac{\partial v}{\partial x} - \frac{\partial u}{\partial y} \tag{12.20}$$

Here, ω is a scalar, which is the z-component of the vorticity vector. Clearly, in a 2-D flow field, the z-component of the vorticity vector is the only nonzero component. Thus, we may say that in a 2-D flow field, vorticity behaves like a scalar.

Now, combine equations 12.11 and 12.20, we have

$$\omega = \frac{\partial v}{\partial x} - \frac{\partial u}{\partial y} = \frac{\partial}{\partial x}\left(-\frac{\partial \psi}{\partial x}\right) - \frac{\partial}{\partial y}\left(\frac{\partial \psi}{\partial y}\right)$$

$$= -\frac{\partial^2 \psi}{\partial x^2} - \frac{\partial^2 \psi}{\partial y^2} \tag{12.21}$$

12.2.3 VORTICITY TRANSPORT EQUATION

Recalling the z-component of the 2-D velocity field using equation 12.20,

$$\omega = \frac{\partial v}{\partial x} - \frac{\partial u}{\partial y} \Rightarrow v\omega = v\frac{\partial v}{\partial x} - v\frac{\partial u}{\partial y}$$

$$\Rightarrow v\frac{\partial u}{\partial y} = v\frac{\partial v}{\partial x} - v\omega = \frac{1}{2}\frac{\partial \left(v^2\right)}{\partial x} - v\omega \tag{12.22}$$

Thus,

$$u\frac{\partial u}{\partial x} + v\frac{\partial u}{\partial y} = \frac{1}{2}\frac{\partial \left(u^2\right)}{\partial x} + \frac{1}{2}\frac{\partial \left(v^2\right)}{\partial x} - v\omega$$

$$= \frac{\partial q}{\partial x} - v\omega \text{ where, } q = \frac{1}{2}\left(u^2 + v^2\right) \tag{12.23}$$

Similarly,

$$\omega = \frac{\partial v}{\partial x} - \frac{\partial u}{\partial y} \Rightarrow u\omega = u\frac{\partial v}{\partial x} - u\frac{\partial u}{\partial y}$$

$$\Rightarrow u\frac{\partial v}{\partial x} = u\frac{\partial u}{\partial y} + u\omega = \frac{1}{2}\frac{\partial \left(u^2\right)}{\partial y} + u\omega \tag{12.24}$$

and

$$u\frac{\partial v}{\partial x} + v\frac{\partial v}{\partial y} = \frac{1}{2}\frac{\partial \left(u^2\right)}{\partial y} + u\omega + v\frac{\partial v}{\partial y} = \frac{\partial q}{\partial y} + u\omega \tag{12.25}$$

Using equations 12.23 and 12.25, we can now rewrite the x-momentum equation 12.9 as

$$\frac{\partial u}{\partial t} + \frac{\partial s}{\partial x} - v\omega = \frac{1}{Re}\left(\frac{\partial^2 u}{\partial x^2} + \frac{\partial^2 u}{\partial y^2}\right) \text{ where, } s = p + q \tag{12.26}$$

Differentiating equation 12.26 with respect to y, we have

$$\frac{\partial}{\partial t}\left(\frac{\partial u}{\partial y}\right) + \frac{\partial^2 s}{\partial x \partial y} - \frac{\partial (v\omega)}{\partial y} = \frac{1}{Re}\left(\frac{\partial^2}{\partial x^2} + \frac{\partial^2}{\partial y^2}\right)\left(\frac{\partial u}{\partial y}\right) \tag{12.27}$$

Similarly, using equations 12.23 and 12.25, we can now rewrite the y-momentum equation 12.10 as

$$\frac{\partial v}{\partial t} + \frac{\partial s}{\partial y} + u\omega = \frac{1}{Re}\left(\frac{\partial^2 v}{\partial x^2} + \frac{\partial^2 v}{\partial y^2}\right) \tag{12.28}$$

Differentiating equation 12.28 with respect to x, we have

$$\frac{\partial}{\partial t}\left(\frac{\partial v}{\partial x}\right) + \frac{\partial^2 s}{\partial x \partial y} + \frac{\partial (u\omega)}{\partial x} = \frac{1}{Re}\left(\frac{\partial^2}{\partial x^2} + \frac{\partial^2}{\partial y^2}\right)\left(\frac{\partial v}{\partial x}\right) \tag{12.29}$$

Now, subtracting equation 12.27 from equation 12.29 and using the expression of vorticity from equation 12.20, we have the following vorticity transport equation

$$\frac{\partial \omega}{\partial t} + \frac{\partial (u\omega)}{\partial x} + \frac{\partial (v\omega)}{\partial y} = \frac{1}{Re}\left(\frac{\partial^2 \omega}{\partial x^2} + \frac{\partial^2 \omega}{\partial y^2}\right) \qquad (12.30)$$

We can now solve equations 12.21 and 12.30 and use equation 12.11 to find the unknowns u and v. Once we know the u and v, pressure p may be easily computed using equations 12.9 and 12.10.

12.2.4 SOLUTION STRATEGY

To solve equations 12.21 and 12.30, we must discretize the equations. We usually use central difference scheme to discretize equation 12.21

$$\frac{\psi_{i-1,j}^{n+1} - 2\psi_{i,j}^{n+1} + \psi_{i+1,j}^{n+1}}{(\Delta x)^2} + \frac{\psi_{i,j-1}^{n+1} - 2\psi_{i,j}^{n+1} + \psi_{i,j+1}^{n+1}}{(\Delta y)^2} = -\omega_{i,j}^{n+1} \qquad (12.31)$$

Here, we have used uniform grids in x and y directions. While Δx is the grid size in x direction, Δy is the grid size in y direction. Further, if we assume both i and j to start from zero, then any quantity $\phi_{i,j} = \phi(x = i\Delta x, y = j\Delta y)$.

Similarly, using central differences for both the convection and diffusion terms and forward difference for the temporal derivative, we have the following implicit discretization for equation 12.30:

$$\frac{\omega_{i,j}^{n+1} - \omega_{i,j}^{n}}{\Delta t} + \frac{(u\omega)_{i+1,j}^{n+1} - (u\omega)_{i-1,j}^{n+1}}{2\Delta x} + \frac{(v\omega)_{i,j+1}^{n+1} - (u\omega)_{i,j-1}^{n+1}}{2\Delta y}$$
$$= \frac{\omega_{i-1,j}^{n+1} - 2\omega_{i,j}^{n+1} + \omega_{i+1,j}^{n+1}}{(Re\Delta x)^2} + \frac{\omega_{i,j-1}^{n+1} - 2\omega_{i,j}^{n+1} + \omega_{i,j+1}^{n+1}}{(Re\Delta y)^2} \qquad (12.32)$$

Here, Δt is the discretized time step. Equation 12.30 may be discretized as explicit or Crank–Nicolson as well. For explicit discretization, the convection and diffusion terms are discretized at the n-th time level. Equation 12.21, however, is always discretized at $(n+1)$-th time level.

The convective terms in equation 12.32 requires the values of u and v that are obtained from the following discretized equations:

$$u_{i,j}^{n+1} = \frac{\psi_{i,j+1}^{n+1} - \psi_{i,j-1}^{n+1}}{2\Delta y} \qquad (12.33)$$

$$v_{i,j}^{n+1} = \frac{\psi_{i-1,j}^{n+1} - \psi_{i+1,j}^{n+1}}{2\Delta x} \qquad (12.34)$$

Equations 12.31–12.34, with suitable boundary conditions, lead to a system of simultaneous linear equations that may be solved using an iterative technique.

Example: The lid-driven cavity problem, shown in Figure 12.3 is a classical problem in fluid mechanics. Here, fluid within a cavity flows within the cavity due to the motion of one boundary of the cavity. Here, all quantities are non-dimensionalized and $Re = 40$.

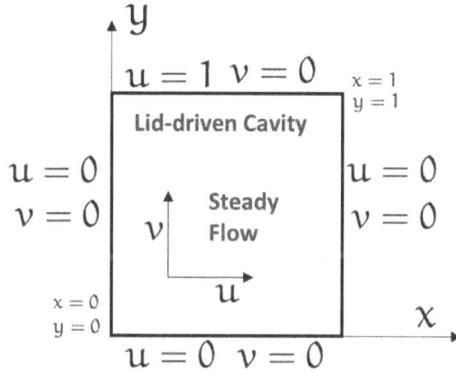

Figure 12.3 Lid-driven cavity.

SOLUTION

Since we are dealing with steady flow here, the governing equations are

$$\frac{\partial^2 \psi}{\partial x^2} + \frac{\partial^2 \psi}{\partial y^2} = -\omega \quad u = \frac{\partial \psi}{\partial y} \quad v = -\frac{\partial \psi}{\partial x} \tag{12.35}$$

$$\frac{\partial (u\omega)}{\partial x} + \frac{\partial (v\omega)}{\partial y} = \frac{1}{Re}\left(\frac{\partial^2 \omega}{\partial x^2} + \frac{\partial^2 \omega}{\partial y^2}\right) \tag{12.36}$$

Since this is a steady-state problem, the discretized equations now assume the following form:

$$\frac{\psi_{i-1,j} - 2\psi_{i,j} + \psi_{i+1,j}}{(\Delta x)^2} + \frac{\psi_{i,j-1} - 2\psi_{i,j} + \psi_{i,j+1}}{(\Delta y)^2} = -\omega_{i,j} \tag{12.37}$$

$$\frac{(u\omega)_{i+1,j} - (u\omega)_{i-1,j}}{2\Delta x} + \frac{(v\omega)_{i,j+1} - (u\omega)_{i,j-1}}{2\Delta y}$$

$$= \frac{\omega_{i-1,j} - 2\omega_{i,j} + \omega_{i+1,j}}{(Re\Delta x)^2} + \frac{\omega_{i,j-1} - 2\omega_{i,j} + \omega_{i,j+1}}{(Re\Delta y)^2} \tag{12.38}$$

$$u_{i,j} = \frac{\psi_{i,j+1} - \psi_{i,j-1}}{2\Delta y} \quad v_{i,j} = \frac{\psi_{i-1,j} - \psi_{i+1,j}}{2\Delta x} \tag{12.39}$$

Since no fluid crosses the cavity boundary, the boundary of the cavity may be considered as a streamline. Thus, the stream function boundary condition is set as constant. The value of the constant can be set arbitrarily; here we set the same as zero.

$$\psi_{boundary} = constant = 0, \text{ set arbitrarily} \qquad (12.40)$$

Similarly, for velocities, we use the no-slip, impermeable boundary conditions

$$x = 0: \quad u = v = 0 \qquad (12.41)$$
$$x = 1: \quad u = v = 0 \qquad (12.42)$$
$$y = 0: \quad u = v = 0 \qquad (12.43)$$
$$y = 1: \quad u = 1, v = 0 \qquad (12.44)$$

For the vorticity boundary condition, we use equation 12.37 along with these no-slip conditions. Let us take an example for $x = 0$ boundary. Since $\psi = $ constant at all boundaries, at $x = 0$ boundary, $\partial^2 \psi / \partial y^2 = 0$. Thus, using equation 12.37,

$$x = 0: \quad \omega_{i,j} = -\frac{\psi_{i-1,j} - 2\psi_{i,j} + \psi_{i+1,j}}{(\Delta x)^2} \qquad (12.45)$$

$$\text{again } v = 0: \quad \Rightarrow \psi_{i+1,j} = \psi_{i-1,j}$$

$$\text{thus at } x = 0: \quad \omega_{i,j} = \frac{2\psi_{i,j} - 2\psi_{i+1,j}}{(\Delta x)^2} \qquad (12.46)$$

Similarly,

$$x = 1: \quad \omega_{i,j} = \frac{2\psi_{i,j} - 2\psi_{i-1,j}}{(\Delta x)^2} \qquad (12.47)$$

$$y = 0: \quad \omega_{i,j} = \frac{2\psi_{i,j} - 2\psi_{i,j+1}}{(\Delta y)^2} \qquad (12.48)$$

$$y = 1: \quad \omega_{i,j} = \frac{2\psi_{i,j} - 2\psi_{i,j-1}}{(\Delta y)^2} - \frac{2u_{i,j}}{\Delta y} \qquad (12.49)$$

This discretization of the governing equations and the boundary conditions develop a system of linear equations that may be solved using the Gauss–Seidel method. A PYTHON program to solve the linear system is given in the course website. The steam function contour for $Re = 100$ is shown in Figure 12.4.

12.3   PRIMITIVE VARIABLE FORMULATION

While the $\psi - \omega$ approach, described earlier, is quite successful for the simulation of several 2-D flow fields, the technique is difficult to replicate in complex geometries as well as in 3-D cases. Thus, for 3-D cases, we directly solve equations 12.1 and 12.2. Such techniques are known as the primitive variable approaches. Needless to

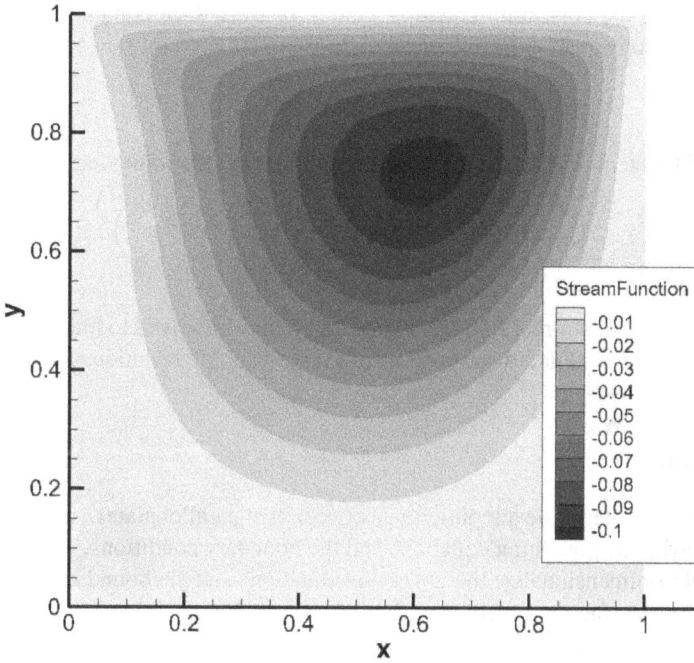

Figure 12.4 Steady stream function contour for the lid-driven cavity at $Re = 100$.

mention that, unlike the $\psi - \omega$ approach that has limited applicability, the primitive variable approach is the general method of solving the equations of fluid mechanics. Consider the following 2-D dimensionless governing equations that we intend to solve using the primitive variable approach:

$$\frac{\partial u}{\partial t} + u\frac{\partial u}{\partial x} + v\frac{\partial u}{\partial y} = -\frac{\partial p}{\partial x} + \frac{1}{Re}\left(\frac{\partial^2 u}{\partial x^2} + \frac{\partial^2 u}{\partial y^2}\right) \tag{12.50}$$

$$\frac{\partial v}{\partial t} + u\frac{\partial v}{\partial x} + v\frac{\partial v}{\partial y} = -\frac{\partial p}{\partial y} + \frac{1}{Re}\left(\frac{\partial^2 v}{\partial x^2} + \frac{\partial^2 v}{\partial y^2}\right) \tag{12.51}$$

$$\frac{\partial u}{\partial x} + \frac{\partial v}{\partial y} = 0 \tag{12.52}$$

The basic idea behind the primitive variable approach is to convert the continuity equation 12.52 to the governing equation of pressure. Applying a divergence operator to both sides of the coordinate-free N-S equation, we have

$$\nabla \cdot \left[\frac{\partial \mathbf{u}}{\partial t} + \nabla \cdot (\mathbf{uu})\right] = \nabla \cdot \left(-\nabla p + \frac{1}{Re}\nabla^2 \mathbf{u}\right) \tag{12.53}$$

Now, using the continuity equation $\nabla \cdot \mathbf{u} = 0$, we reduce equation 12.53 to the pressure Poisson equation as follows:

$$\nabla^2 p = -\nabla \cdot \left[\nabla \cdot (\mathbf{uu}) \right] \tag{12.54}$$

In a 2-D system, the pressure Poisson equation takes the following form:

$$\frac{\partial^2 p}{\partial x^2} + \frac{\partial^2 p}{\partial y^2} = -\frac{\partial^2 \left(u^2 \right)}{\partial x^2} - 2\frac{\partial^2 (uv)}{\partial x \partial y} - \frac{\partial^2 \left(v^2 \right)}{\partial y^2} \tag{12.55}$$

In principle, equations 12.50, 12.51 and 12.55 may be solved to find the unknown $\mathbf{u}$ and p fields. A detailed discussion of primitive-variable techniques is beyond the scope of this book.

PROBLEMS

1. Consider the steady developing flow between two parallel plates.
    a. Write the governing equations and the boundary conditions.
    b. Non-dimensionalize the governing equations and the boundary conditions.
    c. Develop the $\psi - \omega$ equations and the boundary conditions.
    d. Write a PYTHON program to solve the governing equations.

2. For flows where $Re \ll 1$, viscous forces dominate over inertia forces. Such flows are known as creeping flows. Consider creeping flows in a lid-driven cavity.
    a. Write the governing equations and the boundary conditions.
    b. Non-dimensionalize the governing equations and the boundary conditions.
    c. Develop the $\psi - \omega$ equations and the boundary conditions.
    d. Write a PYTHON program to solve the governing equations.

3. For the lid-driven cavity problem, shown in the example in Section 12.1, conduct parametric studies for $Re = 1$, $Re = 100$, $Re = 200$. Discuss, qualitatively, how the flow field changes with the Re.

4. For laminar flow over an infinite parallel plate, we often assume boundary layer (BL) approximation. Under BL approximation, we often solve the Blasius equation to find the laminar boundary flow over a flat surface. In the ODE chapter (Chapter 9), we have solved the Blasius equation. Now solve the complete mass and momentum equations for flow over the flat plate and compare your result with the Blasius solution.

13 Electrochemical Systems

13.1 INTRODUCTION

Generally, a chemical reaction consumes or releases energy in the form of heat when reactants are consumed to generate products. In contrast, an electrochemical reaction releases or consumes at least some amount of energy as electricity. Electrochemical systems utilize electrochemical reactions either to manufacture new products or to generate electricity. A popular application of electrochemical technique is electrolysis of a certain material. Electrolysis consumes electricity to synthesize a product. For instance, high-purity hydrogen and oxygen are often generated through electrolysis of water. Conversely, we may use a fuel cell to combine hydrogen and oxygen to generate electricity, where water is generated as the product.

13.1.1 FUEL CELLS

Figure 13.1 shows the working principle of a typical fuel cell using hydrogen and oxygen. Hydrogen enters the anode where hydrogen is converted to proton and electron via anode half-reaction. While the electron flows through the external circuit toward the cathode, the proton crosses the electrolyte and reaches the cathode. At the cathode, proton and electron combine with the oxygen through the cathode half-reaction. The net effect of the two half-reactions is the flow of electricity through the outer circuit.

Summarizing

$$\text{Anode: } H_2 \rightleftharpoons 2H^+ + 2e \tag{13.1}$$

$$\text{Cathode: } \frac{1}{2}O_2 + 2H^+ + 2e \rightleftharpoons H_2O \tag{13.2}$$

$$\text{Overall: } H_2 + \frac{1}{2}O_2 \rightleftharpoons H_2O + \text{electricity} \tag{13.3}$$

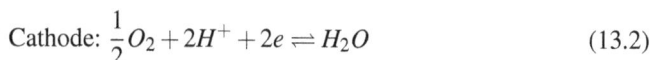

As suggested by Figure 13.1, fuel cell is a steady flow machine. As long as we supply the fuel and the oxidizer, the fuel cell will run and generate current. Under such steady-state condition, the performance of a fuel cell may degrade slowly only due to material degradation.

13.1.2 BATTERIES AND FUEL CELLS

While fuel cell is a futuristic device, the most common electrochemical device that we use everyday is the battery. The most popular industrial battery, especially in the automobile sector, is the lead-acid battery. In a fully charged lead-acid battery, lead and lead-oxide electrodes stay within a sulfuric acid electrolyte. The reactions in a

DOI: 10.1201/9781003049272-13

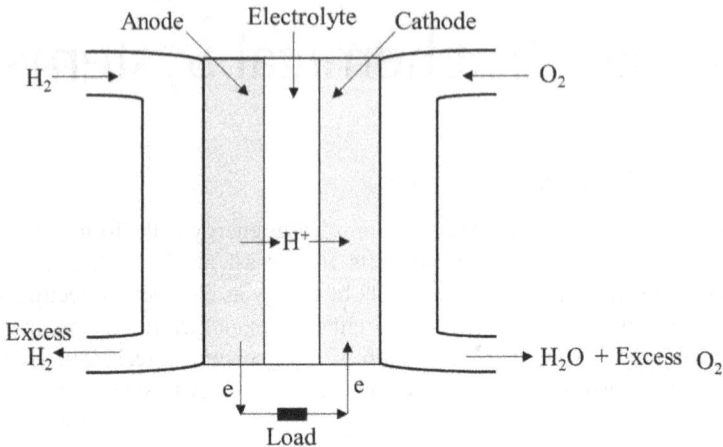

Figure 13.1 Working principle of a fuel cell.

lead-acid battery are

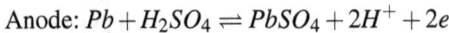

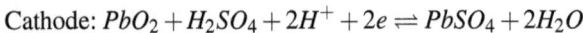

$$\text{Anode: } Pb + H_2SO_4 \rightleftharpoons PbSO_4 + 2H^+ + 2e$$
$$\text{Cathode: } PbO_2 + H_2SO_4 + 2H^+ + 2e \rightleftharpoons PbSO_4 + 2H_2O$$
$$\text{Overall: } Pb + PbO_2 + 2H_2SO_4 \rightleftharpoons 2PbSO_4 + 2H_2O + \text{electricity} \qquad (13.4)$$

As the electrochemical reactions continue, both the electrodes deplete and are replaced by $PbSO_4$. Eventually, a fully-discharged cell will contain $PbSO_4$ and H_2O. The cell may then be recharged to reverse the reactions such that we can have a fresh lead-acid battery. The battery materials, of course, slowly degrade over the charge-discharge cycles.

The working principle of a fuel cell is same as that of a standard battery. Unlike fuel cells, the batteries are, however, not steady flow machines. Each battery comes with a specified amount of reactant. As the reaction proceeds, the reactants are consumed and product accumulates in the battery, eventually stopping the operation. The battery then, depending on the materials, must be either recharged or replaced.

13.2 FUEL CELL THERMODYNAMICS

To analyze a fuel cell from a purely thermodynamic viewpoint, we idealize the fuel cell system as

1. Steady, isothermal flow
2. Changes in potential and kinetic energies between reactant and product are negligible
3. Stoichiometric fuel-oxidizer mixture

We now apply the first and second law to the fuel cell system as follows:

$$\dot{N}_f h_f + \dot{N}_o h_o + \dot{Q} = \dot{N}_p h_p + \dot{W} \tag{13.5}$$

$$\dot{N}_f s_f + \dot{N}_o s_o + \frac{\dot{Q}}{T} + \dot{S}_g = \dot{N}_p s_p \tag{13.6}$$

Here, $\dot{N}_f$ is molar flow rate of fuel, $\dot{N}_o$ is molar flow rate of oxidizer, $\dot{N}_p$ is molar flow rate of product, h_f is enthalpy of fuel per unit mole, h_o is enthalpy of oxidizer per unit mole, h_p is enthalpy of product per unit mole, $\dot{Q}$ is rate of heat input, $\dot{W}$ is rate of work input, s_f is entropy of fuel per unit mole, s_o is entropy of oxidizer per unit mole, s_p is entropy of product per unit mole, $\dot{S}_g$ is rate of entropy generation and T is fuel cell temperature.

Now, combining equations 13.5, and 13.6, we have

$$\dot{W} = \dot{N}_f h_f + \dot{N}_o h_o - \dot{N}_p h_p$$
$$+ T \left(\dot{N}_p s_p - \dot{N}_f s_f - \dot{N}_o s_o - \dot{S}_g \right) \tag{13.7}$$

Therefore, the rate of work output per unit mole of fuel $\dot{w}$ is given by

$$\dot{w} = \frac{\dot{W}}{\dot{N}_f} = h_f + \frac{\dot{N}_o}{\dot{N}_f} h_o - \frac{\dot{N}_p}{\dot{N}_f} h_p$$
$$+ T \left(\frac{\dot{N}_p}{\dot{N}_f} s_p - s_f - \frac{\dot{N}_o}{\dot{N}_f} s_o - \frac{\dot{S}_g}{\dot{N}_f} \right)$$
$$= h_{in} - h_{out} - T \left(s_{in} - s_{out} \right) - \frac{T \dot{S}_g}{\dot{N}_f} \tag{13.8}$$

Here, h_{in} is inlet enthalpy per unit mole of fuel, h_{out} is outlet enthalpy per unit mole of fuel, s_{in} is inlet entropy per unit mole of fuel and s_{out} is outlet entropy per unit mole of fuel.

Now, from equation 13.8

$$\dot{w} = g_{in} - g_{out} - \frac{T \dot{S}_g}{\dot{N}_f} = -\Delta g - \frac{T \dot{S}_g}{\dot{N}_f} \tag{13.9}$$

Here, g is the Gibbs free energy per unit mole of fuel and Δg is the charge in g between outlet and inlet of the fuel cell. Now applying the second law of thermodynamics,

$$\dot{S}_g \geq 0 \Rightarrow \dot{w} \leq -\Delta g \Rightarrow \dot{w}_{max} = -\Delta g \tag{13.10}$$

Since $\dot{S}_g = 0$ indicates reversible process, and $\dot{w}_{max}$ the rate of work output when processes are reversible.

13.2.1   REVERSIBLE VOLTAGE

The previous thermodynamic analysis does not assume anything about the device through which the work, shown in equation 13.10 is produced. We know that the device generates work through an isothermal, chemical reaction. In fuel cell, we generate electrical work through chemical reaction while keeping the cell in isothermal condition. For moving a charge Q through a potential E, we need electrical work W_{elec} as

$$W_{elec} = EQ = NEF \tag{13.11}$$

where,

$\quad\quad$ N is number of moles of electron and F is Faraday's constant;

$$F = 96485\ C/mol \tag{13.12}$$

Now, combining equations 13.9, and 13.12, we have

$$nE^0 F = -\Delta g \Rightarrow E^0 = -\frac{\Delta g}{nF} \tag{13.13}$$

where,

n is number of moles of electron per mole of fuel and E^0 is reversible voltage.

$$\tag{13.14}$$

Thus, the reversible voltage E^0 is the maximum possible voltage that may appear in a fuel cell. Reversible voltage may be realized only at an open-circuit condition.

13.2.2   REVERSIBLE EFFICIENCY

From equation 13.5,

$$\dot{q} = -\frac{\dot{Q}}{\dot{N}_f} = h_f + \frac{\dot{N}_o}{\dot{N}_f} h_o - \frac{\dot{N}_p}{\dot{N}_f} h_p - \dot{w} = h_{in} - h_{out} - \dot{w} \tag{13.15}$$

$$\Rightarrow \dot{q} + \dot{w} = h_{in} - h_{out} = -\Delta h \tag{13.16}$$

Here, $\dot{q}$ is the rate of heat output for each mole of fuel and Δh is the change in enthalpy between outlet and inlet. Thus, from equation 13.16, we find that the heat and work output together from the previous reaction may be $-\Delta h$. We already know that the maxmium electrical work output could be $-\Delta g$. The thermodynamic efficiency or the reversible efficiency ε_{rev} of the fuel cell is

$$\varepsilon_{rev} = \frac{\Delta g}{\Delta h} \tag{13.17}$$

At isothermal condition, we know

$$\Delta g = \Delta h - T\Delta s \tag{13.18}$$

Thus, equation 13.17 may be rewritten as

$$\varepsilon_{rev} = \frac{\Delta g}{\Delta h} = 1 - \frac{T \Delta s}{\Delta h} \tag{13.19}$$

Equation 13.17, or equation 13.19 shows the maximum possible efficiency that a fuel cell may attain.

Example: For an $H_2 - O_2$ fuel cell, find the reversible voltage and efficiency.

SOLUTION

Assuming the product to be in liquid phase

$$H_2 + \frac{1}{2} O_2 \rightleftharpoons H_2 O_{liq} \tag{13.20}$$

Using thermodynamic property data,

$$\Delta g = -237.3 \frac{kJ}{kmol}; \quad \Delta h = -286 \frac{kJ}{kmol}; \quad n = 2 \tag{13.21}$$

Here, $n = 2$ indicates that each H_2 molecule will generate two electrons, as shown in equation 13.1.

Now, using equations 13.13, and 13.17

$$E^0 = -\frac{\Delta g}{nF} = 1.229V \tag{13.22}$$

$$\varepsilon_{rev} = \frac{\Delta g}{\Delta h} = 0.8297 \tag{13.23}$$

It is interesting to note that an $H_2 - O_2$ fuel cell may attain a maximum potential of 1.229 V, irrespective of the shape or size of the cell. Further discussion of fuel cell thermodynamics may be found elsewhere [Larminie et al., 2003].

13.3   CLASSIFICATIONS

Fuel cells are generally classified according to the operating temperature.

1. Low-Temperature Fuel Cell: Operating temperature range is $25 - 100°C$; example, polymer electrolyte membrane fuel cell (PEMFC), direct methanol fuel cell (DMFC).
2. Medium-Temperature Fuel Cell: Operating temperature range is $100 - 300°C$; example, phosphoric acid fuel cell (PAFC).
3. Solid Oxide Fuel Cell: Operating temperature range is $300 - 1000°C$; example, solid oxide fuel cell (SOFC).

The two most important fuel cells are briefly described here.

13.3.1   PEMFC

PEMFC usually operates at ambient temperature typically with H_2 fuel. The electrochemical reactions of PEMFC is shown in equations 13.1–13.3. Since it operates in ambient temperature, the startup time of PEMFC is very low. While PEMFC is particularly suitable for automobile applications, at low temperature, the liquid water generated in PEMFC forms ice and thus stops the operation.

Due to low operating temperature, the electrochemical reactions require high-quality catalyst. Usually in PEMFC, platinum catalyst is used, which increases the cost of PEMFC. PEMFC uses carbon-based electrodes and solid polymer electrolyte.

13.3.2   SOFC

In contrast to PEMFC, SOFC operates at $600-1000°C$. Due to high operating temperature, SOFC requires high startup time, and thus, is not quite suitable for automobile applications. On the other hand, high operating temperature improves the reaction rates in SOFCs. SOFCs are particularly convenient for stationary power plants.

SOFCs use a yttria-stabilized-zirconia electrolyte that conducts O^{-2} ions. SOFCs are fuel-flexible, which means SOFC may use a variety of gaseous and even liquid fuels. For H_2-O_2 combination, the reactions in SOFC are

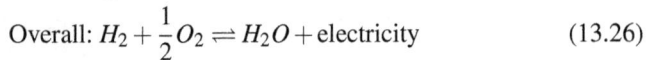

$$\text{Anode: } H_2 + O^{-2} \rightleftharpoons H_2O + 2e \tag{13.24}$$

$$\text{Cathode: } \frac{1}{2}O_2 + 2e \rightleftharpoons O^{-2} \tag{13.25}$$

$$\text{Overall: } H_2 + \frac{1}{2}O_2 \rightleftharpoons H_2O + \text{electricity} \tag{13.26}$$

13.4   LOSSES IN FUEL CELLS

Real fuel cell performance is often described through a curve, also known as the polarization curve. Voltage V in an actual fuel cell is given by

$$V = E^0 - \eta_{activation} - \eta_{Ohmic} - \eta_{concentraton} \tag{13.27}$$

For current $I = 0$, both the forward and reverse reactions proceed at the same rate. To provide a boost to the forward reaction, we use some energy, which is indicated by the activation loss. As the current increases, the electrical resistance in the cell incurs further losses, known as Ohmic losses. Ohmic losses have two parts: resistance to electron flow, and resistance to ion flow. Usually in fuel cell, the resistance to ion flow dominates over the resistance to the electron flow. Finally as the current increases, the reaction sites are either starved of reactants or are flooded by the product water. Thus, at high current condition, the flow of reactant and oxidant to the reaction sites ultimately dictates the fuel cell performance. At this stage, the cell experiences a sudden drop in voltage, as shown in Figure 13.2. A detailed discussion of fuel cell losses may be found elsewhere [O'Hayre et al., 2016].

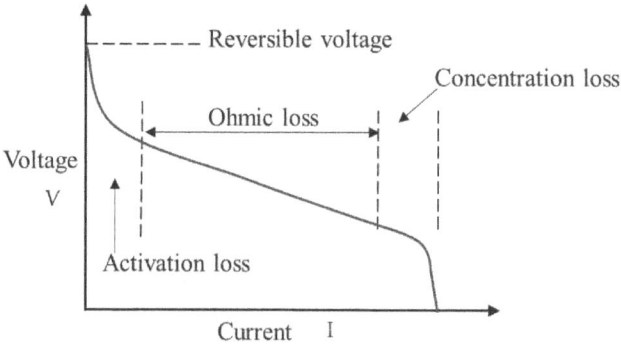

Figure 13.2   Polarization curve of a real fuel cell.

PROBLEMS

1. Consider a fuel cell working with carbon and oxygen. Find the reversible voltage and reversible efficiency of the cell
2. Can the reversible efficiency be more than 1, justify your answer.

REFERENCES

J. Larminie, A. Dicks, and M. S. McDonald. *Fuel Cell Systems Explained*, Volume 2. John Wiley & Sons, UK, 2003.

R. O'Hayre, S. -W. Cha, W. Colella, and F. B. Prinz. *Fuel Cell Fundamentals*. John Wiley & Sons, 2016.

14 Inverse Problems

14.1 INTRODUCTION

A mathematical model of deterministic physical systems usually accepts the *causes* as the input and calculates *effects* as output. Such problems are usually called *direct* or *forward problems*. The *inverse problems*, on the other hand, find causes from the effects. Inverse problems appear in a wide range of engineering applications. Examples include estimation of material properties from experimental data, tomographic image reconstruction as well as recovery of the unknown boundary condition of a mathematical model. Consider the following steady, 1-D heat conduction problem:

$$0 < x < L: \quad \frac{d}{dx}\left(K\frac{dT}{dx}\right) = 0 \tag{14.1}$$

Let us assume the following boundary conditions:

$$T(x=0) = T_1; T(x=L) = T_2 \tag{14.2}$$

We may non-dimensionalize this problem using dimensionless variables θ, z and k, where

$$\theta = \frac{T - T_1}{T_2 - T_1}; z = \frac{x}{L}; k = \frac{K}{K_0} \tag{14.3}$$

Here, K_0 is the value of thermal conductivity K at $T = T_1$. Clearly for a constant conductivity material, $k = 1$. In general, $K = K(T)$ and thus $k = k(\theta)$.

The governing equation 14.1 and the boundary conditions 14.2 thus may be written in dimensionless form as follows:

$$0 < z < 1: \quad \frac{d}{dz}\left(k\frac{d\theta}{dz}\right) = 0 \tag{14.4}$$

Let us assume the following boundary conditions:

$$\theta(z=0) = 0; \theta(z=1) = 1 \tag{14.5}$$

If we now solve this problem, we will have the solution $\theta(z)$. Such a problem is known as a *direct* or forward problem. The *inverse* problems appear if the thermal conductivity k is unknown and we wish to *estimate* k using experimentally obtained values of temperatures at certain z locations. The keyword here is *estimate*, as we will see shortly that exact evaluation of k, in general, may not be possible.

DOI: 10.1201/9781003049272-14

The estimation of k is, however, possible only when the temperature field $\theta(z)$ depends on k. For instance, for a constant property material, the gradient distribution $\theta(z)$ is independent of the thermal conductivity k. In inverse problem jargon, we note that the temperature field is not *sensitive* enough to the unknown k. Clearly, for constant property case, the conductivity k cannot be estimated from the measured temperatures.

For the variable conductivity material, we have $k = k(\theta)$, and thus, equation 14.4 suggests that $\theta(z)$ must be sensitive to $k = k(\theta)$.

The inverse problem of finding $k = k(\theta)$ may now be posed in the following two ways:

1. Parameter estimation problem
2. Function estimation problem

When we do not know the boundary heat flux $q(t, x = L)$ at $x = L$ and wish to *estimate* the same utilizing some experimental data $u_{ij} = ut = t_i, x = x_j, 0 < x_j < L$ available with the domain. The second problem appears when we do now know the property α, and wish to estimate the same from the experimental data, described earlier. The first problem is known as a function estimation problem, while the second one is known as a parameter estimation problem. Each of these inverse problems suffers from several complexities occurred due to the *ill-posed* nature of inverse problems. While a well-posed problem shows existence, uniqueness and stability of solutions, ill-posed problems lack at least one of these three properties [Hadamard, 1923].

For instance, let us say we are about to estimate the unknown heat flux $q(t, x = L)$ at $x = L$ using the experimental data $u_{ij} = ut = t_i, x = x_j, 0 < x_j < L$. Now, if $t_i \approx 0$ and $x_j \approx 0$, the experimental data contain very little information about the unknown function $q(t, x = L)$. Furthermore, all experimental data contain noise and thus we end up using noisy data that are not heavily dependent on the unknown. Needless to say, we may end up with non-unique, unstable estimation. If we measure data u_{ij} close to $x = L$ and for time sufficiently away from the initial time, we should be able to impose uniqueness. Stability may, however, still be an issue. The stability, in an inverse problem, is imposed via appropriate *regularization* technique [Alifanov, 2012]. While there exists a variety of approaches to solve inverse problems, here we will discuss only the conjugate gradient technique. A detailed discussion of various methodologies for inverse thermal problems may be found elsewhere.

14.2 INVERSE HEAT CONDUCTION: CONJUGATE-GRADIENT APPROACH

Consider the following simple 1-D steady-state heat conduction problem:

$$0 < z < 1: \quad \frac{d}{dz}\left[k(\theta)\frac{d\theta}{dz}\right] = 0 \tag{14.6}$$

with boundary conditions

$$\theta\,(z=0)=0 \quad \theta\,(x=L)=1 \tag{14.7}$$

While the direct problem involves the solution of temperature field $\theta\,(z)$, we will discuss the inverse problem of estimating the unknown thermal conductivity $k\,(\theta)$ when we know a few values of measured temperature $Y_i\,(z=z_i)\,,i=1,2,\cdots,m$. The inverse problem, in such a case, leads to the minimization of the following functional $J\,(x)$:

$$J\,(z)=\sum_{i=1}^{m}(\theta_i-Y_i)^2=\int_{z=0}^{1}(\theta-Y)^2\,\delta\,(z-z_i)\,dz \tag{14.8}$$

Here, θ_i is the temperature at $z=z_i$ calculated from the direct problem, Y_i is the measured temperature at $z=z_i$ and δ is the Dirac delta function.

To obtain the value of θ_i, the direct problem is first solved using a reasonable initial guess of thermal conductivity $k\,(\theta)$. In the conjugate gradient method, we iteratively update the initial guess of $k\,(\theta)$, such that

$$k^{(n+1)}=k^{(n)}-\beta^{(n)}P^{(n)} \tag{14.9}$$

Here, $P^{(n)}$ is the descent direction while $\beta^{(n)}$ is the step size for advancing to $(n+1)$-th iteration from n-th iteration.

14.2.1 SENSITIVITY PROBLEM

The sensitivity problem determines the sensitivity of the computed temperature θ to the unknown conductivity k. Let us assume that the thermal conductivity changes from k to $k+\Delta k$ and the temperature changes from θ to $\theta+\Delta\theta$. Thus, using equation 14.3, we have

$$\frac{d}{dz}\left[(k+\Delta k)\frac{d\,(\theta+\Delta\theta)}{dz}\right]=0 \tag{14.10}$$

Assuming both Δk and $\Delta\theta$ are small enough, we set

$$\frac{d}{dz}\left[\Delta k\frac{d\,(\Delta\theta)}{dz}\right]=0 \tag{14.11}$$

Now, combining equations 14.6, 14.10 and 14.11, we have

$$\frac{d}{dz}\left[k\frac{d\,(\Delta\theta)}{dz}\right]+\frac{d}{dz}\left(\Delta k\frac{d\theta}{dz}\right)=0 \tag{14.12}$$

The boundary conditions shown in equation 14.7, however, remain unchanged, and thus, the boundary conditions for this sensitivity problem are given by

$$\Delta\theta\,(z=0)=\Delta\theta\,(z=1)=0 \tag{14.13}$$

14.2.2 ADJOINT PROBLEM

Using the adjoint function $\lambda\,(z)$, we write the functional shown in equation 14.8 as

$$J = \int_{z=0}^{1} \lambda\,(z)\,\frac{d}{dz}\left[k\,(\theta)\,\frac{d\theta}{dz}\right]dz + \int_{z=0}^{1}(\theta - Y)^2\,\delta\,(z - z_i)\,dz \qquad (14.14)$$

As the thermal conductivity changes from k to $k + \Delta k$, the adjoint function changes from $J + \Delta J$, such that

$$J + \Delta J = \int_{z=0}^{1} \lambda\,(z)\,\frac{d}{dz}\left[(k + \Delta k)\,\frac{d\,(\theta + \Delta\theta)}{dz}\right]dz + \int_{z=0}^{1}(\theta + \Delta\theta - Y)^2\,\delta\,(z - z_i)\,dz$$
$$(14.15)$$

Subtracting equation 14.15 from 14.14, and neglecting the higher-order terms,

$$\Delta J = \int_{z=0}^{1} \lambda\,\frac{d}{dz}\left[k\,\frac{d\,(\Delta\theta)}{dz} + \Delta k\,\frac{d\theta}{dz}\right]dz + 2\int_{z=0}^{1}\Delta\theta\,(\theta - Y)\,\delta\,(z - z_i)\,dz$$

$$= \left[\lambda k\,\frac{d\,(\Delta\theta)}{dz}\right]_{z=0}^{1} - \int_{z=0}^{1}k\,\frac{d\lambda}{dz}\,\frac{d\,(\Delta\theta)}{dz}\,dz + \left[\lambda\Delta k\,\frac{d\theta}{dz}\right]_{z=0}^{1} - \int_{z=0}^{1}\Delta k\,\frac{d\lambda}{dz}\,\frac{d\theta}{dz}\,dz$$

$$+ 2\int_{z=0}^{1}\Delta\theta\,(\theta - Y)\,\delta\,(z - z_i)\,dz$$

$$= \left[\lambda k\,\frac{d\,(\Delta\theta)}{dz}\right]_{z=0}^{1} - \left[\Delta\theta k\,\frac{d\lambda}{dz}\right]_{z=0}^{1} + \int_{z=0}^{1}\Delta\theta\,\frac{d}{dz}\left(k\,\frac{d\lambda}{dz}\right)dz + \left[\lambda\Delta k\,\frac{d\theta}{dz}\right]_{z=0}^{1}$$

$$- \int_{z=0}^{1}\Delta k\,\frac{d\lambda}{dz}\,\frac{d\theta}{dz}\,dz + 2\int_{z=0}^{1}\Delta\theta\,(\theta - Y)\,\delta\,(z - z_i)\,dz \qquad (14.16)$$

Using equation 14.13, we simplify equation 14.16 as

$$\Delta J = \int_{z=0}^{1}\left[\frac{d}{dz}\left(k\,\frac{d\lambda}{dz}\right) + 2\,(\theta - Y)\,\delta\,(z - z_i)\right]\Delta\theta\,dz$$

$$+ \left[\lambda k\,\frac{d\,(\Delta\theta)}{dz}\right]_{z=0}^{1} + \left[\lambda\Delta k\,\frac{d\theta}{dz}\right]_{z=0}^{1} - \int_{z=0}^{1}\Delta k\,\frac{d\lambda}{dz}\,\frac{d\theta}{dz}\,dz \qquad (14.17)$$

We now set the governing equation and the boundary conditions for $\lambda\,(z)$ as

$$\frac{d}{dz}\left(k\,\frac{d\lambda}{dz}\right) + 2\,(\theta - Y)\,\delta\,(z - z_i) = 0 \qquad (14.18)$$

$$\lambda\,(z = 0) = \lambda\,(z = 1) = 0 \qquad (14.19)$$

Now, using equations 14.17–14.19,

$$\Delta J = -\int_{z=0}^{1}\Delta k\,\frac{d\lambda}{dz}\,\frac{d\theta}{dz}\,dz \qquad (14.20)$$

For the gradient of the functional being J', we know

$$\Delta J = -\int_{z=0}^{1} J' \Delta k \, dz \tag{14.21}$$

Finally, using equations 14.20 and 14.21, the gradient of the functional is set as

$$J' = -\frac{d\lambda}{dz} \frac{d\theta}{dz} \tag{14.22}$$

Here, the temperature θ is calculated from the direct problem, shown in equations 14.6 and 14.7, while the adjoint λ is calculated from equations 14.18 and 14.19.

14.2.3   DESCENT DIRECTION AND STEP SIZE

Recall the basic procedure of updating the conductivity k in the conjugate gradient method, shown in equation 14.9. We are now ready to find the descent direction P and the step size β.

In the conjugate gradient method, the descent direction is given by

$$P^{(n)} = J'^{(n)} + \nu^{(n)} P^{(n-1)} \tag{14.23}$$

Here, ν is the conjugate coefficient and $\nu^{(0)} = 0$. For $n > 0$, ν is given by

$$\nu^{(n)} = \frac{\int_{z=0}^{1} \left[J'^{(n)} \right]^2 dz}{\int_{z=0}^{1} \left[J'^{(n-1)} \right]^2 dz} \tag{14.24}$$

To find the optimum value of β, we set

$$\frac{\partial J^{(n+1)}}{\partial \beta^{(n)}} = 0 \text{ for } J^{(n)} = \int_{z=0}^{1} \left(\theta \left[k^{(n)} \right] - Y \right)^2 \delta(z - z_i) \, dz \tag{14.25}$$

Here, $\theta \left[k^{(n)} \right]$ indicates θ evaluated with the n-th iteration value of the conductivity k. Now, extending equation 14.25

$$J^{(n+1)} = \int_{z=0}^{1} \left(\theta \left[k^{(n+1)} \right] - Y \right)^2 \delta(z - z_i) \, dz \tag{14.26}$$

Combining equations 14.9 and 14.26,

$$J^{(n+1)} = \int_{z=0}^{1} \left(\theta \left[k^{(n)} - \beta^{(n)} P^{(n)} \right] - Y \right)^2 \delta(z - z_i) \, dz \tag{14.27}$$

Setting $\Delta k = P^{(n)}$ and linearizing using a Taylor series expansion,

$$J^{(n+1)} = \int_{z=0}^{1} \left(\theta - \beta^{(n)} \Delta \theta - Y \right)^2 \delta(z - z_i) \, dz \tag{14.28}$$

Differentiating both sides,

$$\frac{\partial J^{(n+1)}}{\partial \beta^{(n)}} = 0 = -2 \int_{z=0}^{1} \Delta\theta \left(\theta - \beta^{(n)} \Delta\theta - Y \right) \delta (z - z_i) \, dz \qquad (14.29)$$

Simplifying,

$$\beta^{(n)} = \frac{\int_{z=0}^{1} \Delta\theta \, (\theta - Y) \, \delta (z - z_i) \, dz}{\int_{z=0}^{1} (\Delta\theta)^2 \, \delta (z - z_i) \, dz} \qquad (14.30)$$

14.3 REGULARIZATION AND STOPPING CRITERION

The solution of inverse problems requires minimization of the functional J as follows:

$$J(z) = \int_{z=0}^{1} (\theta - Y)^2 \, \delta (z - z_i) \, dz < \varepsilon \qquad (14.31)$$

where, ε is a small, positive, real number, such that $\varepsilon \approx 0$.

J, as shown in equation 14.31, uses experimental data Y. Accuracy of experimental results, therefore, has significant influences on the estimated parameter k. Experimental data always involve random noise. As the inverse algorithm attempts to minimize J, along with the information contained in the data, the noise also creeps in the estimated parameter, and thus, the noise in experimental data may lead to erroneous estimation of parameter. To avoid such a situation, we impose a technique called regularization. Regularization of algorithm is usually achieved using one of the following two techniques: (a) modification of J, or (b) premature stopping of the minimization of J. Some of these techniques are discussed here.

14.3.1 DISCREPANCY PRINCIPLE

Discrepancy principle, initially suggested by Tikhonov [1963], assumes that at $z = z_i$, the difference between the computed and the measured values of temperature is given by

$$|\theta_i - Y_i| = w\sigma \qquad (14.32)$$

Here, σ is the standard deviation, calculated based on repeated measurements, while the weight w is a real positive number indicating the confidence interval of measurements.

The discrepancy principle may be implemented in two ways. First, the minimization of the functional J may be minimized, as shown in equation 14.31, such that

$$\varepsilon \approx w\sigma \qquad (14.33)$$

Thus, iteration stops as soon as the functional J reaches the value $w\sigma$. Alternately, setting $\varepsilon \approx 0$, the functional J may be modified as

$$J(z) = \int_{z=0}^{1} (\theta - Y)^2 \, \delta (z - z_i) \, dz + w\sigma \qquad (14.34)$$

In both these cases, we assume that the standard deviation σ remains independent of location.

14.3.2 ADDITIONAL MEASUREMENT APPROACH

Additional measurement approach, proposed by Özisik and Orlande [2021], constitutes two functionals J_1 and J_2 while attempting to minimize J_1 only. At some point, during the minimization process, J_2 starts to increase, indicating the onset of instability. The iteration stops at this point.

The success of additional measurement technique depends on the amount of experimental data. It is difficult to constitute two functionals, if the amount of data is inadequate. Furthermore, in this method, we assume that arbitrary grouping of measured data in J_1 and J_2 will lead to the same estimated value of the unknown parameter. Such an assumption may not be always justified.

14.3.3 SMOOTHING OF EXPERIMENTAL DATA

If we investigate experimental data in frequency domain, the high-frequency part usually contains noise. If we can remove part of the high-frequency component, using suitable filters, the remaining data may be considered *clean* enough that does not require any further regularization. This concept was applied successfully by Al-Khalidy [1998] for inverse heat conduction problems.

14.4 COMPLETE ALGORITHM

The complete algorithm for inverse estimation is given here. This algorithm may be used for both parameter estimation and function estimation problems.

1. Based on the physical problem, assume a reasonable value of the unknown parameter k.
2. Solve the direct problem, shown in equations 14.6 and 14.7 to calculate θ.
3. Compute the functional J.
4. Check the stopping criterion $J < \varepsilon$. Stop iteration if the stopping criterion is satisfied; otherwise follow the steps below.
5. Solve the adjoint problem, shown in equations 14.18 and 14.19.
6. Compute J' using equation 14.21.
7. Compute the descent direction using equations 14.23 and 14.24.
8. Solve the sensitivity problem, using equations 14.12 and 14.13 while setting $\Delta k = P^{(n)}$.
9. Compute step size β using equation 14.30.
10. Update the unknown parameter k using equation 14.9.
11. Go to step 2 and continue till convergence.

PROBLEMS

Consider steady, 1-D heat conduction problems where we wish to use inverse technique to estimate the unknown thermal conductivity using measured temperatures at few discrete locations. In each of the following cases, formulate the inverse problem, and write the complete algorithm to estimate the unknown thermal conductivity:

1. Thermal conductivity is a second-degree polynomial function of temperature. Assume known temperatures at the boundaries.
2. Thermal conductivity is a unknown function of temperature. Assume known temperatures at the boundaries.
3. Thermal conductivity is a second-degree polynomial function of temperature. Assume known temperature at one boundary and known heat flux at the other.
4. Thermal conductivity is a unknown function of temperature. Assume known temperature at one boundary and known heat flux at the other.

REFERENCES

N. Al-Khalidy. A general space marching algorithm for the solution of two-dimensional boundary inverse heat conduction problems. *Numerical Heat Transfer, Part B*, 34(3):339–360, 1998.

O. M. Alifanov. *Inverse Heat Transfer Problems*. Springer Science & Business Media, 2012.

J. Hadamard. *La notion de différentielle dans l'enseignement*. Hebrew University, 1923.

M. N. Özisik and H. R. Orlande. *Inverse Heat Transfer: Fundamentals and Applications*. CRC Press, 2021.

A. N. Tikhonov. On the solution of ill-posed problems and the method of regularization. In *Doklady Akademii Nauk*, Volume 151, pages 501–504. Russian Academy of Sciences, 1963.

A Thermophysical Properties (Working Fluids)

The thermophysical properties presented below have been adapted from the following reference:

Steven G. Penocello. *Thermal Energy Systems Design and Analysis*. CRC Press, 2019.

DOI: 10.1201/9781003049272-A

Table A.1
Thermophysical Properties of Saturated Water (SI Units)

T °C	P kPa	ρ_f kg/m³	v_g m³/kg	u_f kJ/kg	u_g kJ/kg	h_f kJ/kg	h_g kJ/kg	s_f kJ/kg-K	s_g kJ/kg-K	T °C
0.01	0.61165	999.79	205.99	0.0	2374.9	0.00061178	2500.9	0.0	9.1555	0.01
10	1.2282	999.65	106.30	42.020	2388.6	42.021	2519.2	0.15109	8.8998	10
20	2.3393	998.16	57.757	83.912	2402.3	83.914	2537.4	0.29648	8.6660	20
30	4.2470	995.61	32.878	125.73	2415.9	125.73	2555.5	0.43675	8.4520	30
40	7.3849	992.18	19.515	167.53	2429.4	167.53	2573.5	0.57240	8.2555	40
50	12.352	988.00	12.027	209.33	2442.7	209.34	2591.3	0.70381	8.0748	50
60	19.946	983.16	7.6672	251.16	2455.9	251.18	2608.8	0.83129	7.9081	60
70	31.201	977.73	5.0395	293.03	2468.9	293.07	2626.1	0.95513	7.7540	70
80	47.414	971.77	3.4052	334.96	2481.6	335.01	2643.0	1.0756	7.6111	80
90	70.182	965.30	2.3591	376.97	2494.0	377.04	2659.5	1.1929	7.4781	90
99.974	101.325	958.37	1.6732	418.95	2506.0	419.06	2675.5	1.3069	7.3544	99.974
100	101.42	958.35	1.6718	419.06	2506.0	419.17	2675.6	1.3072	7.3541	100
110	143.38	950.95	1.2093	461.26	2517.7	461.42	2691.1	1.4188	7.2381	110
120	198.67	943.11	0.89121	503.60	2528.9	503.81	2705.9	1.5279	7.1291	120
130	270.28	934.83	0.66800	546.09	2539.5	546.38	2720.1	1.6346	7.0264	130
140	361.54	926.13	0.50845	588.77	2549.6	589.16	2733.4	1.7392	6.9293	140
150	476.16	917.01	0.39245	631.66	2559.1	632.18	2745.9	1.8418	6.8371	150
160	618.23	907.45	0.30678	674.79	2567.8	675.47	2757.4	1.9426	6.7491	160
170	792.19	897.45	0.24259	718.20	2575.7	719.08	2767.9	2.0417	6.6650	170
180	1002.8	887.00	0.19384	761.92	2582.8	763.05	2777.2	2.1392	6.5840	180
190	1255.2	876.08	0.15636	806.00	2589.0	807.43	2785.3	2.2355	6.5059	190
200	1554.9	864.66	0.12721	850.47	2594.2	852.27	2792.0	2.3305	6.4302	200
210	1907.7	852.72	0.10429	895.39	2598.3	897.63	2797.3	2.4245	6.3563	210
220	2319.6	840.22	0.086092	940.82	2601.2	943.58	2800.9	2.5177	6.2840	220
230	2797.1	827.12	0.071503	986.81	2602.9	990.19	2802.9	2.6101	6.2128	230
240	3346.9	813.37	0.059705	1033.4	2603.1	1037.6	2803.0	2.7020	6.1423	240
250	3976.2	798.89	0.050083	1080.8	2601.8	1085.8	2800.9	2.7935	6.0721	250
260	4692.3	783.63	0.042173	1129.0	2598.7	1135.0	2796.6	2.8849	6.0016	260
270	5503.0	767.46	0.035621	1178.1	2593.7	1185.3	2789.7	2.9765	5.9304	270
280	6416.6	750.28	0.030153	1228.3	2586.4	1236.9	2779.9	3.0685	5.8579	280
290	7441.8	731.91	0.025555	1279.9	2576.5	1290.0	2766.7	3.1612	5.7834	290
300	8587.9	712.14	0.021660	1332.9	2563.6	1345.0	2749.6	3.2552	5.7059	300
310	9865.1	690.67	0.018335	1387.9	2547.1	1402.2	2727.9	3.3510	5.6244	310
320	11284.	667.09	0.015471	1445.3	2526.0	1462.2	2700.6	3.4494	5.5372	320
330	12858.	640.77	0.012979	1505.8	2499.2	1525.9	2666.0	3.5518	5.4422	330
340	14601.	610.67	0.010781	1570.6	2464.4	1594.5	2621.8	3.6601	5.3356	340
350	16529.	574.71	0.0088024	1642.1	2418.1	1670.9	2563.6	3.7784	5.2110	350
360	18666.	527.59	0.0069493	1726.3	2351.8	1761.7	2481.5	3.9167	5.0536	360
370	21044.	451.43	0.0049544	1844.1	2230.3	1890.7	2334.5	4.1112	4.8012	370
373.946	22064.	322.00	0.0031056	2015.7	2015.7	2084.3	2084.3	4.4070	4.4070	373.946

Table A.2
Thermophysical Properties of Single-Phase Water (SI Units)

T °C	ρ kg/m³	v m³/kg	u kJ/kg	h kJ/kg	s kJ/kg-K	c_p kJ/kg-K	μ mPa-s	k W/m-K	Pr	T °C
				P = 1 kPa (0.001 MPa)						
50	0.0067072	149.09	2445.3	2594.4	9.2430	1.8761	0.010537	0.020232	0.97716	50
100	0.0058075	172.19	2516.4	2688.6	9.5139	1.8914	0.012336	0.024160	0.96574	100
150	0.0051210	195.27	2588.4	2783.7	9.7531	1.9139	0.014252	0.028482	0.95766	150
200	0.0045797	218.35	2661.7	2880.0	9.9682	1.9403	0.016240	0.033149	0.95056	200
250	0.0041419	241.44	2736.3	2977.7	10.165	1.9692	0.018270	0.038120	0.94381	250
300	0.0037805	264.52	2812.4	3077.0	10.346	1.9996	0.020325	0.043361	0.93729	300
350	0.0034772	287.59	2890.1	3177.7	10.514	2.0312	0.022390	0.048845	0.93107	350
400	0.0032189	310.67	2969.4	3280.1	10.672	2.0636	0.024456	0.054544	0.92525	400
450	0.0029963	333.74	3050.4	3384.1	10.821	2.0969	0.026515	0.060439	0.91991	450
500	0.0028025	356.82	3133.0	3489.8	10.963	2.1309	0.028562	0.066508	0.91510	500
600	0.0024816	402.97	3303.4	3706.3	11.226	2.2006	0.032605	0.079104	0.90704	600
700	0.0022266	449.12	3480.8	3930.0	11.468	2.2716	0.036564	0.092212	0.90074	700
800	0.0020191	495.27	3665.4	4160.7	11.694	2.3424	0.040428	0.10573	0.89562	800
900	0.0018470	541.42	3856.9	4398.4	11.906	2.4114	0.044194	0.11959	0.89114	900
1000	0.0017019	587.58	4055.3	4642.8	12.106	2.4776	0.047860	0.13371	0.88686	1000
1100	0.0015780	633.71	4260.0	4893.8	12.295	2.5403	0.051428	0.14803	0.88250	1100
1200	0.0014708	679.90	4470.9	5150.8	12.476	2.5989	0.054901	0.16252	0.87790	1200
1400	0.0012950	772.20	4909.1	5681.3	12.813	2.7035	0.061575	0.19183	0.86779	1400
1600	0.0011567	864.53	5366.6	6231.1	13.124	2.7922	0.067916	0.22138	0.85658	1600
				P = 100 kPa (0.01 MPa)						
50	988.03	0.0010121	209.32	209.42	0.70377	4.1813	0.54652	0.64062	3.5671	50
100	0.58967	1.6959	2506.2	2675.8	7.3610	2.0766	0.012234	0.024564	1.0342	100
150	0.51636	1.9366	2582.9	2776.6	7.6148	1.9846	0.014192	0.028843	0.97655	150
200	0.46031	2.1724	2658.2	2875.5	7.8356	1.9754	0.016204	0.033436	0.95735	200
250	0.41560	2.4062	2733.9	2974.5	8.0346	1.9893	0.018249	0.038340	0.94690	250
300	0.37895	2.6389	2810.6	3074.5	8.2172	2.0124	0.020313	0.043530	0.93908	300
350	0.34832	2.8709	2888.7	3175.8	8.3866	2.0399	0.022384	0.048975	0.93230	350
400	0.32230	3.1027	2968.3	3278.6	8.5452	2.0698	0.024453	0.054649	0.92617	400
450	0.29992	3.3342	3049.4	3382.8	8.6946	2.1015	0.026515	0.060527	0.92061	450
500	0.28046	3.5656	3132.2	3488.7	8.8361	2.1344	0.028564	0.066586	0.91562	500
600	0.24827	4.0279	3302.8	3705.6	9.0998	2.2029	0.032608	0.079173	0.90727	600
700	0.22272	4.4899	3480.4	3929.4	9.3424	2.2731	0.036568	0.092282	0.90076	700
800	0.20194	4.9520	3665.0	4160.2	9.5681	2.3434	0.040433	0.10581	0.89549	800
900	0.18472	5.4136	3856.6	4398.0	9.7800	2.4122	0.044198	0.11967	0.89091	900
1000	0.17020	5.8754	4055.0	4642.6	9.9800	2.4782	0.047864	0.13380	0.88656	1000
1100	0.15780	6.3371	4259.8	4893.5	10.170	2.5407	0.051432	0.14813	0.88216	1100
1200	0.14708	6.7990	4470.7	5150.6	10.350	2.5993	0.054904	0.16262	0.87754	1200
1400	0.12950	7.7220	4908.9	5681.2	10.688	2.7038	0.061578	0.19194	0.86742	1400
1600	0.11567	8.6453	5366.5	6231.0	10.998	2.7923	0.067918	0.22149	0.85623	1600

Table A.3

Thermophysical Properties of Single-Phase Water (SI Units)

T °C	ρ kg/m³	v m³/kg	u kJ/kg	h kJ/kg	s kJ/kg-K	c_p kJ/kg-K	μ mPa-s	k W/m-K	Pr	T °C
P = 500 kPa (0.05 MPa)										
50	988.21	0.0010119	209.26	209.76	0.70358	4.1804	0.54660	0.64083	3.5657	50
100	958.54	0.0010433	418.94	419.47	1.3069	4.2148	0.28169	0.67744	1.7526	100
150	917.02	0.0010905	631.65	632.19	1.8418	4.3070	0.18262	0.68103	1.1549	150
200	2.3528	0.42503	2643.3	2855.8	7.0610	2.1429	0.016060	0.034648	0.99325	200
250	2.1078	0.47443	2723.8	2961.0	7.2724	2.0788	0.018164	0.039254	0.96191	250
300	1.9135	0.52260	2803.2	3064.6	7.4614	2.0670	0.020265	0.044223	0.94720	300
350	1.7539	0.57016	2883.0	3168.1	7.6346	2.0763	0.022359	0.049512	0.93764	350
400	1.6199	0.61732	2963.7	3272.3	7.7955	2.0957	0.024444	0.055079	0.93005	400
450	1.5056	0.66419	3045.6	3377.7	7.9465	2.1206	0.026516	0.060886	0.92352	450
500	1.4066	0.71093	3129.0	3484.5	8.0892	2.1490	0.028571	0.066902	0.91776	500
600	1.2436	0.80412	3300.4	3702.5	8.3543	2.2121	0.032623	0.079455	0.90823	600
700	1.1149	0.89694	3478.5	3927.0	8.5977	2.2793	0.036585	0.092568	0.90085	700
800	1.0104	0.98971	3663.6	4158.4	8.8240	2.3479	0.040450	0.10612	0.89497	800
900	0.92399	1.0823	3855.4	4396.6	9.0362	2.4155	0.044215	0.12001	0.88997	900
1000	0.85121	1.1748	4054.0	4641.4	9.2364	2.4807	0.047880	0.13416	0.88535	1000
1100	0.78910	1.2673	4259.0	4892.6	9.4263	2.5427	0.051447	0.14852	0.88079	1100
1200	0.73545	1.3597	4470.0	5149.8	9.6071	2.6008	0.054918	0.16304	0.87608	1200
1400	0.64745	1.5445	4908.4	5680.6	9.9448	2.7048	0.061590	0.19238	0.86593	1400
1600	0.57828	1.7293	5366.1	6230.7	10.255	2.7931	0.067928	0.22195	0.85481	1600
P = 1000 kPa (1 MPa)										
50	988.43	0.0010117	209.18	210.19	0.70335	4.1793	0.54670	0.64109	3.5639	50
100	958.77	0.0010430	418.80	419.84	1.3065	4.2136	0.28183	0.67772	1.7522	100
150	917.31	0.0010901	631.41	632.50	1.8412	4.3054	0.18274	0.68137	1.1547	150
200	4.8539	0.20602	2622.2	2828.3	6.6955	2.4281	0.015876	0.036312	1.0616	200
250	4.2965	0.23275	2710.4	2943.1	6.9265	2.2106	0.018058	0.040464	0.98657	250
300	3.8762	0.25798	2793.6	3051.6	7.1246	2.1425	0.020205	0.045122	0.95939	300
350	3.5398	0.28250	2875.7	3158.2	7.3029	2.1248	0.022329	0.050201	0.94512	350
400	3.2615	0.30661	2957.9	3264.5	7.4669	2.1293	0.024433	0.055627	0.93526	400
450	3.0262	0.33045	3040.9	3371.3	7.6200	2.1452	0.026518	0.061343	0.92732	450
500	2.8240	0.35411	3125.0	3479.1	7.7641	2.1677	0.028581	0.067303	0.92052	500
600	2.4931	0.40111	3297.5	3698.6	8.0310	2.2237	0.032642	0.079811	0.90945	600
700	2.2330	0.44783	3476.2	3924.1	8.2755	2.2871	0.036607	0.092929	0.90098	700
800	2.0227	0.49439	3661.7	4156.1	8.5024	2.3534	0.040473	0.10650	0.89433	800
900	1.8490	0.54083	3853.9	4394.8	8.7150	2.4196	0.044237	0.12043	0.88881	900
1000	1.7030	0.58720	4052.7	4639.9	8.9155	2.4839	0.047901	0.13462	0.88385	1000
1100	1.5784	0.63355	4257.9	4891.4	9.1056	2.5452	0.051466	0.14901	0.87908	1100
1200	1.4710	0.67981	4469.0	5148.9	9.2866	2.6028	0.054936	0.16355	0.87427	1200
1400	1.2948	0.77232	4907.7	5680.0	9.6245	2.7062	0.061605	0.19294	0.86408	1400
1600	1.1564	0.86475	5365.5	6230.3	9.9351	2.7940	0.067940	0.22253	0.85305	1600

Table A.4

Thermophysical Properties of Single-Phase Water (SI Units)

T °C	ρ kg/m³	v m³/kg	u kJ/kg	h kJ/kg	s kJ/kg-K	c_p kJ/kg-K	μ mPa-s	k W/m-K	Pr	T °C
				$P = 5000$ kPa (5 MPa)						
50	990.16	0.0010099	208.59	213.64	0.70150	4.1702	0.54751	0.64317	3.5500	50
100	960.63	0.0010410	417.64	422.85	1.3034	4.2045	0.28290	0.67999	1.7493	100
150	919.56	0.0010875	629.55	634.98	1.8368	4.2926	0.18376	0.68408	1.1531	150
200	867.26	0.0011531	847.91	853.68	2.3251	4.4761	0.13546	0.66287	0.91469	200
250	800.09	0.0012499	1079.5	1085.7	2.7910	4.8562	0.10658	0.61802	0.83747	250
300	22.053	0.045345	2699.0	2925.7	6.2110	3.1722	0.019794	0.054298	1.1564	300
350	19.242	0.051970	2809.5	3069.3	6.4516	2.6608	0.022157	0.056664	1.0404	350
400	17.290	0.057837	2907.5	3196.7	6.6483	2.4610	0.024406	0.060565	0.99174	400
450	15.792	0.063323	3000.6	3317.2	6.8210	2.3717	0.026582	0.065367	0.96444	450
500	14.581	0.068582	3091.7	3434.7	6.9781	2.3323	0.028703	0.070782	0.94579	500
600	12.706	0.078703	3273.3	3666.8	7.2605	2.3216	0.032821	0.082832	0.91991	600
700	11.297	0.088519	3457.7	3900.3	7.5136	2.3515	0.036804	0.095933	0.90215	700
800	10.188	0.098155	3646.9	4137.7	7.7458	2.3986	0.040669	0.10969	0.88929	800
900	9.2861	0.10769	3841.8	4380.2	7.9618	2.4528	0.044423	0.12386	0.87967	900
1000	8.5364	0.11715	4042.6	4628.3	8.1648	2.5091	0.048074	0.13832	0.87209	1000
1100	7.9018	0.12655	4249.3	4882.0	8.3566	2.5649	0.051625	0.15295	0.86574	1100
1200	7.3571	0.13592	4461.6	5141.2	8.5388	2.6186	0.055080	0.16770	0.86008	1200
1400	6.4687	0.15459	4902.0	5675.0	8.8784	2.7168	0.061724	0.19738	0.84960	1400
1600	5.7739	0.17319	5361.1	6227.1	9.1900	2.8016	0.068037	0.22712	0.83927	1600
				$P = 10,000$ kPa (10 MPa)						
50	992.31	0.0010077	207.86	217.94	0.69920	4.1592	0.54854	0.64574	3.5331	50
100	962.93	0.0010385	416.23	426.62	1.2996	4.1935	0.28425	0.68279	1.7458	100
150	922.32	0.0010842	627.27	638.11	1.8313	4.2773	0.18502	0.68744	1.1512	150
200	870.94	0.0011482	844.31	855.80	2.3174	4.4491	0.13670	0.66697	0.91191	200
250	805.70	0.0012412	1073.4	1085.8	2.7792	4.7934	0.10799	0.62346	0.83023	250
300	715.29	0.0013980	1329.4	1343.3	3.2488	5.6807	0.086433	0.55506	0.88459	300
350	44.564	0.022440	2699.6	2924.0	5.9459	4.0117	0.022177	0.069107	1.2874	350
400	37.827	0.026436	2833.1	3097.4	6.2141	3.0953	0.024553	0.068716	1.1060	400
450	33.578	0.029781	2944.5	3242.3	6.4219	2.7473	0.026802	0.071561	1.0289	450
500	30.478	0.032811	3047.0	3375.1	6.5995	2.5830	0.028966	0.075923	0.98547	500
600	26.057	0.038377	3242.0	3625.8	6.9045	2.4576	0.033117	0.087077	0.93470	600
700	22.937	0.043598	3434.0	3870.0	7.1693	2.4370	0.037098	0.10000	0.90406	700
800	20.564	0.048629	3628.2	4114.5	7.4085	2.4571	0.040946	0.11390	0.88328	800
900	18.675	0.053548	3826.5	4362.0	7.6290	2.4952	0.044679	0.12834	0.86867	900
1000	17.126	0.058391	4029.9	4613.8	7.8349	2.5411	0.048306	0.14307	0.85798	1000
1100	15.827	0.063183	4238.5	4870.3	8.0288	2.5898	0.051835	0.15797	0.84979	1100
1200	14.719	0.067939	4452.3	5131.7	8.2126	2.6384	0.055269	0.17295	0.84314	1200
1400	12.924	0.077375	4895.0	5668.7	8.5543	2.7301	0.061876	0.20297	0.83229	1400
1600	11.528	0.086745	5355.6	6223.1	8.8671	2.8110	0.068161	0.23288	0.82276	1600

Table A.5
Thermophysical Properties of Single-Phase Water (SI Units)

T °C	ρ kg/m³	v m³/kg	u kJ/kg	h kJ/kg	s kJ/kg-K	c_p kJ/kg-K	μ mPa-s	k W/m-K	Pr	T °C
				P = 50000 kPa (50 MPa)						
50	1008.7	0.00099138	202.46	252.03	0.68100	4.0816	0.55737	0.66526	3.4197	50
100	980.27	0.0010201	405.93	456.94	1.2705	4.1156	0.29480	0.70446	1.7223	100
150	942.70	0.0010600	610.98	661.02	1.7912	4.1741	0.19471	0.71324	1.1395	150
200	896.97	0.0011149	819.45	875.19	2.2628	4.2827	0.14600	0.69787	0.89596	200
250	842.41	0.0011871	1034.2	1093.5	2.7013	4.4666	0.11782	0.66278	0.79399	250
300	776.48	0.0012879	1259.6	1324.0	3.1218	4.7801	0.098674	0.61003	0.77319	300
350	693.25	0.0014425	1503.9	1576.1	3.5431	5.3746	0.083217	0.54007	0.82816	350
400	577.79	0.0017307	1787.8	1874.4	4.0029	6.7899	0.068075	0.44839	1.0309	400
450	402.04	0.0024873	2160.3	2284.7	4.5896	9.5716	0.050974	0.31973	1.5260	450
500	257.07	0.0038900	2528.1	2722.6	5.1762	7.2889	0.040922	0.20590	1.4486	500
600	163.72	0.0061080	2947.1	3252.5	5.8245	4.1028	0.039185	0.15037	1.0692	600
700	129.59	0.0077166	3228.7	3614.6	6.2178	3.2857	0.041453	0.14796	0.92050	700
800	110.22	0.0090728	3472.2	3925.8	6.5225	2.9831	0.044392	0.15755	0.84055	800
900	97.121	0.010296	3702.0	4216.8	6.7819	2.8556	0.047518	0.17090	0.79399	900
1000	87.405	0.011441	3927.3	4499.4	7.0131	2.8041	0.050694	0.18585	0.76486	1000
1100	79.785	0.012534	4152.2	4778.9	7.2244	2.7902	0.053869	0.20147	0.74606	1100
1200	73.583	0.013590	4378.6	5058.1	7.4207	2.7961	0.057017	0.21726	0.73377	1200
1400	63.977	0.015631	4839.2	5620.8	7.7788	2.8344	0.063187	0.24858	0.72047	1400
1600	56.788	0.017609	5312.1	6192.6	8.1015	2.8845	0.069157	0.27902	0.71494	1600
				P = 100000 kPa (100 MPa)						
50	1027.4	0.00097333	196.59	293.92	0.65865	4.0071	0.57007	0.68738	3.3232	50
100	999.76	0.0010002	395.09	495.11	1.2375	4.0385	0.30772	0.72981	1.7028	100
150	964.85	0.0010364	594.29	697.93	1.7475	4.0781	0.20615	0.74335	1.1310	150
200	923.74	0.0010826	795.14	903.40	2.2064	4.1460	0.15651	0.73314	0.88509	200
250	876.66	0.0011407	999.06	1113.1	2.6277	4.2499	0.12813	0.70539	0.77194	250
300	823.17	0.0012148	1207.6	1329.1	3.0219	4.3987	0.10961	0.66281	0.72739	300
350	762.34	0.0013118	1422.8	1554.0	3.3979	4.6070	0.095916	0.60850	0.72618	350
400	692.93	0.0014431	1646.8	1791.1	3.7639	4.8942	0.084622	0.54536	0.75942	400
450	614.16	0.0016282	1881.9	2044.7	4.1271	5.2577	0.074632	0.47655	0.82340	450
500	528.28	0.0018929	2126.9	2316.2	4.4900	5.5688	0.065842	0.40661	0.90175	500
600	374.21	0.0026723	2597.9	2865.1	5.1581	5.1682	0.054338	0.29451	0.95355	600
700	282.04	0.0035456	2976.1	3330.7	5.6639	4.1810	0.050714	0.24292	0.87286	700
800	230.64	0.0043358	3281.7	3715.3	6.0406	3.5764	0.050780	0.23223	0.78201	800
900	198.32	0.0050424	3551.4	4055.6	6.3440	3.2643	0.052275	0.23619	0.72247	900
1000	175.75	0.0056899	3804.0	4373.0	6.6038	3.1013	0.054397	0.24654	0.68426	1000
1100	158.82	0.0062964	4048.8	4678.4	6.8347	3.0158	0.056827	0.25983	0.65957	1100
1200	145.50	0.0068729	4290.3	4977.6	7.0450	2.9731	0.059419	0.27445	0.64368	1200
1400	125.53	0.0079662	4772.5	5569.1	7.4216	2.9516	0.064813	0.30474	0.62775	1400
1600	111.03	0.0090066	5260.0	6160.6	7.7555	2.9673	0.070273	0.33451	0.62337	1600

Table A.6
Thermophysical Properties of Liquids at Saturation (SI Units)

T °C	P_{sat} kPa	ρ kg/m³	h_{fg} kJ/kg	c_p kJ/ kg-K	μ mPa-s	ν mm²/s	k W/m-K	α mm²/s	Pr	T °C
				Water						
5	0.87258	999.92	2489.0	4.2055	1.5183	1.5184	0.56772	0.13501	11.247	5
10	1.2282	999.65	2477.2	4.1955	1.3060	1.3064	0.57871	0.13798	9.4682	10
15	1.7058	999.06	2465.4	4.1888	1.1376	1.1387	0.58874	0.14068	8.0940	15
20	2.3393	998.16	2453.5	4.1844	1.0016	1.0035	0.59795	0.14317	7.0092	20
25	3.1699	997.00	2441.7	4.1816	0.89004	0.89271	0.60646	0.14547	6.1369	25
30	4.2470	995.61	2429.8	4.1801	0.79722	0.80074	0.61434	0.14762	5.4245	30
35	5.6290	993.99	2417.9	4.1795	0.71912	0.72347	0.62165	0.14964	4.8348	35
40	7.3849	992.18	2406.0	4.1796	0.65272	0.65786	0.62844	0.15154	4.3411	40
45	9.5950	990.17	2394.0	4.1804	0.59575	0.60167	0.63474	0.15334	3.9236	45
50	12.352	988.00	2381.9	4.1815	0.54650	0.55314	0.64057	0.15505	3.5674	50
60	19.946	983.16	2357.7	4.1851	0.46602	0.47400	0.65096	0.15820	2.9961	60
70	31.201	977.73	2333.0	4.1902	0.40353	0.41272	0.65972	0.16103	2.5630	70
80	47.414	971.77	2308.0	4.1969	0.35404	0.36432	0.66697	0.16354	2.2278	80
90	70.182	965.30	2282.5	4.2053	0.31417	0.32546	0.67277	0.16573	1.9638	90
100	101.42	958.35	2256.4	4.2157	0.28158	0.29382	0.67721	0.16762	1.7529	100
150	476.16	917.01	2113.7	4.3071	0.18261	0.19914	0.68102	0.17243	1.1549	150
200	1554.9	864.66	1939.7	4.4958	0.13458	0.15565	0.66001	0.16978	0.91675	200
250	3976.2	798.89	1715.2	4.8701	0.10628	0.13304	0.61689	0.15855	0.83908	250
300	8587.9	712.14	1404.6	5.7504	0.085855	0.12056	0.55265	0.13495	0.89334	300
				Ammonia						
−60	21.893	713.62	1441.8	4.3031	0.39129	0.54832	0.75700	0.24652	2.2242	−60
−50	40.836	702.09	1415.9	4.3599	0.32887	0.46841	0.72228	0.23596	1.9851	−50
−40	71.692	690.15	1388.6	4.4137	0.28124	0.40751	0.68811	0.22590	1.8040	−40
−30	119.43	677.83	1359.7	4.4645	0.24407	0.36008	0.65463	0.21632	1.6646	−30
−20	190.08	665.14	1329.1	4.5138	0.21441	0.32235	0.62196	0.20716	1.5560	−20
−10	290.71	652.06	1296.7	4.5636	0.19022	0.29172	0.59014	0.19832	1.4710	−10
0	429.38	638.57	1262.2	4.6165	0.17009	0.26636	0.55920	0.18969	1.4042	0
10	615.05	624.64	1225.5	4.6757	0.15303	0.24499	0.52912	0.18117	1.3523	10
20	857.48	610.20	1186.4	4.7448	0.13832	0.22668	0.49986	0.17265	1.3130	20
30	1167.2	595.17	1144.4	4.8282	0.12545	0.21078	0.47135	0.16403	1.2850	30
40	1555.4	579.44	1099.3	4.9318	0.11404	0.19681	0.44354	0.15521	1.2680	40
50	2034.0	562.86	1050.5	5.0635	0.10379	0.18440	0.41632	0.14607	1.2624	50
60	2615.6	545.24	997.30	5.2351	0.094483	0.17329	0.38959	0.13649	1.2696	60
70	3313.5	526.31	938.90	5.4648	0.085933	0.16328	0.36324	0.12629	1.2928	70
80	4142.0	505.67	873.97	5.7837	0.077979	0.15421	0.33710	0.11526	1.3379	80
90	5116.7	482.75	800.58	6.2501	0.070468	0.14597	0.31102	0.10308	1.4161	90
100	6255.3	456.63	715.63	6.9912	0.063231	0.13847	0.28477	0.089202	1.5523	100
110	7578.3	425.61	613.39	8.3621	0.056028	0.13164	0.25813	0.072528	1.8151	110
120	9112.5	385.49	480.31	11.940	0.048340	0.12540	0.23124	0.050238	2.4961	120

Table A.7
Thermophysical Properties of Liquids at Saturation (SI Units)

T °C	P_{sat} kPa	ρ kg/m³	h_{fg} kJ/kg	c_p kJ/ kg-K	μ mPa-s	ν mm²/s	k W/m-K	α mm²/s	Pr	T °C
					Propane					
−100	2.8994	643.74	480.44	2.0538	0.42569	0.66127	0.16437	0.12432	5.3190	−100
−80	13.049	622.76	462.41	2.1059	0.31590	0.50726	0.15206	0.11595	4.3750	−80
−60	42.693	601.08	443.63	2.1720	0.24359	0.40526	0.13981	0.10709	3.7844	−60
−40	111.12	578.43	423.36	2.2558	0.19255	0.33288	0.12795	0.098058	3.3948	−40
−30	167.83	566.64	412.41	2.3054	0.17232	0.30411	0.12221	0.093552	3.2508	−30
−20	244.52	554.45	400.77	2.3608	0.15473	0.27906	0.11662	0.089096	3.1322	−20
−10	345.28	541.80	388.30	2.4230	0.13928	0.25706	0.11121	0.084712	3.0346	−10
−5	406.04	535.27	381.72	2.4570	0.13223	0.24704	0.10857	0.082551	2.9926	−5
0	474.46	528.59	374.87	2.4932	0.12559	0.23759	0.10597	0.080412	2.9547	0
5	551.12	521.75	367.73	2.5318	0.11930	0.22866	0.10343	0.078295	2.9205	5
10	636.60	514.73	360.28	2.5733	0.11335	0.22021	0.10093	0.076198	2.8899	10
20	836.46	500.06	344.31	2.6662	0.10229	0.20455	0.096073	0.072059	2.8387	20
30	1079.0	484.39	326.70	2.7767	0.092188	0.19032	0.091409	0.067961	2.8004	30
40	1369.4	467.46	307.07	2.9127	0.082844	0.17722	0.086923	0.063839	2.7760	40
50	1713.3	448.87	284.86	3.0893	0.074066	0.16501	0.082598	0.059565	2.7702	50
60	2116.8	427.97	259.23	3.3375	0.065654	0.15341	0.078398	0.054887	2.7950	60
70	2586.8	403.62	228.62	3.7350	0.057364	0.14212	0.074277	0.049271	2.8845	70
80	3131.9	373.29	189.80	4.5445	0.048787	0.13069	0.070213	0.041389	3.1577	80
90	3764.1	328.83	132.77	7.6233	0.038819	0.11805	0.067139	0.026783	4.4077	90
					Pentane					
−80	0.10043	717.29	440.72	1.9990	0.78523	1.0947	0.15551	0.10846	10.093	−80
−40	2.7187	681.58	412.64	2.0792	0.41915	0.61496	0.13791	0.097314	6.3193	−40
−20	9.0283	663.44	398.87	2.1384	0.33262	0.50136	0.12935	0.091172	5.4990	−20
−10	15.191	654.22	391.89	2.1727	0.29997	0.45851	0.12520	0.088078	5.2057	−10
0	24.448	644.87	384.81	2.2099	0.27220	0.42210	0.12113	0.084997	4.9660	0
10	37.835	635.37	377.56	2.2500	0.24825	0.39071	0.11716	0.081953	4.7675	10
20	56.558	625.70	370.11	2.2929	0.22732	0.36331	0.11329	0.078964	4.6010	20
30	81.993	615.82	362.40	2.3385	0.20883	0.33911	0.10951	0.076044	4.4594	30
40	115.67	605.70	354.38	2.3870	0.19232	0.31752	0.10584	0.073207	4.3373	40
50	159.25	595.30	346.01	2.4382	0.17744	0.29806	0.10227	0.070459	4.2303	50
60	214.54	584.59	337.22	2.4925	0.16391	0.28038	0.098799	0.067805	4.1350	60
70	283.46	573.51	327.94	2.5501	0.15150	0.26416	0.095426	0.065249	4.0486	70
80	368.01	562.01	318.11	2.6113	0.14005	0.24919	0.092149	0.062789	3.9687	80
90	470.34	550.01	307.63	2.6770	0.12939	0.23524	0.088964	0.060421	3.8934	90
100	592.65	537.42	296.40	2.7482	0.11940	0.22216	0.085868	0.058139	3.8212	100
120	906.71	510.00	271.14	2.9137	0.10100	0.19803	0.079923	0.053785	3.6819	120
140	1330.5	478.31	240.75	3.1340	0.084073	0.17577	0.074275	0.049548	3.5474	140
160	1887.9	439.49	202.15	3.4903	0.067854	0.15439	0.068887	0.044908	3.4379	160
180	2609.9	385.55	146.99	4.4391	0.051042	0.13239	0.063884	0.037326	3.5468	180

Table A.8
Thermophysical Properties of Liquids at Saturation (SI Units)

T °C	P_{sat} kPa	ρ kg/m³	h_{fg} kJ/kg	c_p kJ/ kg-K	μ mPa-s	ν mm²/s	k W/m-K	α mm²/s	Pr	T °C
				Hexane						
−50	0.20366	721.49	412.78	1.9903	0.81257	1.1262	0.14730	0.10257	10.980	−50
−40	0.46089	712.71	406.50	2.0189	0.68420	0.96000	0.14312	0.099466	9.6516	−40
−30	0.96456	703.92	400.26	2.0493	0.58487	0.83087	0.13912	0.096439	8.6156	−30
−20	1.8855	695.11	394.03	2.0817	0.50631	0.72839	0.13528	0.093491	7.7911	−20
−10	3.4719	686.25	387.79	2.1161	0.44302	0.64556	0.13160	0.090626	7.1233	−10
0	6.0652	677.34	381.52	2.1525	0.39118	0.57753	0.12809	0.087857	6.5735	0
10	10.114	668.35	375.18	2.1909	0.34813	0.52088	0.12472	0.085177	6.1152	10
20	16.187	659.27	368.74	2.2312	0.31192	0.47313	0.12149	0.082595	5.7283	20
30	24.973	650.07	362.17	2.2733	0.28112	0.43245	0.11839	0.080115	5.3979	30
40	37.292	640.74	355.45	2.3172	0.25467	0.39746	0.11542	0.077741	5.1126	40
60	76.424	621.60	341.38	2.4100	0.21169	0.34056	0.10984	0.073325	4.6445	60
80	142.54	601.63	326.24	2.5094	0.17836	0.29647	0.10471	0.069357	4.2745	80
100	246.29	580.60	309.70	2.6161	0.15178	0.26141	0.099987	0.065829	3.9711	100
120	399.76	558.16	291.35	2.7318	0.12999	0.23289	0.095627	0.062714	3.7136	120
140	616.32	533.84	270.63	2.8612	0.11165	0.20915	0.091587	0.059963	3.4879	140
160	910.70	506.85	246.68	3.0142	0.095730	0.18887	0.087815	0.057480	3.2859	160
180	1299.6	475.83	218.02	3.2169	0.081352	0.17097	0.084245	0.055037	3.1065	180
200	1803.4	437.73	181.55	3.5578	0.067525	0.15426	0.080798	0.051882	2.9733	200
220	2450.9	382.16	127.74	4.6726	0.052142	0.13644	0.077711	0.043519	3.1352	220
				Heptane						
−60	0.010397	750.18	419.34	2.0026	1.5345	2.0455	0.14821	0.098656	20.734	−60
−40	0.077005	733.56	405.97	2.0420	0.99774	1.3601	0.14210	0.094868	14.337	−40
−20	0.39420	717.02	393.11	2.0922	0.70470	0.98282	0.13591	0.090599	10.848	−20
0	1.5223	700.45	380.59	2.1526	0.52712	0.75255	0.12973	0.086037	8.7468	0
10	2.7467	692.11	374.39	2.1863	0.46334	0.66946	0.12666	0.083705	7.9978	10
20	4.7222	683.72	368.21	2.2221	0.41085	0.60090	0.12363	0.081374	7.3844	20
30	7.7770	675.27	362.01	2.2597	0.36706	0.54358	0.12063	0.079052	6.8762	30
40	12.326	666.72	355.77	2.2992	0.33007	0.49507	0.11766	0.076757	6.4498	40
60	28.039	649.32	343.04	2.3827	0.27127	0.41778	0.11186	0.072299	5.7785	60
80	57.090	631.36	329.77	2.4717	0.22681	0.35924	0.10624	0.068077	5.2770	80
100	106.23	612.67	315.71	2.5658	0.19210	0.31354	0.10082	0.064136	4.8887	100
120	183.60	593.03	300.58	2.6653	0.16422	0.27693	0.095614	0.060493	4.5778	120
140	298.61	572.12	284.03	2.7714	0.14126	0.24690	0.090607	0.057145	4.3206	140
160	461.74	549.53	265.61	2.8874	0.12184	0.22171	0.085784	0.054064	4.1009	160
180	684.66	524.57	244.66	3.0201	0.10495	0.20007	0.081105	0.051193	3.9082	180
200	980.41	496.14	220.14	3.1853	0.089771	0.18094	0.076509	0.048413	3.7375	200
220	1364.2	462.01	190.09	3.4269	0.075454	0.16332	0.071899	0.045413	3.5963	220
240	1855.1	416.64	149.97	3.9284	0.060780	0.14588	0.067211	0.041064	3.5525	240
260	2479.4	335.09	82.509	7.2271	0.041997	0.12533	0.064864	0.026785	4.6792	260

Table A.9
Thermophysical Properties of Liquids at Saturation (SI Units)

T °C	P_{sat} kPa	ρ kg/m³	h_{fg} kJ/kg	c_p kJ/ kg-K	μ mPa-s	ν mm²/s	k W/m-K	α mm²/s	Pr	T °C
					Octane					
−40	0.012988	750.20	404.85	2.0404	1.4814	1.9747	0.14666	0.095817	20.609	−40
−20	0.000067	734.17	391.98	2.0866	0.98686	1.3442	0.13930	0.090930	14.783	−20
10	0.75269	710.22	373.52	2.1752	0.61684	0.86853	0.12914	0.083592	10.390	10
0	0.38526	718.21	379.58	2.1433	0.71102	0.98999	0.13242	0.086020	11.509	0
10	0.75269	710.22	373.52	2.1752	0.61684	0.86853	0.12914	0.083592	10.390	10
20	1.3916	702.20	367.53	2.2090	0.54141	0.77102	0.12595	0.081198	9.4956	20
40	4.1263	686.04	355.65	2.2821	0.42907	0.62543	0.11987	0.076565	8.1687	40
60	10.450	669.61	343.76	2.3610	0.35007	0.52281	0.11410	0.072172	7.2439	60
80	23.298	652.79	331.67	2.4445	0.29170	0.44686	0.10860	0.068058	6.5659	80
100	46.824	635.45	319.17	2.5318	0.24674	0.38829	0.10335	0.064237	6.0447	100
120	86.423	617.43	306.03	2.6225	0.21086	0.34151	0.098299	0.060708	5.6255	120
140	148.66	598.53	292.02	2.7169	0.18135	0.30300	0.093434	0.057457	5.2734	140
160	241.17	578.48	276.85	2.8161	0.15645	0.27045	0.088722	0.054463	4.9657	160
180	372.53	556.91	260.16	2.9223	0.13494	0.24230	0.084137	0.051699	4.6867	180
200	552.24	533.27	241.43	3.0403	0.11596	0.21745	0.079653	0.049129	4.4262	200
220	790.84	506.68	219.89	3.1805	0.098855	0.19510	0.075248	0.046695	4.1783	220
240	1100.3	475.51	194.21	3.3681	0.083037	0.17463	0.070916	0.044279	3.9438	240
260	1495.1	436.27	161.58	3.6869	0.067831	0.15548	0.066742	0.041493	3.7471	260
280	1995.0	378.16	113.81	4.6731	0.051845	0.13710	0.063640	0.036012	3.8071	280
					Nonane					
−40	0.0023016	765.37	401.57	2.0099	2.3580	3.0808	0.14782	0.096089	32.062	−40
−20	0.018188	749.46	389.10	2.0606	1.3789	1.8399	0.14066	0.091083	20.200	−20
0	0.10056	733.69	377.09	2.1212	0.94012	1.2814	0.13410	0.086167	14.871	0
20	0.42095	717.95	365.44	2.1896	0.69754	0.97158	0.12807	0.081466	11.926	20
40	1.4136	702.14	354.04	2.2639	0.54552	0.77694	0.12249	0.077060	10.082	40
60	3.9784	686.18	342.75	2.3425	0.44174	0.64376	0.11732	0.072986	8.8203	60
80	9.7038	669.98	331.42	2.4245	0.36631	0.54675	0.11250	0.069256	7.8947	80
100	21.057	653.42	319.91	2.5090	0.30880	0.47260	0.10798	0.065864	7.1754	100
120	41.500	636.39	308.02	2.5957	0.26328	0.41372	0.10373	0.062796	6.5883	120
140	75.519	618.75	295.58	2.6843	0.22615	0.36550	0.099707	0.060031	6.0884	140
160	128.59	600.34	282.38	2.7753	0.19509	0.32497	0.095872	0.057543	5.6475	160
180	207.09	580.92	268.17	2.8694	0.16856	0.29015	0.092184	0.055303	5.2467	180
200	318.26	560.20	252.65	2.9685	0.14545	0.25964	0.088598	0.053277	4.8735	200
220	470.16	537.72	235.40	3.0761	0.12497	0.23240	0.085061	0.051424	4.5193	220
240	671.82	512.80	215.80	3.1993	0.10646	0.20760	0.081502	0.049678	4.1788	240
260	933.60	484.23	192.78	3.3544	0.089332	0.18448	0.077827	0.047915	3.8503	260
280	1268.1	449.57	164.28	3.5893	0.072933	0.16223	0.073892	0.045792	3.5427	280
300	1692.7	402.26	124.95	4.1366	0.056076	0.13940	0.069525	0.041783	3.3364	300
320	2238.0	289.18	37.983	25.073	0.031260	0.10810	0.078846	0.010875	9.9406	320

Table A.10

Thermophysical Properties of Liquids at Saturation (SI Units)

T °C	P_{sat} kPa	ρ kg/m³	h_{fg} kJ/kg	c_p kJ/ kg-K	μ mPa-s	ν mm²/s	k W/m-K	α mm²/s	Pr	T °C
					Decane					
−20	0.0039896	761.59	386.87	2.0404	1.9189	2.5196	0.14144	0.091021	27.682	−20
0	0.026319	745.90	375.18	2.1040	1.2731	1.7068	0.13604	0.086682	19.690	0
20	0.12780	730.33	363.83	2.1742	0.91238	1.2493	0.13074	0.082337	15.173	20
40	0.48671	714.78	352.74	2.2494	0.68976	0.96499	0.12557	0.078101	12.356	40
60	1.5248	699.16	341.82	2.3282	0.54215	0.77543	0.12053	0.074047	10.472	60
80	4.0763	683.38	330.95	2.4096	0.43875	0.64203	0.11563	0.070222	9.1428	80
100	9.5701	667.35	320.02	2.4929	0.36304	0.54401	0.11090	0.066658	8.1612	100
120	20.184	650.97	308.87	2.5775	0.30554	0.46935	0.10633	0.063370	7.4065	120
140	38.944	634.13	297.37	2.6632	0.26046	0.41074	0.10195	0.060367	6.8041	140
160	69.756	616.67	285.32	2.7499	0.22415	0.36349	0.097764	0.057651	6.3050	160
180	117.39	598.44	272.55	2.8380	0.19419	0.32449	0.093789	0.055223	5.8761	180
200	187.42	579.21	258.83	2.9284	0.16891	0.29162	0.090036	0.053082	5.4937	200
220	286.20	558.65	243.88	3.0228	0.14712	0.26335	0.086512	0.051229	5.1407	220
240	420.86	536.33	227.30	3.1245	0.12795	0.23857	0.083224	0.049664	4.8038	240
260	599.39	511.52	208.51	3.2397	0.11070	0.21641	0.080176	0.048381	4.4730	260
280	830.94	483.02	186.56	3.3828	0.094727	0.19611	0.077367	0.047349	4.1419	280
300	1126.4	448.49	159.62	3.5932	0.079368	0.17697	0.074807	0.046421	3.8123	300
320	1500.1	402.23	123.34	4.0398	0.063558	0.15801	0.072646	0.044707	3.5345	320
340	1974.8	317.26	57.874	8.5345	0.043216	0.13622	0.075060	0.027721	4.9138	340
					Dodecane					
0	0.0017553	764.28	376.34	2.1375	2.2663	2.9652	0.14146	0.086590	34.245	0
20	0.011684	749.36	364.36	2.1964	1.4866	1.9838	0.13647	0.082916	23.926	20
40	0.057945	734.56	352.90	2.2626	1.0596	1.4425	0.13172	0.079254	18.201	40
60	0.22694	719.77	341.84	2.3340	0.79919	1.1103	0.12715	0.075689	14.670	60
80	0.73371	704.91	331.08	2.4091	0.62760	0.89033	0.12274	0.072276	12.319	80
100	2.0269	689.88	320.52	2.4870	0.50761	0.73580	0.11845	0.069041	10.657	100
120	4.9161	674.60	310.04	2.5666	0.41956	0.62193	0.11427	0.065998	9.4235	120
140	10.699	658.97	299.50	2.6475	0.35230	0.53462	0.11017	0.063151	8.4658	140
160	21.267	642.88	288.77	2.7292	0.29915	0.46533	0.10614	0.060495	7.6921	160
180	39.168	626.21	277.70	2.8118	0.25594	0.40871	0.10217	0.058026	7.0435	180
200	67.635	608.82	266.11	2.8952	0.21992	0.36122	0.098248	0.055738	6.4807	200
220	110.58	590.53	253.82	2.9800	0.18926	0.32049	0.094367	0.053626	5.9765	220
240	172.59	571.10	240.61	3.0669	0.16267	0.28484	0.090527	0.051685	5.5110	240
260	258.91	550.25	226.23	3.1575	0.13925	0.25306	0.086729	0.049919	5.0694	260
280	375.49	527.53	210.32	3.2545	0.11830	0.22425	0.082981	0.048334	4.6396	280
300	529.05	502.35	192.41	3.3630	0.099309	0.19769	0.079303	0.046941	4.2115	300
320	727.34	473.69	171.72	3.4942	0.081853	0.17280	0.075730	0.045753	3.7767	320
340	979.60	439.82	146.96	3.6770	0.065539	0.14901	0.072340	0.044731	3.3313	340
360	1297.7	396.77	115.18	4.0295	0.049828	0.12558	0.069335	0.043367	2.8958	360

Table A.11
Thermophysical Properties of Liquids at Saturation (SI Units)

T °C	P_{sat} kPa	ρ kg/m³	h_{fg} kJ/kg	c_p kJ/ kg-K	μ mPa-s	ν mm²/s	k W/m-K	α mm²/s	Pr	T °C
				Methanol						
−80	0.0025173	887.13	1296.9	2.2124	5.8336	6.5758	0.25568	0.13027	50.479	−80
−60	0.020074	867.24	1276.4	2.2267	2.9632	3.4169	0.24074	0.12467	27.408	−60
−40	0.20096	847.70	1254.6	2.2652	1.7755	2.0945	0.22826	0.11887	17.620	−40
−20	1.0283	828.52	1231.0	2.3224	1.1615	1.4019	0.21794	0.11326	12.378	−20
10	7.4384	800.28	1191.2	2.4496	0.68284	0.85325	0.20571	0.10493	8.1314	10
0	4.0562	809.65	1205.1	2.4011	0.80548	0.99485	0.20941	0.10772	9.2356	0
10	7.4384	800.28	1191.2	2.4496	0.68284	0.85325	0.20571	0.10493	8.1314	10
20	13.032	790.93	1176.6	2.5047	0.58498	0.73962	0.20233	0.10214	7.2416	20
40	35.518	772.10	1145.0	2.6340	0.44145	0.57176	0.19637	0.096555	5.9216	40
60	84.713	752.79	1109.6	2.7880	0.34372	0.45659	0.19119	0.091094	5.0123	60
80	181.11	732.58	1069.2	2.9658	0.27424	0.37435	0.18651	0.085844	4.3608	80
100	353.73	710.95	1022.1	3.1689	0.22275	0.31332	0.18212	0.080837	3.8760	100
120	640.81	687.29	966.15	3.4024	0.18306	0.26635	0.17782	0.076041	3.5027	120
140	1090.2	660.83	898.89	3.6783	0.15129	0.22894	0.17346	0.071360	3.2082	140
160	1759.6	630.42	817.72	4.0224	0.12497	0.19823	0.16893	0.066617	2.9757	160
180	2715.8	594.32	723.80	4.4939	0.10251	0.17248	0.16421	0.061482	2.8053	180
200	4027.0	549.21	621.13	5.2742	0.082795	0.15075	0.15957	0.055089	2.7365	200
220	5787.7	484.87	467.93	7.5369	0.064314	0.13264	0.15743	0.043078	3.0791	220
240	8182.6	310.87	73.259	390.28	0.038619	0.12423	0.26891	0.0022164	56.049	240
				Ethanol						
−80	0.00040787	875.68	1014.2	1.9428	3.8662	4.4150	0.19801	0.11639	37.933	−80
−60	0.0061203	857.77	997.73	1.9811	2.2846	2.6634	0.18843	0.11088	24.020	−60
−40	0.055632	840.37	981.23	2.0479	1.5075	1.7938	0.18060	0.10494	17.094	−40
−20	0.34589	823.28	964.14	2.1414	2.9166	3.5427	0.17418	0.098800	35.857	−20
10	3.1485	797.86	936.24	2.3236	1.4638	1.8347	0.16657	0.089848	20.420	10
0	1.6017	806.34	945.94	2.2571	1.8177	2.2542	0.16888	0.092789	24.294	0
10	3.1485	797.86	936.24	2.3236	1.4638	1.8347	0.16657	0.089848	20.420	10
20	5.8759	789.34	926.01	2.3961	1.1931	1.5115	0.16445	0.086949	17.384	20
40	17.880	772.01	903.52	2.5590	0.81899	1.0609	0.16065	0.081318	13.046	40
60	46.734	753.99	877.53	2.7438	0.58416	0.77476	0.15726	0.076015	10.192	60
80	107.81	734.85	847.02	2.9481	0.43008	0.58527	0.15408	0.071124	8.2289	80
100	224.17	714.02	810.92	3.1739	0.32453	0.45451	0.15096	0.066616	6.8229	100
120	427.30	690.78	768.09	3.4249	0.24908	0.36058	0.14777	0.062460	5.7729	120
140	756.55	664.23	717.34	3.6958	0.19300	0.29057	0.14444	0.058839	4.9383	140
160	1257.6	633.44	657.70	3.9678	0.15017	0.23707	0.14105	0.056118	4.2245	160
180	1980.7	597.60	588.18	4.2485	0.11713	0.19600	0.13779	0.054273	3.6113	180
200	2980.6	554.99	505.72	4.6872	0.091299	0.16450	0.13490	0.051859	3.1722	200
220	4320.5	497.13	397.54	6.0630	0.069103	0.13900	0.13288	0.044087	3.1529	220
240	6094.7	333.83	122.67	63.991	0.037405	0.11205	0.17769	0.0083181	13.470	240

Table A.12
Thermophysical Properties of Liquids at Saturation (SI Units)

T °C	P_{sat} kPa	ρ kg/m³	h_{fg} kJ/kg	c_p kJ/ kg-K	μ mPa-s	ν mm²/s	k W/m-K	α mm²/s	Pr	T °C
					Acetone					
−80	0.015650	897.33	634.81	1.9969	1.6933	1.8870	0.19740	0.11016	17.130	−80
−60	0.13085	876.00	615.66	2.0099	1.0035	1.1455	0.18978	0.10779	10.628	−60
−40	0.72172	854.82	596.79	2.0281	0.68472	0.80102	0.18183	0.10488	7.6375	−40
−20	2.9169	833.58	577.95	2.0540	0.50851	0.61003	0.17366	0.10143	6.0144	−20
10	15.454	801.21	549.13	2.1087	0.35657	0.44504	0.16130	0.095471	4.6615	10
0	9.2991	812.10	558.87	2.0883	0.39775	0.48978	0.16540	0.097527	5.0220	0
10	15.454	801.21	549.13	2.1087	0.35657	0.44504	0.16130	0.095471	4.6615	10
20	24.662	790.19	539.22	2.1311	0.32192	0.40739	0.15720	0.093349	4.3642	20
40	56.582	767.66	518.73	2.1822	0.26705	0.34787	0.14907	0.088987	3.9093	40
60	115.67	744.28	497.07	2.2417	0.22586	0.30346	0.14109	0.084564	3.5886	60
80	215.48	719.79	473.88	2.3102	0.19410	0.26966	0.13332	0.080171	3.3636	80
100	372.30	693.84	448.71	2.3895	0.16904	0.24363	0.12578	0.075864	3.2115	100
110	477.84	680.19	435.21	2.4341	0.15842	0.23291	0.12210	0.073749	3.1581	110
120	604.83	666.00	420.97	2.4828	0.14883	0.22347	0.11850	0.071660	3.1184	120
140	934.13	635.64	389.81	2.5969	0.13213	0.20786	0.11148	0.067534	3.0779	140
160	1383.8	601.79	353.96	2.7457	0.11791	0.19594	0.10471	0.063371	3.0919	160
180	1980.9	562.85	311.35	2.9628	0.10534	0.18715	0.098144	0.058852	3.1800	180
200	2757.9	515.54	257.85	3.3565	0.093515	0.18139	0.091688	0.052986	3.4234	200
220	3757.4	449.93	182.00	4.6149	0.080823	0.17964	0.085422	0.041139	4.3665	220
					Benzene					
10	0.0060742	889.31	444.41	1.6954	0.76992	0.86575	0.14637	0.097079	8.9180	10
20	0.010030	878.84	437.15	1.7204	0.66038	0.75141	0.14289	0.094507	7.9509	20
30	0.015919	868.31	429.95	1.7476	0.57090	0.65749	0.13952	0.091944	7.1510	30
40	0.024388	857.69	422.79	1.7766	0.49715	0.57964	0.13624	0.089410	6.4829	40
50	0.036204	846.96	415.62	1.8071	0.43583	0.51458	0.13305	0.086932	5.9194	50
60	0.052249	836.12	408.41	1.8388	0.38446	0.45982	0.12992	0.084500	5.4416	60
70	0.073517	825.12	401.13	1.8717	0.34109	0.41338	0.12683	0.082125	5.0336	70
80	0.10111	813.97	393.75	1.9054	0.30423	0.37376	0.12378	0.079809	4.6832	80
90	0.13623	802.63	386.23	1.9401	0.27268	0.33973	0.12076	0.077553	4.3806	90
100	0.18016	791.10	378.54	1.9756	0.24551	0.31034	0.11777	0.075356	4.1183	100
110	0.23428	779.33	370.65	2.0120	0.22196	0.28481	0.11480	0.073214	3.8900	110
120	0.30002	767.32	362.51	2.0494	0.20142	0.26250	0.11184	0.071124	3.6908	120
140	0.47249	742.43	345.39	2.1277	0.16752	0.22564	0.10597	0.067083	3.3636	140
160	0.71033	716.12	326.81	2.2127	0.14084	0.19667	0.10013	0.063189	3.1124	160
180	1.0272	687.98	306.31	2.3087	0.11937	0.17350	0.094325	0.059386	2.9216	180
200	1.4378	657.34	283.22	2.4238	0.10170	0.15471	0.088562	0.055586	2.7833	200
220	1.9582	623.05	256.47	2.5756	0.086833	0.13937	0.082853	0.051631	2.6993	220
240	2.6067	582.87	224.10	2.8102	0.073968	0.12690	0.077243	0.047156	2.6911	240
260	3.4064	531.50	181.64	3.3008	0.062322	0.11726	0.071954	0.041014	2.8589	260

Table A.13
Thermophysical Properties of Liquids at Saturation (SI Units)

T °C	P_{sat} kPa	ρ kg/m³	h_{fg} kJ/kg	c_p kJ/ kg-K	μ mPa-s	ν mm²/s	k W/m-K	α mm²/s	Pr	T °C
				Cyclohexane						
10	6.3407	787.96	401.90	1.7893	1.1414	1.4486	0.11984	0.084994	17.044	10
20	10.343	778.60	395.65	1.8357	0.95778	1.2301	0.11772	0.082363	14.935	20
30	16.240	769.15	389.38	1.8826	0.81419	1.0586	0.11559	0.079827	13.261	30
40	24.642	759.61	383.06	1.9299	0.69989	0.92139	0.11345	0.077391	11.906	40
50	36.267	749.95	376.67	1.9777	0.60744	0.80997	0.11132	0.075056	10.791	50
60	51.936	740.17	370.19	2.0260	0.53155	0.71815	0.10920	0.072821	9.8618	60
70	72.566	730.25	363.59	2.0748	0.46845	0.64149	0.10709	0.070682	9.0757	70
80	99.166	720.19	356.85	2.1242	0.41537	0.57676	0.10500	0.068637	8.4031	80
90	132.83	709.96	349.94	2.1745	0.37025	0.52151	0.10294	0.066679	7.8212	90
100	174.73	699.54	342.84	2.2257	0.33154	0.47393	0.10090	0.064806	7.3131	100
110	226.10	688.92	335.50	2.2782	0.29803	0.43261	0.098900	0.063012	6.8654	110
120	288.24	678.07	327.90	2.3323	0.26882	0.39644	0.096936	0.061295	6.4677	120
130	362.51	666.96	319.98	2.3881	0.24316	0.36458	0.095014	0.059653	6.1117	130
140	450.30	655.55	311.72	2.4461	0.22047	0.33632	0.093137	0.058082	5.7904	140
160	672.31	631.60	293.91	2.5703	0.18221	0.28849	0.089538	0.055153	5.2307	160
180	966.43	605.71	273.93	2.7097	0.15116	0.24956	0.086170	0.052501	4.7534	180
200	1345.7	577.00	251.01	2.8735	0.12526	0.21709	0.083070	0.050102	4.3330	200
220	1824.4	544.05	223.93	3.0833	0.10296	0.18924	0.080297	0.047868	3.9534	220
240	2419.0	504.04	190.40	3.4051	0.082864	0.16440	0.078017	0.045457	3.6166	240
				Toluene						
-60	0.0054177	941.08	466.17	1.4936	2.7659	2.9390	0.15229	0.10835	27.126	-60
-40	0.042077	922.37	452.76	1.5274	1.5984	1.7329	0.14761	0.10477	16.540	-40
-20	0.22511	903.83	440.01	1.5723	1.0598	1.1725	0.14253	0.10030	11.691	-20
0	0.90575	885.35	427.75	1.6257	0.76681	0.86611	0.13720	0.095323	9.0860	0
20	2.9189	866.82	415.81	1.6855	0.58660	0.67674	0.13171	0.090150	7.5067	20
40	7.8923	848.12	404.01	1.7501	0.46600	0.54945	0.12615	0.084992	6.4648	40
60	18.540	829.15	392.16	1.8184	0.38047	0.45887	0.12065	0.080020	5.7344	60
80	38.868	809.79	380.07	1.8896	0.31717	0.39166	0.11522	0.075300	5.2014	80
100	74.246	789.93	367.57	1.9632	0.26869	0.34015	0.10995	0.070901	4.7975	100
120	131.37	769.40	354.43	2.0390	0.23046	0.29953	0.10489	0.066860	4.4800	120
140	218.12	748.04	340.46	2.1175	0.19950	0.26670	0.10010	0.063196	4.2203	140
160	343.44	725.61	325.40	2.1993	0.17383	0.23957	0.095618	0.059918	3.9983	160
180	517.19	701.83	308.96	2.2859	0.15207	0.21667	0.091490	0.057027	3.7995	180
200	750.13	676.28	290.75	2.3803	0.13323	0.19700	0.087750	0.054512	3.6138	200
220	1054.0	648.38	270.24	2.4874	0.11658	0.17980	0.084424	0.052347	3.4347	220
240	1441.7	617.21	246.61	2.6175	0.10155	0.16454	0.081531	0.050466	3.2603	240
260	1928.2	581.17	218.52	2.7945	0.087639	0.15080	0.079082	0.048693	3.0969	260
280	2531.6	537.13	183.35	3.0901	0.074250	0.13823	0.077109	0.046457	2.9755	280
300	3275.7	476.60	134.38	3.8919	0.060238	0.12639	0.076019	0.040984	3.0839	300

Table A.14
Thermophysical Properties of Liquids at Saturation (SI Units)

T °C	P_{sat} kPa	ρ kg/m³	h_{fg} kJ/kg	c_p kJ/ kg-K	μ mPa-s	ν mm²/s	k W/m-K	α mm²/s	Pr	T °C
				R22						
−100	2.0102	1571.3	268.26	1.0612	0.89602	0.57023	0.14312	0.085832	6.6436	−100
−90	4.8130	1544.9	262.53	1.0612	0.73051	0.47286	0.13778	0.084043	5.6264	−90
−80	10.372	1518.2	256.84	1.0624	0.60936	0.40137	0.13259	0.082203	4.8826	−80
−70	20.469	1491.2	251.12	1.0655	0.51759	0.34711	0.12752	0.080264	4.3245	−70
−60	37.505	1463.7	245.33	1.0710	0.44603	0.30474	0.12258	0.078197	3.8970	−60
−50	64.530	1435.6	239.39	1.0793	0.38879	0.27082	0.11774	0.075990	3.5639	−50
−40	105.23	1406.8	233.24	1.0905	0.34197	0.24308	0.11299	0.073646	3.3007	−40
−30	163.89	1377.2	226.81	1.1049	0.30289	0.21994	0.10836	0.071209	3.0886	−30
−20	245.31	1346.5	220.02	1.1227	0.26971	0.20030	0.10378	0.068654	2.9175	−20
−10	354.79	1314.7	212.79	1.1439	0.24106	0.18336	0.099250	0.065992	2.7785	−10
0	497.99	1281.5	205.05	1.1692	0.21598	0.16853	0.094743	0.063230	2.6654	0
10	680.95	1246.7	196.69	1.1993	0.19371	0.15538	0.090247	0.060359	2.5743	10
20	910.02	1209.9	187.60	1.2356	0.17370	0.14356	0.085742	0.057353	2.5031	20
30	1191.9	1170.7	177.64	1.2807	0.15547	0.13280	0.081205	0.054161	2.4520	30
40	1533.6	1128.5	166.60	1.3389	0.13867	0.12288	0.076604	0.050696	2.4239	40
50	1942.7	1082.3	154.19	1.4191	0.12296	0.11361	0.071900	0.046812	2.4269	50
60	2427.5	1030.4	139.94	1.5392	0.10799	0.10481	0.067040	0.042270	2.4795	60
70	2997.4	969.74	123.00	1.7434	0.093339	0.096252	0.061948	0.036643	2.6268	70
80	3663.8	893.74	101.57	2.1814	0.078270	0.087576	0.056583	0.029024	3.0174	80
				R32						
−110	1.4525	1363.8	438.66	1.5647	0.66291	0.48607	0.23063	0.10808	4.4975	−110
−100	3.8130	1339.0	429.48	1.5600	0.55456	0.41415	0.22432	0.10739	3.8564	−100
−90	8.8687	1313.9	420.20	1.5586	0.47122	0.35864	0.21740	0.10616	3.3782	−90
−80	18.654	1288.4	410.71	1.5606	0.40512	0.31444	0.20999	0.10444	3.0108	−80
−70	36.067	1262.4	400.92	1.5663	0.35146	0.27842	0.20222	0.10227	2.7223	−70
−60	64.955	1235.7	390.73	1.5758	0.30715	0.24856	0.19421	0.099731	2.4923	−60
−50	110.14	1208.4	380.06	1.5895	0.27004	0.22347	0.18604	0.096860	2.3072	−50
−40	177.41	1180.2	368.79	1.6077	0.23861	0.20219	0.17779	0.093707	2.1576	−40
−30	273.44	1151.0	356.83	1.6311	0.21173	0.18396	0.16954	0.090311	2.0370	−30
−20	405.75	1120.6	344.03	1.6607	0.18852	0.16824	0.16135	0.086700	1.9405	−20
−10	582.63	1088.8	330.25	1.6980	0.16830	0.15458	0.15324	0.082891	1.8648	−10
0	813.10	1055.3	315.30	1.7450	0.15049	0.14261	0.14525	0.078879	1.8079	0
10	1106.9	1019.7	298.92	1.8056	0.13463	0.13204	0.13741	0.074632	1.7692	10
20	1474.6	981.38	280.78	1.8859	0.12033	0.12262	0.12970	0.070075	1.7498	20
30	1927.5	939.62	260.41	1.9973	0.10723	0.11412	0.12210	0.065062	1.7541	30
40	2478.3	893.04	237.09	2.1629	0.094988	0.10637	0.11458	0.059318	1.7931	40
50	3141.2	839.26	209.62	2.4385	0.083217	0.099156	0.10703	0.052300	1.8959	50
60	3933.2	773.31	175.51	3.0007	0.071369	0.092290	0.099377	0.042826	2.1550	60
70	4876.8	680.93	127.78	4.8653	0.058172	0.085431	0.092046	0.027784	3.0748	70

Table A.15
Thermophysical Properties of Liquids at Saturation (SI Units)

T °C	P_{sat} kPa	ρ kg/m³	h_{fg} kJ/kg	c_p kJ/ kg-K	μ mPa-s	ν mm²/s	k W/m-K	α mm²/s	Pr	T °C
				R123						
-100	0.011610	1754.5	220.49	0.92605	3.7133	2.1164	0.11303	0.069566	30.423	-100
-80	0.12548	1709.6	212.07	0.92359	2.0932	1.2244	0.10738	0.068005	18.004	-80
-60	0.80750	1665.1	204.21	0.93199	1.3834	0.83082	0.10200	0.065730	12.640	-60
-40	3.5752	1620.0	196.63	0.94805	0.98641	0.60891	0.096071	0.062555	9.7340	-40
-30	6.7480	1597.0	192.87	0.95776	0.84796	0.53097	0.092972	0.060783	8.7354	-30
-20	11.997	1573.8	189.11	0.96816	0.73539	0.46728	0.089848	0.058969	7.9242	-20
-10	20.247	1550.1	185.30	0.97903	0.64240	0.41441	0.086742	0.057156	7.2506	-10
0	32.645	1526.1	181.44	0.99023	0.56459	0.36995	0.083686	0.055377	6.6807	0
10	50.567	1501.6	177.49	1.0017	0.49878	0.33216	0.080707	0.053654	6.1909	10
20	75.610	1476.6	173.44	1.0135	0.44260	0.29974	0.077821	0.051998	5.7645	20
30	109.58	1451.0	169.27	1.0257	0.39427	0.27172	0.075037	0.050416	5.3896	30
40	154.47	1424.8	164.94	1.0385	0.35240	0.24733	0.072359	0.048906	5.0574	40
60	285.89	1370.0	155.73	1.0663	0.28386	0.20720	0.067311	0.046078	4.4967	60
80	489.09	1311.2	145.54	1.0996	0.23054	0.17582	0.062621	0.043431	4.0484	80
100	785.53	1246.9	134.01	1.1433	0.18809	0.15084	0.058190	0.040818	3.6955	100
120	1199.0	1174.4	120.53	1.2072	0.15336	0.13059	0.053884	0.038006	3.4360	120
140	1756.3	1088.3	104.02	1.3178	0.12382	0.11378	0.049522	0.034532	3.2949	140
160	2490.1	975.68	81.886	1.5836	0.096802	0.099215	0.044842	0.029023	3.4185	160
180	3450.6	765.91	40.596	4.5486	0.064292	0.083942	0.039715	0.011400	7.3635	180
				R125						
-100	3.0883	1688.7	189.96	1.0351	1.1336	0.67133	0.11573	0.066213	10.139	-100
-90	7.2856	1656.2	185.18	1.0450	0.89075	0.53782	0.11103	0.064151	8.3836	-90
-80	15.468	1623.4	180.37	1.0581	0.72062	0.44389	0.10630	0.061885	7.1729	-80
-70	30.076	1589.9	175.47	1.0736	0.59524	0.37438	0.10157	0.059500	6.2921	-70
-60	54.318	1555.7	170.41	1.0912	0.49932	0.32097	0.096846	0.057052	5.6259	-60
-50	92.164	1520.5	165.14	1.1107	0.42375	0.27870	0.092188	0.054591	5.1052	-50
-40	148.30	1484.0	159.59	1.1323	0.36278	0.24446	0.087570	0.052114	4.6908	-40
-30	228.06	1446.1	153.71	1.1565	0.31258	0.21615	0.083016	0.049637	4.3546	-30
-20	337.33	1406.4	147.41	1.1840	0.27051	0.19233	0.078538	0.047162	4.0782	-20
-10	482.52	1364.5	140.60	1.2161	0.23469	0.17199	0.074145	0.044681	3.8494	-10
-5	570.72	1342.6	136.97	1.2344	0.21867	0.16287	0.071982	0.043432	3.7501	-5
0	670.52	1319.8	133.16	1.2547	0.20372	0.15435	0.069840	0.042174	3.6599	0
5	782.88	1296.2	129.14	1.2773	0.18970	0.14636	0.067720	0.040902	3.5782	5
10	908.75	1271.5	124.90	1.3029	0.17651	0.13882	0.065619	0.039610	3.5048	10
20	1205.2	1218.3	115.57	1.3666	0.15221	0.12494	0.061473	0.036923	3.3837	20
30	1568.5	1158.4	104.81	1.4575	0.13008	0.11230	0.057382	0.033988	3.3041	30
40	2008.5	1088.4	92.024	1.6052	0.10940	0.10052	0.053317	0.030518	3.2937	40
50	2536.8	1001.1	75.924	1.9102	0.089205	0.089107	0.049261	0.025760	3.4591	50
60	3170.3	872.09	52.112	3.1392	0.067051	0.076885	0.045865	0.016753	4.5892	60

Table A.16
Thermophysical Properties of Liquids at Saturation (SI Units)

T °C	P_{sat} kPa	ρ kg/m³	h_{fg} kJ/kg	c_p kJ/ kg-K	μ mPa-s	ν mm²/s	k W/m-K	α mm²/s	Pr	T °C
				R134a						
−100	0.55940	1582.4	261.49	1.1842	1.8824	1.1896	0.14323	0.076431	15.564	−100
−80	3.6719	1529.0	249.67	1.1981	1.0203	0.66728	0.13154	0.071806	9.2927	−80
−70	7.9814	1501.9	243.82	1.2096	0.80910	0.53873	0.12603	0.069373	7.7656	−70
−60	15.906	1474.3	237.95	1.2230	0.66051	0.44800	0.12071	0.066943	6.6923	−60
−50	29.451	1446.3	231.98	1.2381	0.55089	0.38089	0.11557	0.064541	5.9016	−50
−40	51.209	1417.7	225.86	1.2546	0.46703	0.32943	0.11059	0.062175	5.2983	−40
−30	84.378	1388.4	219.53	1.2729	0.40095	0.28879	0.10576	0.059846	4.8255	−30
−20	132.73	1358.3	212.92	1.2930	0.34758	0.25590	0.10107	0.057547	4.4469	−20
−10	200.60	1327.1	205.97	1.3156	0.30355	0.22873	0.096491	0.055267	4.1386	−10
0	292.80	1294.8	198.60	1.3410	0.26653	0.20585	0.092013	0.052992	3.8845	0
10	414.61	1261.0	190.74	1.3704	0.23487	0.18626	0.087618	0.050705	3.6734	10
20	571.71	1225.3	182.28	1.4049	0.20737	0.16923	0.083284	0.048381	3.4979	20
30	770.20	1187.5	173.10	1.4465	0.18313	0.15422	0.078992	0.045989	3.3533	30
40	1016.6	1146.7	163.02	1.4984	0.16145	0.14079	0.074716	0.043483	3.2378	40
50	1317.9	1102.3	151.81	1.5661	0.14177	0.12861	0.070427	0.040795	3.1527	50
60	1681.8	1052.9	139.12	1.6602	0.12361	0.11741	0.066091	0.037811	3.1051	60
70	2116.8	996.25	124.37	1.8039	0.10651	0.10691	0.061672	0.034316	3.1153	70
80	2633.2	928.24	106.42	2.0648	0.089846	0.096791	0.057147	0.029815	3.2464	80
100	3972.4	651.18	34.385	17.592	0.047429	0.072835	0.058884	0.0051403	14.169	100
				R227ea						
−100	0.28879	1803.4	164.26	0.95223	1.8003	0.99829	0.090519	0.052712	18.939	−100
−90	0.82377	1774.7	160.44	0.96521	1.4397	0.81121	0.088723	0.051794	15.662	−90
−80	2.0645	1745.7	156.67	0.97846	1.1758	0.67357	0.086772	0.050802	13.259	−80
−70	4.6422	1716.0	152.92	0.99209	0.97631	0.56894	0.084685	0.049743	11.438	−70
−60	9.5257	1685.8	149.15	1.0062	0.82125	0.48717	0.082480	0.048623	10.019	−60
−50	18.087	1654.8	145.35	1.0211	0.69788	0.42172	0.080159	0.047438	8.8900	−50
−40	32.140	1623.1	141.46	1.0369	0.59775	0.36827	0.077740	0.046191	7.9729	−40
−30	53.960	1590.5	137.47	1.0539	0.51512	0.32387	0.075236	0.044886	7.2154	−30
−20	86.272	1556.9	133.33	1.0722	0.44597	0.28645	0.072659	0.043527	6.5808	−20
−10	132.23	1522.1	129.00	1.0921	0.38741	0.25452	0.070021	0.042121	6.0425	−10
0	195.36	1486.0	124.44	1.1141	0.33731	0.22699	0.067335	0.040672	5.5810	0
10	279.57	1448.2	119.58	1.1386	0.29406	0.20305	0.064613	0.039186	5.1817	10
20	389.08	1408.4	114.37	1.1663	0.25640	0.18205	0.061868	0.037665	4.8334	20
30	528.42	1366.2	108.72	1.1983	0.22334	0.16347	0.059114	0.036109	4.5271	30
40	702.45	1320.9	102.52	1.2364	0.19403	0.14690	0.056364	0.034514	4.2563	40
50	916.39	1271.4	95.639	1.2837	0.16778	0.13196	0.053634	0.032861	4.0158	50
60	1175.9	1216.5	87.846	1.3463	0.14393	0.11832	0.050941	0.031104	3.8039	60
80	1858.3	1078.2	67.813	1.5960	0.10076	0.093454	0.045838	0.026638	3.5083	80
100	2821.6	786.76	26.802	7.1411	0.051620	0.065611	0.048251	0.0085882	7.6397	100

Table A.17
Thermophysical Properties of Ethylene Glycol–Water Solutions (SI Units)

Conc mass %	T_{freeze} °C	T °C	ρ kg/m³	c_p kJ/kg-K	μ mPa-s	ν mm²/s	k W/m-K	α mm²/s	Pr	T °C
10	−3.357	0	1013.6	4.0374	2.3449	2.3134	0.52328	0.12787	18.092	0
		20	1010.6	4.0459	1.2745	1.2612	0.55289	0.13523	9.3265	20
		40	1004.3	4.0619	0.80135	0.79789	0.57833	0.14176	5.6283	40
		60	995.23	4.0831	0.56338	0.56608	0.59961	0.14756	3.8363	60
		80	983.55	4.1071	0.42808	0.43524	0.61678	0.15269	2.8505	80
		100	969.62	4.1316	0.33982	0.35047	0.62987	0.15723	2.2290	100
		120	953.75	4.1543	0.27241	0.28562	0.63890	0.16125	1.7712	120
		140	936.25	4.1727	0.21315	0.22766	0.64390	0.16482	1.3813	140
		160	917.44	4.1847	0.15735	0.17151	0.64490	0.16798	1.0211	160
		180	897.63	4.1878	0.10594	0.11802	0.64194	0.17077	0.69113	180
20	−7.949	0	1029.1	3.8608	3.1776	3.0878	0.48414	0.12186	25.340	0
		20	1024.1	3.8962	1.6624	1.6233	0.50766	0.12723	12.759	20
		40	1016.4	3.9328	1.0133	0.99691	0.52908	0.13236	7.5318	40
		60	1006.2	3.9693	0.69688	0.69258	0.54832	0.13729	5.0448	60
		80	993.76	4.0048	0.52381	0.52710	0.56533	0.14205	3.7106	80
		100	979.32	4.0379	0.41676	0.42556	0.58004	0.14668	2.9012	100
		120	963.11	4.0675	0.33995	0.35297	0.59241	0.15122	2.3341	120
		140	945.40	4.0926	0.27534	0.29124	0.60235	0.15568	1.8708	140
		160	926.42	4.1118	0.21448	0.23152	0.60982	0.16009	1.4462	160
		180	906.43	4.1240	0.15563	0.17169	0.61475	0.16445	1.0440	180
30	−14.58	0	1045.0	3.6581	4.2976	4.1126	0.44592	0.11665	35.255	0
		20	1038.0	3.7183	2.1664	2.0870	0.46490	0.12045	17.327	20
		40	1028.8	3.7754	1.2856	1.2496	0.48303	0.12436	10.048	40
		60	1017.5	3.8287	0.86604	0.85118	0.50018	0.12840	6.6293	60
		80	1004.3	3.8777	0.63884	0.63613	0.51624	0.13257	4.7986	80
		100	989.41	3.9215	0.49766	0.50299	0.53106	0.13687	3.6749	100
		120	973.15	3.9596	0.39488	0.40577	0.54452	0.14131	2.8715	120
		140	955.71	3.9913	0.30780	0.32207	0.55649	0.14589	2.2076	140
		160	937.30	4.0158	0.22732	0.24253	0.56684	0.15059	1.6105	160
		180	918.16	4.0326	0.15342	0.16709	0.57544	0.15542	1.0751	180
40	−23.81	0	1060.4	3.4342	5.8107	5.4798	0.40990	0.11256	48.682	0
		20	1051.9	3.5190	2.8191	2.6801	0.42530	0.11490	23.326	20
		40	1041.4	3.5978	1.6351	1.5702	0.44050	0.11757	13.355	40
		60	1029.1	3.6697	1.0821	1.0515	0.45535	0.12057	8.7209	60
		80	1015.4	3.7338	0.77997	0.76814	0.46966	0.12388	6.2008	80
		100	1000.5	3.7893	0.58440	0.58413	0.48326	0.12748	4.5823	100
		120	984.55	3.8353	0.43445	0.44126	0.49598	0.13135	3.3594	120
		140	967.93	3.8708	0.30586	0.31599	0.50764	0.13549	2.3322	140
		160	950.85	3.8952	0.19464	0.20470	0.51806	0.13988	1.4634	160
		180	933.57	3.9073	0.10686	0.11447	0.52708	0.14449	0.79220	180

Table A.18
Thermophysical Properties of Propylene Glycol–Water Solutions (SI Units)

Conc mass %	T_{freeze} °C	T °C	ρ kg/m³	c_p kJ/kg-K	μ mPa-s	ν mm²/s	k W/m-K	α mm²/s	Pr	T °C
10	−2.867	0	1009.3	4.0590	2.7446	2.7192	0.51549	0.12583	21.611	0
		20	1006.2	4.0763	1.4323	1.4234	0.54381	0.13258	10.736	20
		40	999.42	4.1011	0.88380	0.88431	0.56879	0.13877	6.3724	40
		60	989.81	4.1303	0.61366	0.61998	0.59018	0.14436	4.2947	60
		80	978.22	4.1610	0.45625	0.46641	0.60775	0.14931	3.1238	80
		100	965.50	4.1901	0.34567	0.35802	0.62128	0.15357	2.3313	100
		120	952.50	4.2147	0.25395	0.26661	0.63053	0.15706	1.6975	120
		140	940.08	4.2317	0.17216	0.18314	0.63528	0.15969	1.1468	140
		160	929.07	4.2380	0.10249	0.11032	0.63528	0.16134	0.68374	160
		180	920.33	4.2308	0.050990	0.055403	0.63031	0.16188	0.34226	180
20	−7.173	0	1020.1	3.9359	4.3124	4.2275	0.47112	0.11734	36.028	0
		20	1014.8	3.9768	2.0301	2.0005	0.49221	0.12197	16.402	20
		40	1006.4	4.0190	1.1693	1.1619	0.51228	0.12666	9.1737	40
		60	995.59	4.0612	0.77776	0.78120	0.53100	0.13133	5.9484	60
		80	983.18	4.1016	0.56382	0.57347	0.54808	0.13591	4.2195	80
		100	969.86	4.1389	0.42044	0.43350	0.56320	0.14030	3.0897	100
		120	956.38	4.1714	0.30437	0.31826	0.57607	0.14440	2.2040	120
		140	943.47	4.1976	0.20190	0.21400	0.58636	0.14806	1.4453	140
		160	931.86	4.2159	0.11581	0.12428	0.59376	0.15114	0.82230	160
		180	922.28	4.2248	0.054220	0.058789	0.59798	0.15347	0.38307	180
30	−12.79	0	1031.6	3.8026	7.1171	6.8994	0.42845	0.10923	63.166	0
		20	1023.8	3.8570	2.9650	2.8961	0.44443	0.11255	25.732	20
		40	1013.4	3.9104	1.5731	1.5523	0.46056	0.11622	13.357	40
		60	1001.2	3.9623	0.99458	0.99343	0.47657	0.12014	8.2692	60
		80	987.60	4.0122	0.70106	0.70987	0.49215	0.12420	5.7153	80
		100	973.42	4.0595	0.51549	0.52957	0.50703	0.12831	4.1273	100
		120	959.27	4.1036	0.36996	0.38567	0.52089	0.13232	2.9145	120
		140	945.79	4.1441	0.24246	0.25636	0.53346	0.13611	1.8835	140
		160	933.65	4.1803	0.13577	0.14542	0.54445	0.13950	1.0425	160
		180	923.48	4.2118	0.060780	0.065816	0.55356	0.14232	0.46245	180
40	−20.57	0	1042.4	3.6416	11.879	11.397	0.38784	0.10218	111.54	0
		20	1032.3	3.7067	4.3838	4.2467	0.40026	0.10461	40.597	20
		40	1020.1	3.7708	2.1408	2.0987	0.41321	0.10743	19.536	40
		60	1006.3	3.8339	1.2827	1.2746	0.42651	0.11055	11.530	60
		80	991.63	3.8958	0.87425	0.88163	0.43998	0.11389	7.7411	80
		100	976.62	3.9567	0.62846	0.64351	0.45345	0.11735	5.4838	100
		120	961.88	4.0164	0.44177	0.45928	0.46673	0.12081	3.8016	120
		140	948.00	4.0749	0.28154	0.29698	0.47964	0.12416	2.3919	140
		160	935.59	4.1324	0.15082	0.16120	0.49201	0.12726	1.2667	160
		180	925.25	4.1886	0.062967	0.068054	0.50367	0.12996	0.52365	180

Table A.19
Thermophysical Properties of Methanol–Water Solutions (SI Units)

Conc mass %	T_{freeze} °C	T °C	ρ kg/m³	c_p kJ/kg-K	μ mPa-s	ν mm²/s	k W/m-K	α mm²/s	Pr	T °C
10	−6.548	−5	984.45	4.2513	3.1070	3.1561	0.50195	0.11993	26.315	−5
		0	984.39	4.2408	2.5348	2.5750	0.50955	0.12206	21.097	0
		5	984.05	4.2301	2.1026	2.1367	0.51700	0.12420	17.204	5
		10	983.46	4.2192	1.7709	1.8007	0.52430	0.12636	14.251	10
		15	982.62	4.2083	1.5122	1.5389	0.53144	0.12852	11.974	15
		20	981.54	4.1975	1.3073	1.3319	0.53840	0.13068	10.192	20
		25	980.23	4.1871	1.1426	1.1657	0.54519	0.13283	8.7754	25
		30	978.70	4.1771	1.0083	1.0302	0.55179	0.13497	7.6326	30
		35	976.96	4.1677	0.89694	0.91809	0.55819	0.13709	6.6969	35
		40	975.02	4.1590	0.80327	0.82385	0.56439	0.13918	5.9193	40
20	−15.09	−15	974.91	4.0064	6.4493	6.6153	0.44563	0.11409	57.983	−15
		−10	974.30	4.0296	5.0298	5.1625	0.45117	0.11492	44.923	−10
		−5	973.51	4.0502	3.9961	4.1048	0.45665	0.11582	35.442	−5
		0	972.52	4.0681	3.2301	3.3214	0.46209	0.11680	28.437	0
		5	971.35	4.0832	2.6531	2.7313	0.46748	0.11787	23.173	5
		10	970.00	4.0953	2.2115	2.2798	0.47283	0.11903	19.154	10
		15	968.48	4.1043	1.8684	1.9292	0.47813	0.12029	16.038	15
		20	966.79	4.1101	1.5979	1.6528	0.48339	0.12165	13.586	20
		30	962.90	4.1114	1.2063	1.2527	0.49378	0.12473	10.044	30
		40	958.37	4.0980	0.94179	0.98270	0.50402	0.12833	7.6574	40
30	−25.69	−25	968.53	3.6331	12.405	12.808	0.39857	0.11327	113.08	−25
		−20	967.11	3.6871	9.3562	9.6744	0.40231	0.11283	85.747	−20
		−15	965.56	3.7381	7.1951	7.4518	0.40604	0.11249	66.241	−15
		−10	963.90	3.7859	5.6353	5.8463	0.40975	0.11228	52.068	−10
		−5	962.12	3.8301	4.4897	4.6665	0.41345	0.11220	41.593	−5
		0	960.22	3.8705	3.6346	3.7851	0.41715	0.11224	33.723	0
		10	956.08	3.9382	2.4870	2.6013	0.42458	0.11276	23.068	10
		20	951.48	3.9865	1.7888	1.8800	0.43210	0.11392	16.503	20
		30	946.42	4.0127	1.3399	1.4157	0.43976	0.11580	12.226	30
		40	940.90	4.0144	1.0355	1.1006	0.44762	0.11851	9.2868	40
40	−38.70	−35	962.74	3.2421	20.860	21.668	0.35960	0.11521	188.07	−35
		−30	960.50	3.3122	15.400	16.034	0.36196	0.11378	140.92	−30
		−25	958.19	3.3800	11.597	12.103	0.36430	0.11249	107.60	−25
		−20	955.83	3.4451	8.8989	9.3101	0.36663	0.11134	83.622	−20
		−10	950.91	3.5659	5.5182	5.8030	0.37127	0.10949	53.001	−10
		0	945.73	3.6712	3.6395	3.8484	0.37595	0.10828	35.541	0
		10	940.29	3.7578	2.5306	2.6913	0.38074	0.10775	24.976	10
		20	934.57	3.8224	1.8385	1.9673	0.38571	0.10797	18.220	20
		30	928.56	3.8618	1.3834	1.4898	0.39093	0.10902	13.666	30
		40	922.27	3.8728	1.0684	1.1585	0.39646	0.11100	10.437	40

Table A.20

Thermophysical Properties of Ethanol–Water Solutions (SI Units)

Conc mass %	T_{freeze} °C	T °C	ρ kg/m³	c_p kJ/kg-K	μ mPa-s	ν mm²/s	k W/m-K	α mm²/s	Pr	T °C
10	−4.388	0	984.99	4.4017	3.3170	3.3676	0.50270	0.11595	29.044	0
		5	984.67	4.3764	2.6376	2.6787	0.50927	0.11818	22.667	5
		10	984.02	4.3506	2.1531	2.1881	0.51577	0.12048	18.162	10
		15	983.07	4.3257	1.7975	1.8284	0.52219	0.12280	14.890	15
		20	981.84	4.3033	1.5287	1.5570	0.52852	0.12509	12.447	20
		25	980.37	4.2850	1.3194	1.3459	0.53476	0.12730	10.572	25
		30	978.67	4.2722	1.1513	1.1764	0.54089	0.12937	9.0939	30
		35	976.77	4.2666	1.0118	1.0359	0.54689	0.13123	7.8939	35
		40	974.70	4.2696	0.89217	0.91532	0.55276	0.13282	6.8914	40
		45	972.49	4.2829	0.78623	0.80846	0.55848	0.13408	6.0295	45
20	−11.13	−10	977.60	4.3827	9.5644	9.7836	0.43871	0.10240	95.547	−10
		−5	976.83	4.3774	6.9767	7.1422	0.44324	0.10366	68.902	−5
		0	975.77	4.3694	5.2493	5.3796	0.44782	0.10504	51.217	0
		5	974.42	4.3595	4.0621	4.1687	0.45243	0.10650	39.142	5
		10	972.82	4.3488	3.2237	3.3137	0.45706	0.10803	30.673	10
		15	970.98	4.3382	2.6161	2.6943	0.46169	0.10960	24.582	15
		20	968.92	4.3287	2.1648	2.2342	0.46631	0.11118	20.095	20
		25	966.65	4.3211	1.8212	1.8841	0.47091	0.11274	16.712	25
		30	964.20	4.3165	1.5534	1.6111	0.47546	0.11424	14.102	30
		40	958.81	4.3200	1.1640	1.2140	0.48440	0.11695	10.381	40
30	−20.14	−20	973.10	4.0715	25.213	25.910	0.38499	0.097171	266.65	−20
		−15	971.36	4.0998	17.304	17.814	0.38802	0.097434	182.83	−15
		−10	969.42	4.1242	12.258	12.644	0.39113	0.097830	129.25	−10
		−5	967.28	4.1451	8.9429	9.2453	0.39431	0.098343	94.012	−5
		0	964.96	4.1632	6.7046	6.9480	0.39754	0.098956	70.213	0
		5	962.46	4.1789	5.1538	5.3548	0.40081	0.099655	53.733	5
		10	959.79	4.1926	4.0530	4.2227	0.40411	0.10042	42.049	10
		20	954.01	4.2161	2.6599	2.7881	0.41071	0.10211	27.305	20
		30	947.72	4.2376	1.8618	1.9645	0.41722	0.10389	18.910	30
		40	940.99	4.2610	1.3652	1.4508	0.42353	0.10563	13.735	40
40	−29.53	−25	965.29	3.6450	35.834	37.122	0.34154	0.097071	382.43	−25
		−20	962.30	3.7020	24.550	25.512	0.34361	0.096455	264.50	−20
		−15	959.22	3.7551	17.312	18.048	0.34574	0.095987	188.03	−15
		−10	956.04	3.8045	12.542	13.118	0.34791	0.095653	137.14	−10
		−5	952.78	3.8503	9.3164	9.7782	0.35012	0.095441	102.45	−5
		0	949.43	3.8926	7.0833	7.4605	0.35235	0.095339	78.252	0
		10	942.51	3.9672	4.3572	4.6229	0.35681	0.095427	48.445	10
		20	935.32	4.0293	2.8757	3.0746	0.36120	0.095843	32.080	20
		30	927.89	4.0799	2.0062	2.1621	0.36541	0.096525	22.399	30
		40	920.26	4.1199	1.4573	1.5836	0.36934	0.097417	16.256	40

Table A.21
Thermophysical Properties of Calcium Chloride–Water Solutions (SI Units)

Conc mass %	T_{freeze} °C	T °C	ρ kg/m³	c_p kJ/kg-K	μ mPa-s	ν mm²/s	k W/m-K	α mm²/s	Pr	T °C
5	−2.361	0	1043.1	3.8720	1.9790	1.8971	0.55904	0.13841	13.707	0
		5	1041.7	3.8734	1.6917	1.6240	0.56753	0.14066	11.546	5
		10	1040.8	3.8764	1.4659	1.4084	0.57601	0.14276	9.8652	10
		15	1040.2	3.8809	1.2855	1.2358	0.58444	0.14477	8.5364	15
		20	1039.6	3.8866	1.1392	1.0958	0.59281	0.14671	7.4687	20
		25	1038.8	3.8932	1.0185	0.98048	0.60108	0.14863	6.5967	25
		30	1037.4	3.9005	0.91727	0.88422	0.60924	0.15057	5.8725	30
		35	1035.2	3.9083	0.83087	0.80265	0.61726	0.15257	5.2608	35
		40	1031.9	3.9164	0.75577	0.73243	0.62512	0.15469	4.7349	40
		45	1027.2	3.9244	0.68928	0.67102	0.63280	0.15698	4.2747	45
10	−5.840	−5	1089.8	3.5602	2.6193	2.4034	0.54737	0.14108	17.036	−5
		0	1088.3	3.5688	2.2273	2.0466	0.55523	0.14296	14.316	0
		5	1087.4	3.5781	1.9163	1.7623	0.56312	0.14473	12.176	5
		10	1086.8	3.5879	1.6668	1.5336	0.57102	0.14644	10.473	10
		15	1086.4	3.5981	1.4642	1.3478	0.57894	0.14811	9.1002	15
		20	1085.9	3.6086	1.2981	1.1954	0.58687	0.14977	7.9817	20
		25	1085.0	3.6192	1.1603	1.0693	0.59480	0.15147	7.0599	25
		30	1083.6	3.6298	1.0447	0.96412	0.60273	0.15324	6.2916	30
		35	1081.4	3.6403	0.94679	0.87553	0.61065	0.15512	5.6441	35
		40	1078.1	3.6505	0.86283	0.80030	0.61857	0.15716	5.0921	40
15	−11.05	−10	1137.8	3.2651	3.5905	3.1557	0.53586	0.14424	21.878	−10
		−5	1136.3	3.2792	3.0425	2.6774	0.54333	0.14581	18.362	−5
		0	1135.4	3.2929	2.6052	2.2945	0.55080	0.14732	15.575	0
		5	1134.8	3.3063	2.2529	1.9852	0.55829	0.14880	13.342	5
		10	1134.4	3.3193	1.9664	1.7335	0.56581	0.15027	11.536	10
		15	1133.9	3.3319	1.7315	1.5270	0.57335	0.15176	10.062	15
		20	1133.2	3.3440	1.5371	1.3564	0.58093	0.15330	8.8479	20
		25	1132.1	3.3556	1.3749	1.2145	0.58856	0.15493	7.8389	25
		30	1130.4	3.3667	1.2384	1.0956	0.59623	0.15667	6.9930	30
		40	1124.6	3.3868	1.0238	0.91041	0.61175	0.16062	5.6681	40
20	−18.26	−15	1188.4	3.0056	5.1425	4.3271	0.52391	0.14667	29.502	−15
		−10	1186.9	3.0232	4.3363	3.6534	0.53117	0.14803	24.680	−10
		−5	1185.7	3.0399	3.6963	3.1173	0.53841	0.14937	20.870	−5
		0	1184.8	3.0558	3.1830	2.6866	0.54564	0.15071	17.826	0
		5	1183.9	3.0709	2.7673	2.3374	0.55287	0.15207	15.371	5
		10	1183.0	3.0851	2.4273	2.0519	0.56011	0.15347	13.370	10
		15	1181.9	3.0985	2.1468	1.8164	0.56737	0.15493	11.724	15
		20	1180.6	3.1110	1.9131	1.6205	0.57466	0.15647	10.357	20
		30	1176.5	3.1336	1.5504	1.3178	0.58936	0.15986	8.2431	30
		40	1169.8	3.1527	1.2852	1.0986	0.60431	0.16385	6.7047	40

Table A.22
Thermophysical Properties of Magnesium Chloride–Water Solutions (SI Units)

Conc mass %	T_{freeze} °C	T °C	ρ kg/m³	c_p kJ/kg-K	μ mPa-s	ν mm²/s	k W/m-K	α mm²/s	Pr	T °C
5	−2.999	0	1043.8	3.8744	2.2022	2.1099	0.55226	0.13656	15.450	0
		5	1043.2	3.8786	1.8527	1.7760	0.56160	0.13880	12.795	5
		10	1042.3	3.8833	1.5858	1.5214	0.57057	0.14096	10.793	10
		15	1041.3	3.8885	1.3774	1.3228	0.57921	0.14305	9.2467	15
		20	1040.0	3.8939	1.2107	1.1641	0.58757	0.14509	8.0231	20
		25	1038.6	3.8993	1.0741	1.0342	0.59570	0.14710	7.0309	25
		30	1037.0	3.9047	0.95932	0.92513	0.60364	0.14908	6.2054	30
		35	1035.2	3.9098	0.86025	0.83096	0.61145	0.15106	5.5007	35
		40	1033.5	3.9145	0.77246	0.74746	0.61915	0.15305	4.8837	40
		45	1031.6	3.9185	0.69275	0.67153	0.62682	0.15506	4.3306	45
10	−8.283	−5	1089.9	3.5633	3.3696	3.0916	0.53090	0.13670	22.616	−5
		0	1089.0	3.5731	2.7921	2.5640	0.54025	0.13884	18.467	0
		5	1087.9	3.5826	2.3499	2.1601	0.54949	0.14099	15.321	5
		10	1086.6	3.5917	2.0031	1.8435	0.55860	0.14313	12.880	10
		15	1085.1	3.6004	1.7246	1.5893	0.56755	0.14527	10.941	15
		20	1083.5	3.6088	1.4955	1.3803	0.57633	0.14740	9.3643	20
		25	1081.7	3.6169	1.3025	1.2041	0.58492	0.14951	8.0539	25
		30	1079.7	3.6247	1.1361	1.0522	0.59330	0.15159	6.9410	30
		35	1077.6	3.6323	0.98975	0.91844	0.60144	0.15365	5.9774	35
		40	1075.4	3.6395	0.85876	0.79853	0.60933	0.15568	5.1294	40
15	−16.79	−15	1137.4	3.2601	6.2748	5.5168	0.50140	0.13522	40.799	−15
		−10	1136.6	3.2732	5.1741	4.5521	0.50986	0.13704	33.217	−10
		−5	1135.7	3.2858	4.3202	3.8042	0.51850	0.13895	27.378	−5
		0	1134.5	3.2981	3.6455	3.2133	0.52726	0.14091	22.803	0
		5	1133.2	3.3100	3.1027	2.7380	0.53609	0.14293	19.157	5
		10	1131.7	3.3218	2.6581	2.3488	0.54495	0.14497	16.203	10
		15	1130.0	3.3333	2.2877	2.0245	0.55377	0.14702	13.771	15
		20	1128.2	3.3448	1.9742	1.7499	0.56251	0.14907	11.739	20
		30	1124.1	3.3677	1.4700	1.3077	0.57951	0.15308	8.5428	30
		40	1119.6	3.3910	1.0802	0.96486	0.59555	0.15686	6.1509	40
20	−28.71	−25	1187.5	2.9861	12.758	10.744	0.47376	0.13360	80.416	−25
		−20	1187.1	2.9996	10.421	8.7787	0.48133	0.13517	64.945	−20
		−15	1186.3	3.0133	8.5696	7.2235	0.48918	0.13684	52.787	−15
		−10	1185.3	3.0269	7.0933	5.9844	0.49727	0.13860	43.178	−10
		−5	1184.0	3.0407	5.9089	4.9906	0.50554	0.14042	35.541	−5
		0	1182.5	3.0545	4.9529	4.1885	0.51395	0.14229	29.436	0
		10	1178.9	3.0822	3.5428	3.0051	0.53098	0.14612	20.565	10
		20	1174.9	3.1100	2.5924	2.2065	0.54796	0.14996	14.713	20
		30	1170.6	3.1376	1.9379	1.6554	0.56448	0.15368	10.771	30
		40	1166.4	3.1651	1.4777	1.2669	0.58015	0.15714	8.0620	40

Table A.23
Thermophysical Properties of Sodium Chloride–Water Solutions (SI Units)

Conc mass %	T_{freeze} °C	T °C	ρ kg/m³	c_p kJ/kg-K	μ mPa-s	ν mm²/s	k W/m-K	α mm²/s	Pr	T °C
5	−3.054	0	1038.1	3.9121	1.8918	1.8223	0.55832	0.13747	13.256	0
		5	1037.4	3.9157	1.6223	1.5638	0.56748	0.13970	11.194	5
		10	1036.4	3.9196	1.4046	1.3552	0.57641	0.14189	9.5512	10
		15	1035.3	3.9237	1.2279	1.1860	0.58511	0.14404	8.2339	15
		20	1034.0	3.9281	1.0837	1.0481	0.59357	0.14615	7.1718	20
		25	1032.5	3.9327	0.96578	0.93542	0.60180	0.14822	6.3112	25
		30	1030.7	3.9376	0.86897	0.84306	0.60980	0.15025	5.6111	30
		35	1028.8	3.9428	0.78943	0.76730	0.61757	0.15224	5.0400	35
		40	1026.7	3.9483	0.72409	0.70523	0.62511	0.15420	4.5735	40
		45	1024.5	3.9540	0.67059	0.65457	0.63241	0.15612	4.1926	45
10	−6.553	−5	1078.0	3.6780	2.4367	2.2604	0.54719	0.13801	16.378	−5
		0	1076.8	3.6883	2.0715	1.9238	0.55567	0.13992	13.750	0
		5	1075.4	3.6979	1.7784	1.6537	0.56406	0.14184	11.659	5
		10	1073.9	3.7068	1.5417	1.4356	0.57236	0.14378	9.9847	10
		15	1072.3	3.7150	1.3498	1.2587	0.58058	0.14574	8.6369	15
		20	1070.6	3.7226	1.1933	1.1146	0.58871	0.14772	7.5457	20
		25	1068.7	3.7294	1.0654	0.99689	0.59675	0.14972	6.6583	25
		30	1066.7	3.7356	0.96053	0.90047	0.60470	0.15175	5.9339	30
		35	1064.6	3.7411	0.87452	0.82148	0.61256	0.15381	5.3410	35
		40	1062.3	3.7459	0.80405	0.75689	0.62033	0.15589	4.8553	40
15	−10.90	−10	1119.6	3.4952	3.2795	2.9292	0.53646	0.13709	21.367	−10
		−5	1118.0	3.5075	2.7510	2.4607	0.54432	0.13881	17.727	−5
		0	1116.2	3.5188	2.3318	2.0890	0.55219	0.14059	14.859	0
		5	1114.4	3.5289	1.9972	1.7921	0.56006	0.14241	12.584	5
		10	1112.5	3.5380	1.7286	1.5537	0.56793	0.14429	10.768	10
		15	1110.5	3.5460	1.5118	1.3613	0.57581	0.14622	9.3097	15
		20	1108.5	3.5529	1.3360	1.2053	0.58370	0.14821	8.1320	20
		25	1106.3	3.5588	1.1930	1.0784	0.59159	0.15026	7.1769	25
		30	1104.0	3.5635	1.0766	0.97510	0.59948	0.15237	6.3995	30
		40	1099.3	3.5699	0.90447	0.82279	0.61528	0.15679	5.2478	40
20	−16.46	−15	1162.7	3.3533	4.6494	3.9989	0.52535	0.13475	29.677	−15
		−10	1160.7	3.3642	3.8326	3.3019	0.53274	0.13643	24.203	−10
		−5	1158.7	3.3742	3.1960	2.7582	0.54019	0.13816	19.963	−5
		0	1156.7	3.3833	2.6960	2.3308	0.54768	0.13995	16.655	0
		5	1154.5	3.3915	2.3006	1.9927	0.55521	0.14179	14.053	5
		10	1152.3	3.3988	1.9860	1.7234	0.56280	0.14370	11.993	10
		15	1150.1	3.4052	1.7342	1.5079	0.57043	0.14566	10.352	15
		20	1147.8	3.4106	1.5319	1.3347	0.57810	0.14768	9.0379	20
		30	1142.9	3.4189	1.2375	1.0827	0.59359	0.15191	7.1272	30
		40	1137.8	3.4234	1.0467	0.91998	0.60927	0.15642	5.8816	40

Table A.24
Thermophysical Properties of Dry Air (SI Units)

T °C	ρ kg/m³	v m³/kg	u kJ/kg	h kJ/kg	s kJ/kg-K	c_p kJ/kg-K	μ mPa-s	k W/m-K	Pr	T °C
				P = 1 kPa (0.001 MPa)						
0	0.012751	78.425	195.14	273.57	8.0991	1.0039	0.017241	0.023970	0.72210	0
10	0.012301	81.294	202.31	283.61	8.1352	1.0043	0.017740	0.024723	0.72064	10
20	0.011881	84.168	209.49	293.65	8.1700	1.0047	0.018232	0.025467	0.71928	20
30	0.011489	87.040	216.66	303.70	8.2037	1.0052	0.018717	0.026202	0.71803	30
40	0.011122	89.912	223.85	313.76	8.2364	1.0057	0.019195	0.026929	0.71687	40
50	0.010778	92.782	231.04	323.82	8.2680	1.0063	0.019667	0.027649	0.71581	50
60	0.010454	95.657	238.23	333.88	8.2987	1.0070	0.020132	0.028361	0.71485	60
70	0.010150	98.522	245.43	343.96	8.3285	1.0078	0.020592	0.029066	0.71397	70
80	0.0098624	101.40	252.64	354.04	8.3574	1.0086	0.021046	0.029763	0.71319	80
90	0.0095908	104.27	259.86	364.13	8.3856	1.0095	0.021494	0.030454	0.71250	90
100	0.0093338	107.14	267.09	374.23	8.4130	1.0105	0.021937	0.031139	0.71190	100
150	0.0082309	121.49	303.41	424.90	8.5405	1.0167	0.024076	0.034469	0.71010	150
200	0.0073611	135.85	340.08	475.93	8.6544	1.0247	0.026103	0.037664	0.71013	200
250	0.0066576	150.20	377.19	527.39	8.7578	1.0342	0.028034	0.040743	0.71163	250
300	0.0060768	164.56	414.81	579.37	8.8527	1.0450	0.029883	0.043721	0.71424	300
350	0.0055892	178.92	452.99	631.91	8.9406	1.0566	0.031658	0.046612	0.71760	350
400	0.0051740	193.27	491.76	685.03	9.0226	1.0685	0.033370	0.049425	0.72142	400
600	0.0039889	250.70	652.80	903.49	9.3065	1.1153	0.039708	0.060071	0.73722	600
800	0.0032455	308.12	822.50	1130.6	9.5405	1.1544	0.045453	0.070012	0.74948	800
				P = 101.325 kPa (1 atm)						
0	1.2927	0.77357	194.91	273.29	6.7722	1.0059	0.017258	0.024009	0.72307	0
10	1.2469	0.80199	202.09	283.35	6.8084	1.0061	0.017756	0.024760	0.72151	10
20	1.2043	0.83036	209.28	293.41	6.8433	1.0064	0.018247	0.025503	0.72007	20
30	1.1644	0.85881	216.46	303.48	6.8771	1.0067	0.018731	0.026237	0.71875	30
40	1.1272	0.88715	223.66	313.55	6.9098	1.0072	0.019209	0.026963	0.71752	40
50	1.0922	0.91558	230.85	323.62	6.9414	1.0077	0.019680	0.027681	0.71641	50
60	1.0594	0.94393	238.06	333.70	6.9721	1.0083	0.020145	0.028392	0.71539	60
70	1.0284	0.97238	245.27	343.79	7.0020	1.0089	0.020604	0.029096	0.71448	70
80	0.99926	1.0007	252.48	353.88	7.0310	1.0097	0.021058	0.029793	0.71366	80
90	0.97170	1.0291	259.71	363.98	7.0592	1.0105	0.021505	0.030483	0.71293	90
100	0.94563	1.0575	266.94	374.09	7.0866	1.0115	0.021948	0.031167	0.71229	100
150	0.83378	1.1994	303.28	424.81	7.2142	1.0174	0.024085	0.034494	0.71038	150
200	0.74562	1.3412	339.97	475.86	7.3282	1.0252	0.026111	0.037686	0.71033	200
250	0.67433	1.4830	377.09	527.35	7.4317	1.0347	0.028042	0.040763	0.71177	250
300	0.61549	1.6247	414.73	579.35	7.5266	1.0454	0.029889	0.043740	0.71434	300
350	0.56611	1.7664	452.92	631.90	7.6145	1.0568	0.031664	0.046629	0.71767	350
400	0.52406	1.9082	491.70	685.04	7.6965	1.0688	0.033375	0.049441	0.72147	400
600	0.40403	2.4751	652.75	903.54	7.9804	1.1154	0.039712	0.060083	0.73723	600
800	0.32875	3.0418	822.46	1130.7	8.2145	1.1545	0.045456	0.070021	0.74948	800

Table A.25
Thermophysical Properties of Dry Air (SI Units)

T °C	ρ kg/m³	v m³/kg	u kJ/kg	h kJ/kg	s kJ/ kg-K	c_p kJ/ kg-K	μ mPa-s	k W/m-K	Pr	T °C
					P = 500 kPa (0.05 MPa)					
0	6.3933	0.15641	193.99	272.20	6.3106	1.0140	0.017324	0.024169	0.72679	0
10	6.1638	0.16224	201.22	282.34	6.3470	1.0135	0.017820	0.024915	0.72487	10
20	5.9505	0.16805	208.44	292.47	6.3822	1.0132	0.018308	0.025651	0.72312	20
30	5.7516	0.17386	215.67	302.60	6.4162	1.0130	0.018790	0.026380	0.72151	30
40	5.5658	0.17967	222.90	312.73	6.4490	1.0129	0.019265	0.027102	0.72004	40
50	5.3917	0.18547	230.12	322.86	6.4809	1.0130	0.019734	0.027816	0.71871	50
60	5.2283	0.19127	237.36	332.99	6.5117	1.0132	0.020198	0.028522	0.71750	60
70	5.0746	0.19706	244.59	343.12	6.5417	1.0136	0.020655	0.029222	0.71641	70
80	4.9298	0.20285	251.84	353.26	6.5708	1.0140	0.021106	0.029915	0.71543	80
90	4.7931	0.20863	259.09	363.41	6.5992	1.0146	0.021553	0.030602	0.71457	90
100	4.6638	0.21442	266.35	373.55	6.6267	1.0153	0.021994	0.031282	0.71380	100
150	4.1101	0.24330	302.78	424.43	6.7547	1.0202	0.024124	0.034595	0.71141	150
200	3.6745	0.27215	339.54	475.61	6.8690	1.0274	0.026146	0.037777	0.71106	200
250	3.3228	0.30095	376.72	527.20	6.9726	1.0364	0.028072	0.040844	0.71230	250
300	3.0326	0.32975	414.40	579.27	7.0677	1.0467	0.029916	0.043814	0.71472	300
350	2.7892	0.35853	452.63	631.89	7.1557	1.0580	0.031688	0.046697	0.71796	350
400	2.5821	0.38728	491.44	685.08	7.2378	1.0697	0.033397	0.049504	0.72168	400
600	1.9910	0.50226	652.58	903.72	7.5219	1.1159	0.039728	0.060131	0.73728	600
800	1.6202	0.61721	822.34	1130.9	7.7561	1.1548	0.045468	0.070060	0.74947	800
					P = 1000 kPa (1 MPa)					
0	12.821	0.077997	192.84	270.84	6.1073	1.0241	0.017412	0.024386	0.73125	0
10	12.353	0.080952	200.12	281.07	6.1441	1.0228	0.017904	0.025122	0.72888	10
20	11.920	0.083893	207.40	291.29	6.1796	1.0217	0.018388	0.025851	0.72674	20
30	11.516	0.086836	214.67	301.50	6.2139	1.0208	0.018867	0.026572	0.72479	30
40	11.140	0.089767	221.94	311.71	6.2470	1.0202	0.019339	0.027287	0.72303	40
50	10.788	0.092696	229.21	321.91	6.2790	1.0197	0.019806	0.027994	0.72143	50
60	10.458	0.095621	236.48	332.10	6.3101	1.0195	0.020266	0.028695	0.71999	60
70	10.148	0.098542	243.76	342.30	6.3403	1.0194	0.020721	0.029389	0.71870	70
80	9.8561	0.10146	251.03	352.49	6.3696	1.0194	0.021170	0.030077	0.71754	80
90	9.5808	0.10438	258.31	362.69	6.3980	1.0196	0.021614	0.030759	0.71650	90
100	9.3208	0.10729	265.60	372.88	6.4257	1.0200	0.022053	0.031434	0.71559	100
150	8.2092	0.12181	302.15	423.97	6.5542	1.0237	0.024175	0.034728	0.71264	150
200	7.3368	0.13630	339.00	475.30	6.6688	1.0301	0.026190	0.037894	0.71192	200
250	6.6333	0.15075	376.25	527.01	6.7727	1.0385	0.028111	0.040950	0.71292	250
300	6.0537	0.16519	413.99	579.18	6.8680	1.0485	0.029951	0.043910	0.71517	300
350	5.5677	0.17961	452.26	631.87	6.9561	1.0594	0.031720	0.046784	0.71828	350
400	5.1542	0.19402	491.11	685.13	7.0383	1.0709	0.033426	0.049585	0.72192	400
600	3.9749	0.25158	652.36	903.94	7.3227	1.1166	0.039749	0.060192	0.73733	600
800	3.2354	0.30908	822.19	1131.3	7.5569	1.1552	0.045484	0.070110	0.74945	800

Table A.26
Thermophysical Properties of Dry Air (SI Units)

T °C	ρ kg/m³	v m³/kg	u kJ/kg	h kJ/kg	s kJ/ kg-K	c_p kJ/ kg-K	μ mPa-s	k W/m-K	Pr	T °C
				P = 5000 kPa (5 MPa)						
0	65.195	0.015339	183.56	260.26	5.6122	1.1079	0.018284	0.026560	0.76270	0
10	62.539	0.015990	191.34	271.29	5.6519	1.0985	0.018731	0.027191	0.75670	10
20	60.116	0.016635	199.06	282.23	5.6899	1.0905	0.019175	0.027824	0.75150	20
30	57.893	0.017273	206.73	293.10	5.7263	1.0836	0.019616	0.028458	0.74695	30
40	55.844	0.017907	214.37	303.91	5.7614	1.0778	0.020054	0.029093	0.74295	40
50	53.949	0.018536	221.98	314.66	5.7952	1.0728	0.020489	0.029727	0.73942	50
60	52.189	0.019161	229.56	325.37	5.8278	1.0685	0.020921	0.030361	0.73629	60
70	50.550	0.019782	237.12	336.03	5.8594	1.0649	0.021349	0.030993	0.73351	70
80	49.018	0.020401	244.66	346.67	5.8899	1.0618	0.021773	0.031623	0.73104	80
90	47.583	0.021016	252.19	357.27	5.9195	1.0591	0.022194	0.032251	0.72885	90
100	46.235	0.021629	259.71	367.85	5.9483	1.0569	0.022612	0.032877	0.72691	100
150	40.552	0.024660	297.22	420.51	6.0807	1.0509	0.024645	0.035963	0.72016	150
200	36.165	0.027651	334.78	473.04	6.1980	1.0509	0.026594	0.038974	0.71708	200
250	32.663	0.030616	372.59	525.67	6.3038	1.0550	0.028464	0.041910	0.71652	250
300	29.796	0.033562	410.78	578.58	6.4004	1.0618	0.030263	0.044772	0.71771	300
350	27.403	0.036492	449.42	631.88	6.4895	1.0704	0.031999	0.047568	0.72006	350
400	25.372	0.039414	488.57	685.64	6.5725	1.0801	0.033678	0.050302	0.72314	400
600	19.597	0.051028	650.67	905.81	6.8586	1.1216	0.039929	0.060729	0.73743	600
800	15.978	0.062586	821.01	1133.9	7.0937	1.1583	0.045621	0.070539	0.74915	800
				P = 10,000 kPa (10 MPa)						
0	131.34	0.0076138	172.15	248.29	5.3747	1.2099	0.019810	0.029882	0.80213	0
10	125.47	0.0079700	180.58	260.28	5.4178	1.1896	0.020156	0.030327	0.79062	10
20	120.19	0.0083202	188.89	272.09	5.4588	1.1725	0.020512	0.030797	0.78091	20
30	115.40	0.0086655	197.09	283.74	5.4979	1.1579	0.020876	0.031287	0.77264	30
40	111.04	0.0090058	205.20	295.26	5.5352	1.1455	0.021245	0.031791	0.76552	40
50	107.04	0.0093423	213.23	306.66	5.5711	1.1348	0.021618	0.032307	0.75937	50
60	103.35	0.0096759	221.21	317.96	5.6055	1.1256	0.021994	0.032833	0.75400	60
70	99.949	0.010005	229.12	329.17	5.6387	1.1176	0.022372	0.033367	0.74931	70
80	96.788	0.010332	237.00	340.31	5.6707	1.1106	0.022750	0.033906	0.74519	80
90	93.844	0.010656	244.83	351.39	5.7016	1.1046	0.023129	0.034451	0.74156	90
100	91.093	0.010978	252.63	362.41	5.7315	1.0993	0.023507	0.034999	0.73835	100
150	79.629	0.012558	291.30	416.88	5.8686	1.0819	0.025385	0.037767	0.72719	150
200	70.903	0.014104	329.73	470.77	5.9889	1.0747	0.027221	0.040543	0.72156	200
250	64.000	0.015625	368.21	524.46	6.0968	1.0738	0.029006	0.043296	0.71940	250
300	58.380	0.017129	406.93	578.22	6.1949	1.0771	0.030739	0.046013	0.71953	300
350	53.704	0.018621	446.00	632.21	6.2853	1.0830	0.032422	0.048690	0.72116	350
400	49.746	0.020102	485.52	686.55	6.3691	1.0907	0.034058	0.051326	0.72373	400
600	38.510	0.025967	648.63	908.30	6.6574	1.1275	0.040193	0.061484	0.73704	600
800	31.469	0.031777	819.57	1137.3	6.8934	1.1620	0.045819	0.071133	0.74847	800

Table A.27
Thermophysical Properties of Saturated R134a (SI Units)

T °C	P kPa	ρ_f kg/m³	v_g m³/kg	u_f kJ/kg	u_g kJ/kg	h_f kJ/kg	h_g kJ/kg	s_f kJ/kg-K	s_g kJ/kg-K	T °C
−103.3	0.38956	1591.1	35.496	71.455	321.11	71.455	334.94	0.41262	1.9639	−103.3
−95	0.93899	1569.1	15.435	81.287	325.29	81.288	339.78	0.46913	1.9201	−95
−90	1.5241	1555.8	9.7698	87.225	327.87	87.226	342.76	0.50201	1.8972	−90
85	2.3990	1542.5	6.3707	93.180	330.49	93.182	345.77	0.53409	1.8766	−85
−80	3.6719	1529.0	4.2682	99.158	333.15	99.161	348.83	0.56544	1.8580	−80
−75	5.4777	1515.5	2.9312	105.16	335.85	105.17	351.91	0.59613	1.8414	−75
−70	7.9814	1501.9	2.0590	111.19	338.59	111.20	355.02	0.62619	1.8264	−70
−65	11.380	1488.2	1.4765	117.26	341.35	117.26	358.16	0.65568	1.8130	−65
−60	15.906	1474.3	1.0790	123.35	344.15	123.36	361.31	0.68462	1.8010	−60
−55	21.828	1460.4	0.80236	129.48	346.96	129.50	364.48	0.71305	1.7902	−55
−50	29.451	1446.3	0.60620	135.65	349.80	135.67	367.65	0.74101	1.7806	−50
−45	39.117	1432.1	0.46473	141.86	352.65	141.89	370.83	0.76852	1.7720	−45
−40	51.209	1417.7	0.36108	148.11	355.51	148.14	374.00	0.79561	1.7643	−40
−35	66.144	1403.1	0.28402	154.40	358.38	154.44	377.17	0.82230	1.7575	−35
−30	84.378	1388.4	0.22594	160.73	361.25	160.79	380.32	0.84863	1.7515	−30
−25	106.40	1373.4	0.18162	167.11	364.12	167.19	383.45	0.87460	1.7461	−25
−20	132.73	1358.3	0.14739	173.54	366.99	173.64	386.55	0.90025	1.7413	−20
−15	163.94	1342.8	0.12067	180.02	369.85	180.14	389.63	0.92559	1.7371	−15
−10	200.60	1327.1	0.099590	186.55	372.69	186.70	392.66	0.95065	1.7334	−10
−5	243.34	1311.1	0.082801	193.13	375.51	193.32	395.66	0.97544	1.7300	−5
0	292.80	1294.8	0.069309	199.77	378.31	200.00	398.60	1.0000	1.7271	0
5	349.66	1278.1	0.058374	206.48	381.08	206.75	401.49	1.0243	1.7245	5
10	414.61	1261.0	0.049442	213.25	383.82	213.58	404.32	1.0485	1.7221	10
15	488.37	1243.4	0.042090	220.09	386.52	220.48	407.07	1.0724	1.7200	15
20	571.71	1225.3	0.035997	227.00	389.17	227.47	409.75	1.0962	1.7180	20
25	665.38	1206.7	0.030912	233.99	391.77	234.55	412.33	1.1199	1.7162	25
30	770.20	1187.5	0.026642	241.07	394.30	241.72	414.82	1.1435	1.7145	30
35	886.98	1167.5	0.023033	248.25	396.76	249.01	417.19	1.1670	1.7128	35
40	1016.6	1146.7	0.019966	255.52	399.13	256.41	419.43	1.1905	1.7111	40
45	1159.9	1125.1	0.017344	262.91	401.40	263.94	421.52	1.2139	1.7092	45
50	1317.9	1102.3	0.015089	270.43	403.55	271.62	423.44	1.2375	1.7072	50
55	1491.5	1078.3	0.013140	278.09	405.55	279.47	425.15	1.2611	1.7050	55
60	1681.8	1052.9	0.011444	285.91	407.38	287.50	426.63	1.2848	1.7024	60
65	1889.8	1025.6	0.0099604	293.92	408.99	295.76	427.82	1.3088	1.6993	65
70	2116.8	996.25	0.0086527	302.16	410.33	304.28	428.65	1.3332	1.6956	70
75	2364.1	964.09	0.0074910	310.68	411.32	313.13	429.03	1.3580	1.6909	75
80	2633.2	928.24	0.0064483	319.55	411.83	322.39	428.81	1.3836	1.6850	80
85	2925.8	887.16	0.0054990	328.93	411.67	332.22	427.76	1.4104	1.6771	85
90	3244.2	837.83	0.0046134	339.06	410.45	342.93	425.42	1.4390	1.6662	90
101.06	4059.3	511.90	0.0019535	381.71	381.71	389.64	389.64	1.5621	1.5621	101.06

Table A.28
Thermophysical Properties of Saturated R134a (SI Units)

T °C	P kPa	c_{pf} kJ/kg-K	c_{pg} kJ/kg-K	μ_f mPa-s	μ_g mPa-s	k_f W/m-K	k_g W/m-K	Pr_f	PR_g	T °C
−103.3	0.38956	1.1838	0.58530	2.1536	0.0068294	0.14524	0.0030801	17.553	1.2978	−103.3
−95	0.93899	1.1861	0.60524	1.5720	0.0071588	0.14022	0.0037452	13.298	1.1569	−95
−90	1.5241	1.1892	0.61730	1.3410	0.0073562	0.13727	0.0041459	11.618	1.0953	−90
−85	2.3990	1.1933	0.62943	1.1624	0.0075528	0.13438	0.0045469	10.322	1.0456	−85
−80	3.6719	1.1981	0.64165	1.0203	0.0077484	0.13154	0.0049479	9.2927	1.0048	−80
−75	5.4777	1.2036	0.65401	0.90473	0.0079430	0.12876	0.0053493	8.4569	0.97112	−75
−70	7.9814	1.2096	0.66654	0.80910	0.0081364	0.12603	0.0057509	7.7656	0.94303	−70
−65	11.380	1.2161	0.67932	0.72879	0.0083286	0.12335	0.0061529	7.1854	0.91952	−65
−60	15.906	1.2230	0.69239	0.66051	0.0085194	0.12071	0.0065554	6.6923	0.89982	−60
−55	21.828	1.2304	0.70582	0.60182	0.0087089	0.11812	0.0069586	6.2688	0.88336	−55
−50	101.325	1.2381	0.71969	0.55089	0.0088970	0.11557	0.0073625	5.9016	0.86968	−50.000
−45	39.117	1.2462	0.73406	0.50633	0.0090837	0.11306	0.0077675	5.5807	0.85845	−45
−40	51.209	1.2546	0.74900	0.46703	0.0092690	0.11059	0.0081736	5.2983	0.84939	−40
−35	66.144	1.2635	0.76458	0.43213	0.0094531	0.10816	0.0085812	5.0482	0.84228	−35
−30	84.378	1.2729	0.78087	0.40095	0.0096361	0.10576	0.0089906	4.8255	0.83693	−30
−25	106.40	1.2827	0.79792	0.37292	0.0098181	0.10340	0.0094022	4.6261	0.83322	−25
−20	132.73	1.2930	0.81580	0.34758	0.0099995	0.10107	0.0098164	4.4469	0.83101	−20
−15	163.94	1.3040	0.83458	0.32456	0.010181	0.098767	0.010234	4.2851	0.83022	−15
−10	200.60	1.3156	0.85435	0.30355	0.010362	0.096491	0.010655	4.1386	0.83079	−10
−5	243.34	1.3279	0.87520	0.28428	0.010543	0.094241	0.011082	4.0056	0.83267	−5
0	292.80	1.3410	0.89723	0.26653	0.010726	0.092013	0.011514	3.8845	0.83584	0
5	349.66	1.3552	0.92059	0.25011	0.010911	0.089806	0.011954	3.7741	0.84031	5
10	414.61	1.3704	0.94546	0.23487	0.011099	0.087618	0.012402	3.6734	0.84612	10
15	488.37	1.3869	0.97206	0.22066	0.011291	0.085444	0.012862	3.5815	0.85335	15
20	571.71	1.4049	1.0007	0.20737	0.011488	0.083284	0.013335	3.4979	0.86211	20
25	665.38	1.4246	1.0316	0.19489	0.011693	0.081134	0.013825	3.4219	0.87256	25
30	770.20	1.4465	1.0655	0.18313	0.011907	0.078992	0.014336	3.3533	0.88493	30
35	886.98	1.4709	1.1028	0.17200	0.012132	0.076853	0.014874	3.2920	0.89953	35
40	1016.6	1.4984	1.1445	0.16145	0.012373	0.074716	0.015446	3.2378	0.91679	40
45	1159.9	1.5298	1.1917	0.15139	0.012633	0.072575	0.016062	3.1912	0.93728	45
50	1317.9	1.5661	1.2461	0.14177	0.012917	0.070427	0.016734	3.1527	0.96180	50
55	1491.5	1.6089	1.3099	0.13253	0.013232	0.068267	0.017481	3.1234	0.99154	55
60	1681.8	1.6602	1.3868	0.12361	0.013587	0.066091	0.018326	3.1051	1.0282	60
65	1889.8	1.7234	1.4822	0.11496	0.013996	0.063894	0.019305	3.1007	1.0745	65
70	2116.8	1.8039	1.6051	0.10651	0.014475	0.061672	0.020471	3.1153	1.1350	70
75	2364.1	1.9115	1.7714	0.098169	0.015053	0.059421	0.021903	3.1579	1.2174	75
80	2633.2	2.0648	2.0122	0.089846	0.015773	0.057147	0.023735	3.2464	1.3372	80
85	2925.8	2.3064	2.3971	0.081372	0.016711	0.054880	0.026205	3.4198	1.5287	85
90	3244.2	2.7559	3.1207	0.072450	0.018023	0.052755	0.029819	3.7848	1.8862	90
101.06	4059.3	−	−	−	−	−	−	−	−	101.06

Table A.29
Thermophysical Properties of Single-Phase R134a (SI Units)

T °C	ρ kg/m³	v m³/kg	u kJ/kg	h kJ/kg	s kJ/kg-K	c_p kJ/kg-K	μ mPa-s	k W/m-K	Pr	T °C
				P = 10 kPa (0.01 MPa)						
−100	1582.4	0.00063195	75.360	75.366	0.43539	1.1842	1.8827	0.14323	15.566	−100
−80	1529.0	0.00065402	99.157	99.163	0.56544	1.1981	1.0204	0.13155	9.2932	−80
−60	0.58001	1.7241	344.41	361.65	1.8400	0.68454	0.0085303	0.0065526	0.89115	−60
−40	0.52903	1.8903	356.77	375.67	1.9029	0.71847	0.0093211	0.0081541	0.82130	−40
−20	0.48655	2.0553	369.85	390.40	1.9635	0.75455	0.010105	0.0097557	0.78156	−20
0	0.45051	2.2197	383.66	405.85	2.0222	0.79068	0.010883	0.011357	0.75764	0
20	0.41950	2.3838	398.19	422.02	2.0793	0.82629	0.011655	0.012959	0.74314	20
40	0.39253	2.5476	413.42	438.90	2.1350	0.86123	0.012422	0.014561	0.73474	40
60	0.36884	2.7112	429.36	456.47	2.1894	0.89548	0.013185	0.016163	0.73048	60
80	0.34786	2.8747	445.97	474.72	2.2425	0.92905	0.013942	0.017764	0.72916	80
100	0.32916	3.0380	463.25	493.63	2.2946	0.96198	0.014696	0.019366	0.72998	100
120	0.31236	3.2014	481.18	513.19	2.3457	0.99429	0.015445	0.020968	0.73239	120
140	0.29721	3.3646	499.75	533.39	2.3958	1.0260	0.016190	0.022570	0.73600	140
160	0.28346	3.5278	518.95	554.23	2.4450	1.0572	0.016931	0.024172	0.74054	160
180	0.27092	3.6911	538.77	575.68	2.4934	1.0879	0.017669	0.025774	0.74579	180
200	0.25945	3.8543	559.20	597.74	2.5411	1.1181	0.018403	0.027376	0.75161	200
220	0.24892	4.0174	580.23	620.40	2.5880	1.1478	0.019133	0.028977	0.75787	220
240	0.23921	4.1804	601.85	643.65	2.6342	1.1771	0.019860	0.030579	0.76448	240
260	0.23022	4.3437	624.05	667.48	2.6798	1.2060	0.020583	0.032181	0.77137	260
				P = 50 kPa (0.05 MPa)						
−100	1582.5	0.00063191	75.353	75.384	0.43535	1.1842	1.8838	0.14324	15.574	−100
−80	1529.1	0.00065398	99.148	99.181	0.56539	1.1981	1.0208	0.13156	9.2960	−80
−60	1474.4	0.00067824	123.34	123.38	0.68458	1.2230	0.66074	0.12072	6.6936	−60
−40	2.7023	0.37006	355.55	374.05	1.7665	0.74790	0.0092706	0.0081729	0.84835	−40
−20	2.4699	0.40487	368.97	389.21	1.8288	0.77053	0.010071	0.0097718	0.79410	−20
0	2.2780	0.43898	382.98	404.92	1.8885	0.80114	0.010860	0.011372	0.76508	0
20	2.1157	0.47266	397.64	421.27	1.9463	0.83373	0.011641	0.012972	0.74815	20
40	1.9759	0.50610	412.97	438.28	2.0024	0.86677	0.012414	0.014573	0.73837	40
60	1.8541	0.53935	428.98	455.94	2.0571	0.89974	0.013181	0.016174	0.73326	60
80	1.7469	0.57244	445.64	474.26	2.1105	0.93240	0.013943	0.017775	0.73138	80
100	1.6516	0.60547	462.96	493.24	2.1627	0.96466	0.014699	0.019376	0.73181	100
120	1.5664	0.63841	480.93	512.85	2.2139	0.99647	0.015451	0.020977	0.73394	120
140	1.4897	0.67128	499.53	533.09	2.2641	1.0278	0.016198	0.022579	0.73736	140
160	1.4202	0.70413	518.75	553.96	2.3134	1.0587	0.016941	0.024180	0.74175	160
180	1.3569	0.73697	538.59	575.44	2.3619	1.0892	0.017680	0.025782	0.74689	180
200	1.2992	0.76970	559.03	597.52	2.4096	1.1192	0.018415	0.027383	0.75263	200
220	1.2461	0.80250	580.08	620.20	2.4565	1.1488	0.019146	0.028985	0.75882	220
240	1.1973	0.83521	601.71	643.47	2.5028	1.1779	0.019874	0.030586	0.76538	240
260	1.1521	0.86798	623.92	667.32	2.5484	1.2067	0.020598	0.032188	0.77223	260

Table A.30
Thermophysical Properties of Single-Phase R134a (SI Units)

T °C	P kg/m³	v m³/kg	u kj/kg	h kj/kg	s kJ/ kg-K	c_p kJ/ kg-K	μ mPa-s	k W/m-K	Pr	T °C
					P = 100 kPa (0.1 MPa)					
−100	1582.5	0.00063191	75.344	75.407	0.43529	1.1842	1.8852	0.14326	15.583	−100
−80	1529.2	0.00065394	99.137	99.202	0.56533	1.1980	1.0214	0.13158	9.2995	−80
−60	1474.5	0.00067820	123.33	123.40	0.68451	1.2229	0.66109	0.12074	6.6956	−60
−40	1417.8	0.00070532	148.09	148.16	0.79554	1.2545	0.46727	0.11061	5.2996	−40
−20	5.0401	0.19841	367.81	387.65	1.7677	0.79505	0.010028	0.0097966	0.81382	−20
0	4.6232	0.21630	382.10	403.73	1.8288	0.81542	0.010832	0.011392	0.77529	0
20	4.2784	0.23373	396.94	420.31	1.8874	0.84352	0.011623	0.012990	0.75474	20
40	3.9860	0.25088	412.40	437.49	1.9441	0.87396	0.012404	0.014589	0.74307	40
60	3.7336	0.26784	428.49	455.28	1.9991	0.90521	0.013177	0.016189	0.73683	60
80	3.5130	0.28466	445.23	473.70	2.0528	0.93667	0.013944	0.017789	0.73421	80
100	3.3181	0.30138	462.61	492.74	2.1053	0.96806	0.014704	0.019389	0.73414	100
120	3.1444	0.31803	480.61	512.42	2.1566	0.99923	0.015458	0.020990	0.73592	120
140	2.9885	0.33462	499.25	532.71	2.2070	1.0301	0.016208	0.022590	0.73907	140
160	2.8476	0.35117	518.50	553.62	2.2564	1.0606	0.016953	0.024191	0.74328	160
180	2.7198	0.36767	538.36	575.13	2.3049	1.0908	0.017694	0.025792	0.74828	180
200	2.6031	0.38416	558.83	597.25	2.3527	1.1205	0.018430	0.027393	0.75391	200
220	2.4961	0.40062	579.89	619.95	2.3997	1.1499	0.019163	0.028994	0.76002	220
240	2.3977	0.41707	601.53	643.24	2.4460	1.1790	0.019892	0.030595	0.76651	240
260	2.3069	0.43348	623.76	667.11	2.4916	1.2076	0.020617	0.032197	0.77331	260
					P = 500 kPa (0.5 MPa)					
−100	1583.0	0.00063171	75.271	75.587	0.43487	1.1839	1.8967	0.14338	15.660	−100
−80	1529.8	0.00065368	99.048	99.375	0.56487	1.1976	1.0260	0.13172	9.3278	−80
−60	1475.2	0.00067787	123.22	123.56	0.68400	1.2223	0.66384	0.12090	6.7113	−60
−40	1418.7	0.00070487	147.96	148.31	0.79496	1.2536	0.46928	0.11080	5.3099	−40
−20	1359.4	0.00073562	173.38	173.75	0.89964	1.2918	0.34910	0.10126	4.4535	−20
0	1295.6	0.00077184	199.66	200.05	0.99959	1.3399	0.26730	0.092142	3.8869	0
20	23.744	0.042116	390.55	411.61	1.7339	0.96352	0.011504	0.013250	0.83654	20
40	21.526	0.046455	407.40	430.63	1.7967	0.94679	0.012349	0.014788	0.79066	40
60	19.808	0.050485	424.40	449.64	1.8555	0.95652	0.013169	0.016353	0.77026	60
80	18.406	0.054330	441.78	468.95	1.9118	0.97514	0.013970	0.017931	0.75974	80
100	17.225	0.058055	459.65	488.68	1.9661	0.99794	0.014758	0.019516	0.75464	100
120	16.211	0.061687	478.04	508.88	2.0189	1.0230	0.015535	0.021105	0.75301	120
140	15.324	0.065257	496.98	529.60	2.0703	1.0494	0.016303	0.022697	0.75376	140
160	14.540	0.068776	516.47	550.86	2.1205	1.0765	0.017062	0.024290	0.75622	160
180	13.840	0.072254	536.54	572.67	2.1697	1.1041	0.017816	0.025885	0.75992	180
200	13.209	0.075706	557.17	595.03	2.2180	1.1318	0.018563	0.027480	0.76456	200
220	12.637	0.079133	578.38	617.94	2.2654	1.1596	0.019306	0.029077	0.76992	220
240	12.116	0.082535	600.14	641.41	2.3121	1.1873	0.020043	0.030674	0.77583	240
260	11.638	0.085925	622.47	665.43	2.3580	1.2149	0.020776	0.032271	0.78217	260

Table A.31
Thermophysical Properties of Single-Phase R134a (SI Units)

T °C	ρ kg/m³	v m³/kg	u kJ/kg	h kJ/kg	s kJ/ kg-K	c_p kJ/ kg-K	μ mPa-s	k W/m-K	Pr	T °C
				P = 1000 kPa (1 MPa)						
−100	1583.6	0.00063147	75.180	75.811	0.43435	1.1835	1.9111	0.14354	15.757	−100
−80	1530.5	0.00065338	98.937	99.591	0.56130	1.1971	1.0317	0.13190	9.3634	−80
−60	1476.1	0.00067746	123.09	123.76	0.68337	1.2216	0.66729	0.12110	6.7310	−60
−40	1419.9	0.00070427	147.79	148.50	0.79425	1.2526	0.47179	0.11102	5.3228	−40
−20	1360.8	0.00073486	173.18	173.91	0.89881	1.2901	0.35116	0.10152	4.4625	−20
0	1297.5	0.00077071	199.39	200.16	0.99860	1.3372	0.26915	0.092451	3.8929	0
20	1227.7	0.00081453	226.69	227.50	1.0952	1.4006	0.20893	0.083611	3.4999	20
40	49.004	0.020406	399.45	419.86	1.7135	1.1340	0.012370	0.015410	0.91026	40
60	43.350	0.023068	418.46	441.53	1.7806	1.0535	0.013232	0.016749	0.83229	60
80	39.372	0.025399	437.00	462.40	1.8414	1.0387	0.014065	0.018226	0.80156	80
100	36.295	0.027552	455.65	483.21	1.8988	1.0438	0.014878	0.019755	0.78613	100
120	33.792	0.029593	474.62	504.21	1.9536	1.0579	0.015674	0.021307	0.77820	120
140	31.692	0.031554	494.00	525.55	2.0065	1.0768	0.016458	0.022873	0.77477	140
160	29.889	0.033457	513.84	547.30	2.0579	1.0986	0.017231	0.024448	0.77430	160
180	28.314	0.035318	534.19	569.51	2.1080	1.1222	0.017996	0.026028	0.77590	180
200	26.922	0.037144	555.05	592.20	2.1570	1.1470	0.018753	0.027612	0.77897	200
220	25.678	0.038944	576.45	615.39	2.2050	1.1724	0.019504	0.029199	0.78315	220
240	24.557	0.040722	598.37	639.10	2.2522	1.1983	0.020250	0.030788	0.78815	240
260	23.539	0.042483	620.84	663.32	2.2985	1.2244	0.020990	0.032378	0.79379	260
				P = 4000 kPa (4 MPa)						
−100	1587.2	0.00063004	74.643	77.163	0.43123	1.1813	2.0024	0.14449	16.371	−100
−80	1534.9	0.00065151	98.286	100.89	0.56090	1.1941	1.0671	0.13295	9.5843	−80
−60	1481.5	0.00067499	122.29	124.99	0.67962	1.2174	0.68828	0.12228	6.8525	−60
−40	1426.6	0.00070097	146.82	149.62	0.79005	1.2465	0.48695	0.11235	5.4027	−40
−20	1369.3	0.00073030	171.97	174.89	0.89400	1.2810	0.36351	0.10304	4.5191	−20
0	1308.6	0.00076418	197.86	200.91	0.99291	1.3227	0.28014	0.094232	3.9322	0
20	1242.9	0.00080457	224.65	227.87	1.0881	1.3756	0.21952	0.085770	3.5207	20
40	1169.4	0.00085514	252.65	256.07	1.1812	1.4485	0.17257	0.077471	3.2266	40
60	1082.4	0.00092387	282.39	286.09	1.2740	1.5642	0.13383	0.069049	3.0317	60
80	967.24	0.0010339	315.32	319.46	1.3713	1.8145	0.098907	0.059942	2.9940	80
100	677.84	0.0014753	364.80	370.70	1.5117	9.4035	0.050372	0.054130	8.7507	100
120	200.28	0.0049930	445.13	465.10	1.7616	1.6841	0.019036	0.026830	1.1949	120
140	165.99	0.0060245	471.28	495.38	1.8368	1.4047	0.019013	0.026158	1.0210	140
160	146.16	0.0068418	495.07	522.43	1.9008	1.3151	0.019427	0.026800	0.95331	160
180	132.37	0.0075546	518.11	548.33	1.9592	1.2798	0.019992	0.027871	0.91801	180
200	121.91	0.0082028	540.97	573.78	2.0142	1.2682	0.020626	0.029134	0.89781	200
220	113.54	0.0088075	563.91	599.14	2.0667	1.2692	0.021295	0.030500	0.88616	220
240	106.60	0.0093809	587.07	624.60	2.1173	1.2776	0.021986	0.031925	0.87988	240
260	100.71	0.0099295	610.55	650.27	2.1664	1.2908	0.022690	0.033386	0.87726	260

Table A.32
Thermophysical Properties of Metals (SI Units)

T °C	ρ kg/m³	c kJ/kg-K	k W/m-K	ρ kg/m³	c kJ/kg-K	k W/m-K	ρ kg/m³	c kJ/kg-K	k W/m-K
	Aluminum			Brass			Bronze		
30	2701	0.9044	236.1	8530	0.3781	116.7	8799	0.4213	52.00
90	2690	0.9320	238.5	8530	0.3886	129.3	8771	0.4453	52.00
150	2678	0.9587	239.1	8544	0.3985	138.4	8742	0.4698	52.81
210	2665	0.9839	236.7	8581	0.4075	142.0	8712	0.4953	54.91
270	2652	1.0090	234.3	8618	0.4165	145.6	8682	0.5208	57.01
330	2640	1.0350	231.8	8655	0.4255	149.2	8652	0.5463	59.11
	Chromium			Copper			Iron		
30	7160	0.4501	93.61	8932	0.3854	400.7	7869	0.4484	79.86
90	7151	0.4711	91.93	8904	0.3926	395.9	7851	0.4742	73.44
150	7141	0.4907	89.72	8876	0.3993	391.4	7833	0.4966	67.79
210	7129	0.5081	86.66	8847	0.4053	387.2	7813	0.5137	63.35
270	7118	0.5255	83.60	8818	0.4113	383.0	7793	0.5308	58.91
330	7106	0.5426	80.55	8789	0.4173	378.8	7772	0.5491	54.52
	Iron-Armco			AISI 1010 Carbon Steel			AISI 302 Stainless Steel		
30	7867	0.4484	72.48	7831	0.4354	63.74	8052	0.4825	15.27
90	7815	0.4742	68.28	7814	0.4618	60.62	8031	0.5008	16.53
150	7782	0.4966	64.24	7797	0.4874	57.55	8008	0.5174	17.61
210	7782	0.5137	60.46	7777	0.5117	54.58	7983	0.5315	18.42
270	7781	0.5308	56.68	7757	0.5360	51.61	7958	0.5456	19.23
330	7779	0.5491	52.93	7737	0.5610	48.65	7932	0.5594	20.04
	AISI 304 Stainless Steel			AISI 316 Stainless Steel			AISI 347 Stainless Steel		
30	7899	0.4782	14.95	8237	0.4691	13.46	7975	0.4826	14.32
90	7877	0.5010	15.97	8214	0.4907	14.54	7954	0.5014	15.24
150	7854	0.5199	16.97	8190	0.5093	15.56	7931	0.5183	16.16
210	7829	0.5325	17.93	8164	0.5231	16.49	7907	0.5321	17.09
270	7805	0.5451	18.89	8139	0.5369	17.42	7882	0.5459	18.02
330	7780	0.5574	19.84	8113	0.5504	18.35	7857	0.5594	18.95
	Lead			Nickel			Inconel X-750		
30	11337	0.1291	35.26	8899	0.4453	90.37	8509	0.4401	11.76
90	11277	0.1309	34.48	8876	0.4699	84.07	8490	0.4605	12.84
150	11214	0.1332	33.70	8853	0.4974	78.51	8470	0.4773	13.91
210	11147	0.1362	32.92	8829	0.5295	74.13	8448	0.4884	14.96
270	11081	0.1392	32.14	8805	0.5616	69.75	8426	0.4995	16.01
330	11014	0.1422	31.36	8781	0.5910	65.63	8405	0.5106	17.06
	Silver			Titanium			Zirconium		
30	10498	0.2351	248.9	4500	0.5229	21.85	6570	0.2787	22.67
90	10462	0.2375	426.5	4492	0.5403	20.95	6563	0.2919	22.01
150	10424	0.2403	423.6	4485	0.5556	20.28	6556	0.3025	21.50
210	10386	0.2436	419.6	4477	0.5676	19.98	6548	0.3091	21.23
270	10348	0.2469	415.7	4469	0.5796	19.68	6540	0.3157	20.96
330	10309	0.2502	411.7	4461	0.5917	19.40	6533	0.3223	20.71

B Thermophysical Properties (Exergy Calculation)

The thermophysical properties for exergy calculations have been adapted from the following reference:

A. Bejan, G. Tsatsaronis, and M. Moran. *Thermal Design and Optimization*. John Wiley & Sons, 1996.

DOI: 10.1201/9781003049272-B

Table B.1

Property Data Table as a Function of Temperature for Various Substances at 1 Bar

1. At $T_{ref} = 298.15$ K ($25°$ C), $p_{ref} = 1$ bar

Substance	Formula	$\bar{c}_p^\circ$	$\bar{h}^\circ$	$\bar{s}^\circ$	$\bar{g}^\circ$
Carbon (graphite)	C(s)	8.53	0	5.740	−1711
Sulfur (rhombic)	S(s)	22.77	0	32.058	−9558
Nitrogen	N_2(g)	28.49	0	191.610	−57128
Oxygen	O_2(g)	28.92	0	205.146	−61164
Hydrogen	H_2(g)	29.13	0	130.679	−38961
Carbon monoxide	CO(g)	28.54	−110528	197.648	−169457
Carbon dioxide	CO_2(g)	35.91	−393521	213.794	−457264
Water	H_2O(g)	31.96	−241856	188.824	−298153
Water	H_2O(l)	75.79	−285829	69.948	−306685
Methane	CH_4(g)	35.05	−74872	186.251	−130403
Sulfur dioxide	SO_2(g)	39.59	−296833	284.094	−370803
Hydrogen sulfide	H_2S(g)	33.06	−20501	205.757	−81847
.monia	NH_3(g)	35.59	−46111	192.451	−103491

2. For $298.15 < T \leq T_{max}$, $p_{ref} = 1$ bar, with $y = 10^{-3}T$

$$\bar{c}_p^\circ = a + by + cy^{-2} + dy^2 \tag{1}$$

$$\bar{h}^\circ = 10^3 \left[H^+ + ay + \frac{b}{2} y^2 - cy^{-1} + \frac{d}{3} y^3 \right] \tag{2}$$

$$\bar{s}^\circ = S^+ + a \ln T + by - \frac{c}{2} y^{-2} + \frac{d}{2} y^2 \tag{3}$$

$$\bar{g}^\circ = \bar{h}^\circ - T\bar{s}^\circ \tag{4}$$

The constants H^+, S^+, a, b, c, and d required by Equations (1)–(4) are given for selected substances in the table below. The maximum temperature, T_{max}, is 1100 K for C(s), 368 K for S(s), 500 K for $H_2O(l)$, 2000 K for $CH_4(g)$, $SO_2(g)$ and $H_2S(g)$, 1500 K for $NH_3(g)$, and 3000 K for all remaining substances. To evaluate the absolute entropy at states where the pressure p differs from $p_{ref} = 1$ bar, Equations 2.71 and 2.72 should be applied, as appropriate. The same caution applies when using absolute entropy data from other references. Also, owing to different data sources and roundoff, the absolute entropy values from other references may differ slightly from those of the current table.

Substance	Formula	H^+	S^+	a	b	c	d
Carbon (graphite)	C(s)	−2.101	−6.540	0.109	38.940	−0.146	−17.385
Sulfur (rhombic)	S(s)	−5.242	−59.014	14.795	24.075	0.071	0
Nitrogen[b]	$N_2(g)$	−9.982	16.203	30.418	2.544	−0.238	0
Oxygen	$O_2(g)$	−9.589	36.116	29.154	6.477	−0.184	−1.017
Hydrogen	$H_2(g)$	−7.823	−22.966	26.882	3.586	0.105	0
Carbon monoxide	CO(g)	−120.809	18.937	30.962	2.439	−0.280	0
Carbon dioxide	$CO_2(g)$	−413.886	−87.078	51.128	4.368	−1.469	0
Water	$H_2O(g)$	−253.871	−11.750	34.376	7.841	−0.423	0
Water	$H_2O(l)$	−289.932	−67.147	20.355	109.198	2.033	0
Methane	$CH_4(g)$	−81.242	96.731	11.933	77.647	0.142	−18.414
Sulfur dioxide	$SO_2(g)$	−315.422	−43.725	49.936	4.766	−1.046	0
Hydrogen sulfide	$H_2S(g)$	−32.887	1.142	34.911	10.686	−0.448	0
Ammonia	$NH_3(g)$	−60.244	−29.402	37.321	18.661	−0.649	0

Special Note: Since the reference state used in the *steam tables* differs from that of the present table, care must be exercised to ensure that steam table data are used consistently with values from this table. Although the reference states differ, each of these sources must yield the same values for the *changes* in property values between any two states. Thus, for specific enthalpy and entropy we have

$$h(T, p) - h(25°C, 1\ \text{bar}) = h^*(T, p) - h^*(25°C, 1\ \text{bar}) \tag{5}$$

$$s(T, p) - s(25°C, 1\ \text{bar}) = s^*(T, p) - s^*(25°C, 1\ \text{bar}) \tag{6}$$

where the terms on the left side are obtained from the present table and the terms on the right side denoted by superscript * are obtained from the steam tables.

Liquid water is a special case of interest. Thus at 25°C we have from the present table, $h = -15866.2$ kJ/kg ($\bar{h} = -285,829$ kJ/kmol), $s = 3.88276$ kJ/kg·K ($\bar{s} = 69.948$ kJ/kmol·K). Applying Equations 2.42c and 2.42d at 25°C, 1 bar we have from the steam tables, $h^* = 104.85$ kJ/kg, $s^* = 0.3670$ kJ/kg·K. Inserting values in Equations (5), (6)

$$h(T, p) = h^*(T, p) - 15,971\ \text{kJ/kg} \tag{7a}$$

$$s(T, p) = s^*(T, p) + 3.51576\ \text{kJ/kg·K} \tag{8a}$$

On a molar basis

$$\bar{h}(T, p) = \bar{h}^*(T, p) - 287718\ \text{kJ/kmol} \tag{7b}$$

$$\bar{s}(T, p) = \bar{s}^*(T, p) + 63.3365\ \text{kJ/kmol·K} \tag{8b}$$

Using these expressions, steam table values of the specific enthalpy and entropy of liquid water can be made consistent with enthalpy and entropy data obtained from the present table.

[a]$\bar{c}_p^°$, $\bar{s}^°$, and S^+ in kJ/kmol·K; $\bar{h}^°$, $\bar{g}^°$, and H^+ in kJ/kmol; T in Kelvin.
[b]Table values for nitrogen are shown as reported in the source given below. Corrected values for H^+, S^+, a, b, c, d are, respectively, −7.069, 51.539, 24.229, 10.521, 0.180, −2.315.

Source: O. Knacke, O. Kubaschewski, and K. Hesselmann, *Thermochemical Properties of Inorganic Substances*, 2nd ed., Springer-Verlag, Berlin, 1991.

Figure B.1 Constant values for equations.

Table B.2
Standard Chemical Exergy for Different Substances

Substance	Formula	Model I[a]	Model II[b]
Nitrogen	$N_2(g)$	639	720
Oxygen	$O_2(g)$	3951	3970
Carbon dioxide	$CO_2(g)$	14176	19870
Water	$H_2O(g)$	8636	9500
Water	$H_2O(l)$	45	900
Carbon (graphite)	$C(s)$	404589	410260
Hydrogen	$H_2(g)$	235249	236100
Sulfur	$S(s)$	598158	609600
Carbon monoxide	$CO(g)$	269412	275100
Sulfur dioxide	$SO_2(g)$	301939	313400
Nitrogen monoxide	$NO(g)$	88851	88900
Nitrogen dioxide	$NO_2(g)$	55565	55600
Hydrogen peroxide	$H_2O_2(g)$	133587	—
Hydrogen sulfide	H_2S	799890	812000
Ammonia	$NH_3(g)$	336684	337900
Oxygen	$O(g)$	231968	233700
Hydrogen	$H(g)$	320822	331300
Nitrogen	$N(g)$	453821	—
Methane	$CH_4(g)$	824348	831650
Acetylene	$C_2H_2(g)$	—	1265800
Ethylene	$C_2H_4(g)$	—	1361100
Ethane	$C_2H_6(g)$	1482033	1495840
Propylene	$C_3H_6(g)$	—	2003900
Propane	$C_3H_8(g)$	—	2154000
n-Butane	$C_4H_{10}(g)$	—	2805800
n-Pentane	$C_5H_{12}(g)$	—	3463300
Benzene	$C_6H_6(g)$	—	3303600
Octane	$C_8H_{18}(l)$	—	5413100
Methanol	$CH_3OH(g)$	715069	722300
Methanol	$CH_3OH(l)$	710747	718000
Ethyl alcohol	$C_2H_5OH(g)$	1348328	1363900
Ethyl alcohol	$C_2H_5OH(l)$	1342086	1375700

C Thermophysical Properties (Emissivity)

The radiation properties provided in Table C.1 have been adapted from the following reference:

Y. Jaluria *Design and Optimization of Thermal Systems*. CRC Press, 2007.

DOI: 10.1201/9781003049272-C

Table C.1
Emissivities ε_n of the Radiation in the Direction of the Normal to the Surface and ε of the Total Hemispherical Radiation for Various Materials for Temperature T

Surface	T, °C	ε_n	ε
Gold, polished	130	0.018	
	400	0.022	
Silver	20	0.020	
Copper, polished	20	0.030	
Lightly oxidized	20	0.037	
Scraped	20	0.070	
Black oxidized	20	0.78	
Oxidized	131	0.76	0.725
Aluminum, bright rolled	170	0.039	0.049
	500	0.050	
Aluminum paint	100	0.20–0.40	
Silumin, cast polished	150	0.186	
Nickel, bright matte	100	0.041	0.046
Polished	100	0.045	0.053
Manganin, bright rolled	118	0.048	0.057
Chrome, polished	150	0.058	0.071
Iron, bright etched	150	0.128	0.158
Bright abrased	20	0.24	
Red rusted	20	0.61	
Hot rolled	20	0.77	
	130	0.60	
Hot cast	100	0.80	
Heavily rusted	20	0.85	
Heat-resistant oxidized	80	0.613	
	200	0.639	
Zinc, gray oxidized	20	0.23–0.28	
Lead, gray oxidized	20	0.28	
Bismuth, bright	80	0.340	0.366
Corundum, emery rough	80	0.855	0.84
Clay, fired	70	0.91	0.86
Lacquer, white	100	0.925	
Red lead	100	0.93	
Enamel, lacquer	20	0.85–0.95	
Lacquer, black matte	80	0.970	
Bakelite lacquer	80	0.935	
Brick, mortar, plaster	20	0.93	
Porcelain	20	0.92–0.94	
Glass	90	0.940	0.876
Ice, smooth, water	0	0.966	0.918
Rough crystals	0	0.985	
Waterglass	20	0.96	
Paper	95	0.92	0.89
Wood, beech	70	0.935	0.91
Tarpaper	20	0.93	

D Standard Pipe Dimension

The thermophysical properties presented here have been adapted from the following reference:

Steven G. Penocello *Thermal Energy Systems Design and Analysis*. CRC Press, 2019.

DOI: 10.1201/9781003049272-D

Table D.1
Standard Pipe Dimension

Nominal Diameter	Outside Diameter			Schedule		Inside Diameter		
	in	ft	cm	Steel (Iron)	SS	in	ft	cm
1/8	0.405	0.033750	1.0287		10S	0.307	0.025583	0.77978
				40 (std)	40S	0.269	0.022417	0.68326
				80 (xs)	80S	0.215	0.017917	0.54610
1/4	0.540	0.045000	1.3716		10S	0.410	0.034167	1.0414
				40	40S	0.364	0.030333	0.92456
				80	80S	0.302	0.025167	0.76708
3/8	0.675	0.056250	1.7145	10	10S	0.545	0.045417	1.3843
				40 (std)	40S	0.493	0.041083	1.2522
				80 (xs)	80S	0.423	0.035250	1.0744
1/2	0.840	0.070000	2.1336		5S	0.710	0.059167	1.8034
					10S	0.674	0.056167	1.7120
				40 (std)	40S	0.622	0.051833	1.5799
				80 (xs)	80S	0.546	0.045500	1.3868
				160		0.466	0.038833	1.1836
				(xxs)		0.252	0.021000	0.64008
3/4	1.050	0.087500	2.6670		5S	0.920	0.076667	2.3368
					10S	0.884	0.073667	2.2454
				40 (std)	40S	0.824	0.068667	2.0930
				80 (xs)	80S	0.742	0.061833	1.8847
				160		0.612	0.051000	1.5545
				(xxs)		0.434	0.036167	1.1024
1	1.315	0.10958	3.3401		5S	1.185	0.098750	3.0099
					10S	1.097	0.091417	2.7864
				40 (std)	40S	1.049	0.087417	2.6645
				80 (xs)	80S	0.957	0.079750	2.4308
				160		0.815	0.067917	2.0701
				(xxs)		0.599	0.049917	1.5215

(Continued)

Nominal Diameter	Outside Diameter			Schedule		Inside Diameter		
	in	ft	cm	Steel (Iron)	SS	in	ft	cm
1 1/4	1.660	0.13833	4.2164		5S	1.530	0.12750	3.8862
					10S	1.442	0.12017	3.6627
				40 (std)	40S	1.380	0.11500	3.5052
				80 (xs)	80S	1.278	0.10650	3.2461
				160		1.160	0.096667	2.9464
				(xxs)		0.896	0.074667	2.2758
1 1/2	1.900	0.15833	4.8260		5S	1.770	0.14750	4.4958
					10S	1.682	0.14017	4.2723
				40 (std)	40S	1.610	0.13417	4.0894
				80 (xs)	80S	1.500	0.12500	3.8100
				160		1.338	0.11150	3.3985
				(xxs)		1.100	0.091667	2.7940
2	2.375	0.19792	6.0325		5S	2.245	0.18708	5.7023
					10S	2.157	0.17975	5.4788
				40 (std)	40S	2.067	0.17225	5.2502
				80 (xs)	80S	1.939	0.16158	4.9251
				160		1.687	0.14058	4.2850
				(xxs)		1.503	0.12525	3.8176
2 1/2	2.875	0.23958	7.3025		5S	2.709	0.22575	6.8809
					10S	2.635	0.21958	6.6929
				40 (std)	40S	2.469	0.20575	6.2713
				80 (xs)	80S	2.323	0.19358	5.9004
				160		2.125	0.17708	5.3975
				(xxs)		1.771	0.14758	4.4983
3	3.500	0.29167	8.8900		5S	3.334	0.27783	8.4684
					10S	3.260	0.27167	8.2804
				40 (std)	40S	3.068	0.25567	7.7927
				80 (xs)	80S	2.900	0.24167	7.3660
				160		2.624	0.21867	6.6650
				(xxs)		2.300	0.19167	5.8420
3 1/2	4.000	0.33333	10.160		5S	3.834	0.31950	9.7384
					10S	3.760	0.31333	9.5504
				40 (std)	40S	3.548	0.29567	9.0119
				80 (xs)	80S	3.364	0.28033	8.5446
4	4.500	0.37500	11.430		5S	4.334	0.36117	11.008
					10S	4.260	0.35500	10.820
				40 (std)	40S	4.026	0.33550	10.226
				80 (xs)	80S	3.826	0.31883	9.7180
				120		3.624	0.30200	9.2050
				160		3.438	0.28650	8.7325
				(xxs)		3.152	0.26267	8.0061

(Continued)

Nominal Diameter	Outside Diameter			Schedule		Inside Diameter		
	in	ft	cm	Steel (Iron)	SS	in	ft	cm
5	5.563	0.46358	14.130		5S	5.345	0.44542	13.576
					10S	5.295	0.44125	13.449
				40 (std)	40S	5.047	0.42058	12.819
				80 (xs)	80S	4.813	0.40108	12.225
				120		4.563	0.38025	11.590
				160		4.313	0.35942	10.955
				(xxs)		4.063	0.33858	10.320
6	6.625	0.55208	16.828	5	5S	6.407	0.53392	16.274
				10	10S	6.357	0.52975	16.147
				40 (std)	40S	6.065	0.50542	15.405
				80 (xs)	80S	5.761	0.48008	14.633
				120		5.501	0.45842	13.973
				160		5.187	0.43225	13.175
				xxs		4.897	0.40808	12.438
8	8.625	0.71875	21.908		5S	8.407	0.70058	21.354
					10S	8.329	0.69408	21.156
				20		8.125	0.67708	20.638
				30		8.071	0.67258	20.500
				40 (std)	40S	7.981	0.66508	20.272
				60		7.813	0.65108	19.845
				80 (xs)	80S	7.625	0.63542	19.368
				100		7.437	0.61975	18.890
				120		7.187	0.59892	18.255
				140		7.001	0.58342	17.783
				(xxs)		6.875	0.57292	17.463
				160		6.813	0.56775	17.305
10	10.750	0.89583	27.305		5S	10.482	0.87350	26.624
					10S	10.420	0.86833	26.467
				20		10.250	0.85417	26.035
				30		10.136	0.84467	25.745
				40 (std)	40S	10.020	0.83500	25.451
				60 (std)	80S	9.750	0.81250	24.765
				80		9.562	0.79683	24.287
				100		9.312	0.77600	23.652
				120		9.062	0.75517	23.017
				140 (xxs)		8.750	0.72917	22.225
				160		8.500	0.70833	21.590

(Continued)

Nominal Diameter	Outside Diameter			Schedule		Inside Diameter		
	in	ft	cm	Steel (Iron)	SS	in	ft	cm
12	12.75	1.0625	32.385		5S	12.438	1.0365	31.593
					10S	12.390	1.0325	31.471
				20		12.250	1.0208	31.115
				30		12.090	1.0075	30.709
				(std)	40S	12.000	1.0000	30.480
				40		11.983	0.99858	30.437
				(xs)	80S	11.750	0.97917	29.845
				60		11.626	0.96883	29.530
				80		11.374	0.94783	28.890
				100		11.062	0.92183	28.097
				120 (xxs)		10.750	0.89583	27.305
				140		10.500	0.87500	26.670
				160		10.126	0.84383	25.720
14	14	1.1667	35.560		5S	13.688	1.1407	34.768
					10S	13.624	1.1353	34.605
				10		13.500	1.1250	34.290
				20		13.376	1.1147	33.975
				30 (std)		13.250	1.1042	33.655
				40		13.124	1.0937	33.335
				(xs)		13.000	1.0833	33.020
				60		12.812	1.0677	32.542
				80		12.500	1.0417	31.750
				100		12.124	1.0103	30.795
				120		11.812	0.98433	30.002
				140		11.500	0.95833	29.210
				160		11.188	0.93233	28.418
16	16	1.3333	40.640		5S	15.670	1.3058	39.802
					10S	15.624	1.3020	39.685
				10		15.500	1.2917	39.370
				20		15.376	1.2813	39.055
				30 (std)		15.250	1.2708	38.735
				40 (xs)		15.000	1.2500	38.100
				60		14.688	1.2240	37.308
				80		14.312	1.1927	36.352
				100		13.938	1.1615	35.403
				120		13.562	1.1302	34.447
				140		13.124	1.0937	33.335
				160		12.182	1.0152	30.942

(*Continued*)

Nominal	Outside Diameter			Schedule		Inside Diameter		
Diameter	in	ft	cm	Steel (Iron)	SS	in	ft	cm
18	18	1.5000	45.720		5S	17.670	1.4725	44.882
					10S	17.624	1.4687	44.765
				10		17.500	1.4583	44.450
				20		17.376	1.4480	44.135
				(std)		17.250	1.4375	43.815
				30		17.124	1.4270	43.495
				(xs)		17.000	1.4167	43.180
				40		16.876	1.4063	42.865
				60		16.500	1.3750	41.910
				80		16.124	1.3437	40.955
				100		15.688	1.3073	39.848
				120		15.250	1.2708	38.735
				140		14.876	1.2397	37.785
				160		14.438	1.2032	36.673
20	20	1.6667	50.800		5S	19.624	1.6353	49.845
					10S	19.564	1.6303	49.693
				10		19.500	1.6250	49.530
				20 (std)		19.250	1.6042	48.895
				30 (xs)		19.000	1.5833	48.260
				40		18.812	1.5677	47.782
				60		18.376	1.5313	46.675
				80		17.938	1.4948	45.563
				100		17.438	1.4532	44.293
				120		17.000	1.4167	43.180
				140		16.500	1.3750	41.910
				160		16.062	1.3385	40.797
22	22	1.8333	55.880		5S	21.624	1.8020	54.925
					10S	21.564	1.7970	54.773
				10		21.500	1.7917	54.610
				20 (std)		21.250	1.7708	53.975
				30 (xs)		21.000	1.7500	53.340
				60		20.250	1.6875	51.435
				80		19.750	1.6458	50.165
				100		19.250	1.6042	48.895
				120		18.750	1.5625	47.625
				140		18.250	1.5208	46.355
				160		17.750	1.4792	45.085

(Continued)

Nominal Diameter	Outside Diameter			Schedule		Inside Diameter		
	in	ft	cm	Steel (Iron)	SS	in	ft	cm
24	24	2.0000	60.960		5S	23.564	1.9637	59.853
				10	10S	23.500	1.9583	59.690
				20 (std)		23.250	1.9375	59.055
				(xs)		23.000	1.9167	58.420
				30		22.876	1.9063	58.105
				40		22.624	1.8853	57.465
				60		22.062	1.8385	56.037
				80		21.562	1.7968	54.767
				100		20.938	1.7448	53.183
				120		20.376	1.6980	51.755
				140		19.876	1.6563	50.485
				160		19.312	1.6093	49.052
26	26	2.1667	66.040	10		25.376	2.1147	64.455
				(std)		25.250	2.1042	64.135
				20 (xs)		25.000	2.0833	63.500
28	28	2.3333	71.120	10		27.376	2.2813	69.535
				(std)		27.250	2.2708	69.215
				20 (xs)		27.000	2.2500	68.580
				30		26.750	2.2292	67.945
30	30	2.5000	76.200		5S	29.500	2.4583	74.930
				10	10S	29.376	2.4480	74.615
				(std)		29.250	2.4375	74.295
				20 (xs)		29.000	2.4167	73.660
				30		28.750	2.3958	73.025
32	32	2.6667	81.280	10		31.376	2.6147	79.695
				(std)		31.250	2.6042	79.375
				20 (xs)		31.000	2.5833	78.740
				30		30.750	2.5625	78.105
				40		30.624	2.5520	77.785
34	34	2.8333	86.360	10		33.376	2.7813	84.775
				(std)		33.250	2.7708	84.455
				20 (xs)		33.000	2.7500	83.820
				30		32.750	2.7292	83.185
				40		32.624	2.7187	82.865
36	36	3.0000	91.440	10		35.376	2.9480	89.855
				(std)		35.250	2.9375	89.535
				20 (xs)		35.000	2.9167	88.900
				30		34.750	2.8958	88.265
				40		34.500	2.8750	87.630

E Pump Performance Curve

(SOURCE: GOULD'S MANUAL)

DOI: 10.1201/9781003049272-E

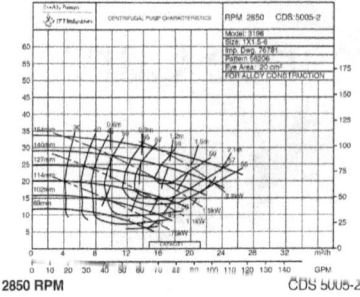

STX
1 x 1½-6
AA

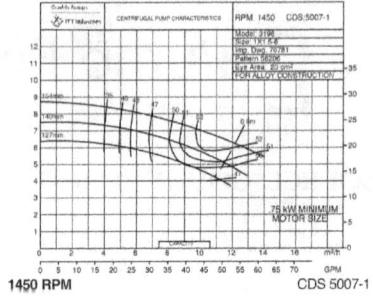

2850 RPM   CDS 5005-2

1450 RPM   CDS 5007-1

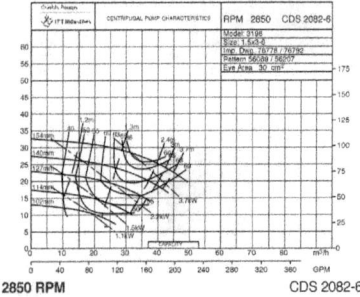

STX
1½ x 3-6
AB

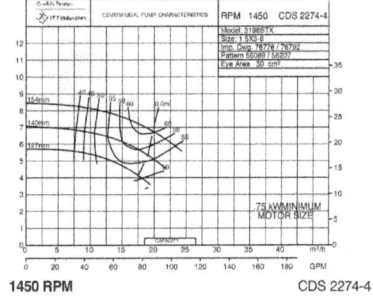

2850 RPM   CDS 2082-6

1450 RPM   CDS 2274-4

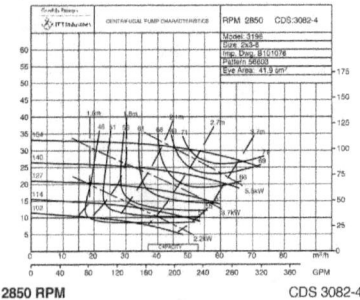

STX
2 x 3-6

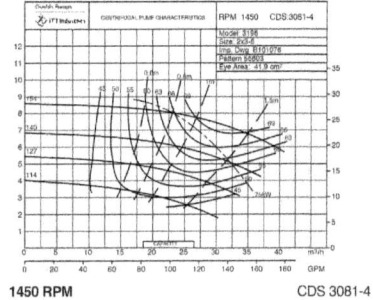

2850 RPM   CDS 3082-4

1450 RPM   CDS 3081-4

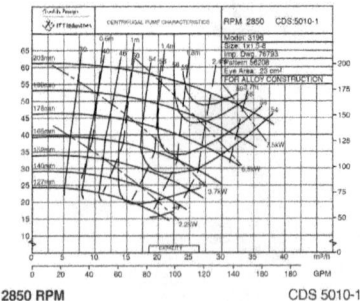

STX
1 x 1½-8
AA

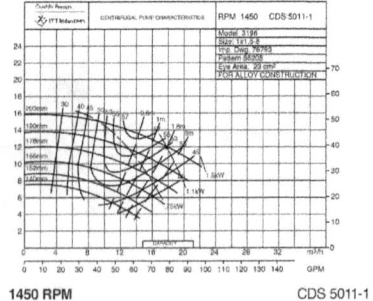

2850 RPM   CDS 5010-1

1450 RPM   CDS 5011-1

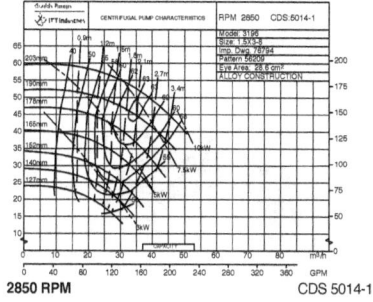

2850 RPM CDS 5014-1

STX
1½ x 3-8
AB

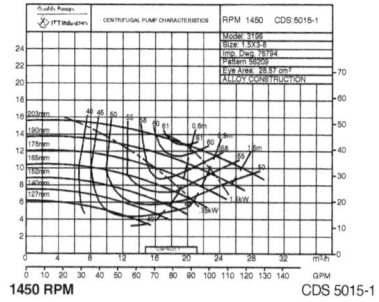

1450 RPM CDS 5015-1

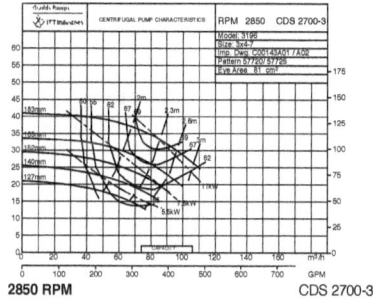

2850 RPM CDS 2700-3

MTX
3 x 4-7
A70

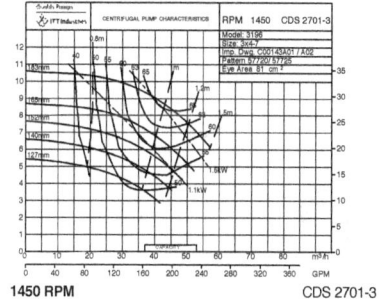

1450 RPM CDS 2701-3

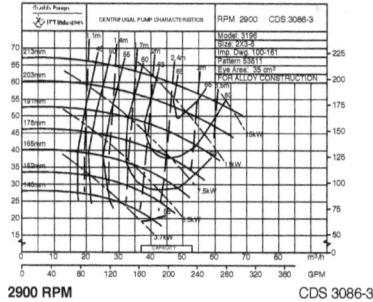

2900 RPM CDS 3086-3

MTX
2 x 3-8
A60

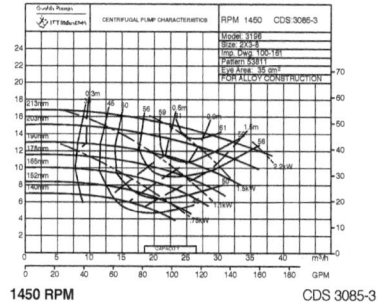

1450 RPM CDS 3085-3

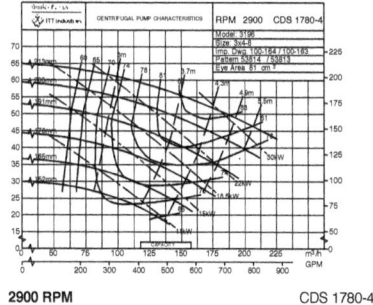

2900 RPM CDS 1780-4

MTX
3 x 4-8
A70

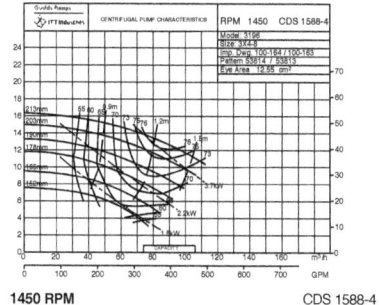

1450 RPM CDS 1588-4

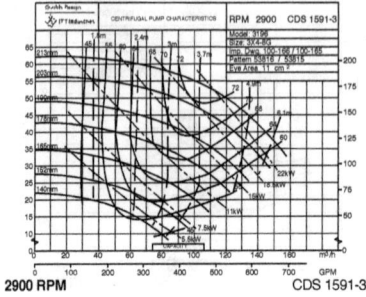

2900 RPM   CDS 1591-3

MTX
**3 x 4-8G
A70**

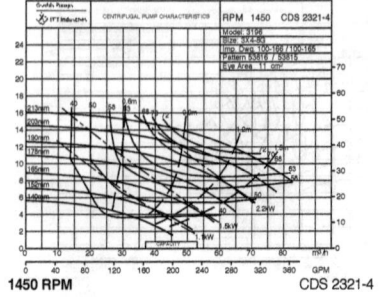

1450 RPM   CDS 2321-4

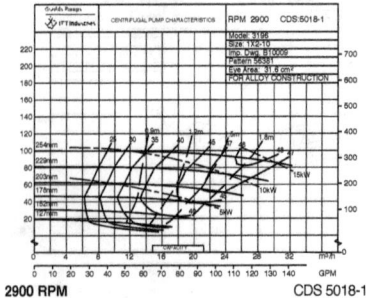

2900 RPM   CDS 5018-1

MTX
**1 x 2-10
A05**

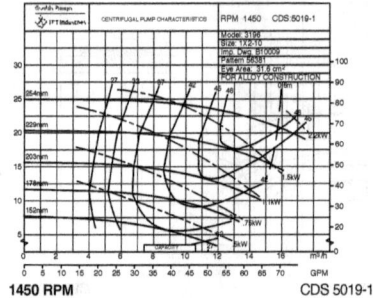

1450 RPM   CDS 5019-1

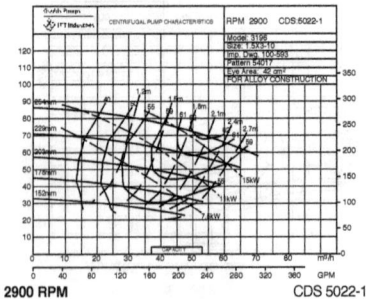

2900 RPM   CDS 5022-1

MTX
**1½ x 3-10
A50**

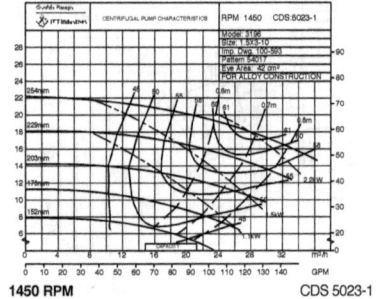

1450 RPM   CDS 5023-1

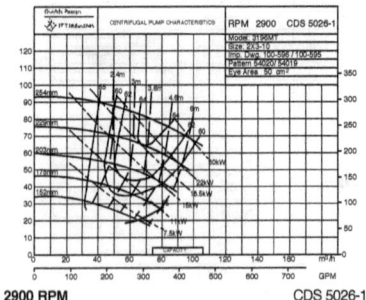

2900 RPM   CDS 5026-1

MTX
**2 x 3-10
A60**

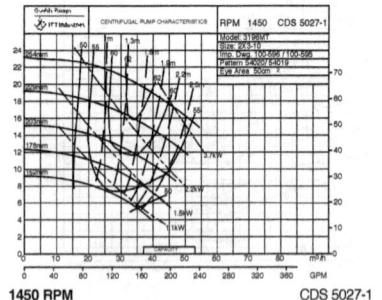

1450 RPM   CDS 5027-1

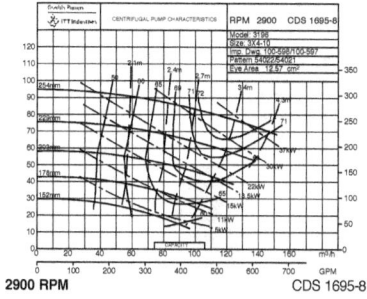

2900 RPM   CDS 1695-8

MTX
3 x 4-10
A70

1450 RPM   CDS 2241-6

1450 RPM   CDS 2825-5

MTX
3 x 4-10H
A40

950 RPM   CDS 2991-4

2965 RPM   CDS 5380

MTX
4 x 6-10G
A80

1475 RPM   CDS 5382

1450 RPM   CDS 4034-3

MTX
4 X 6-10H
A80

1450 RPM   CDS 4029-4

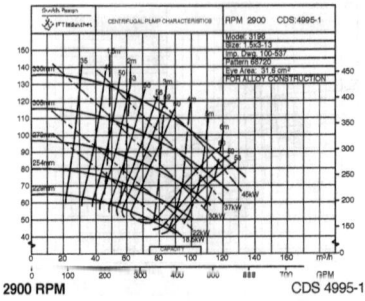

MTX/LTX
1½ x 3-13
A20

2900 RPM   CDS 4995-1

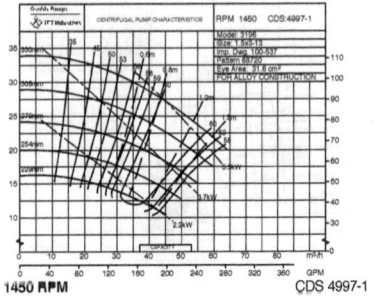

1450 RPM   CDS 4997-1

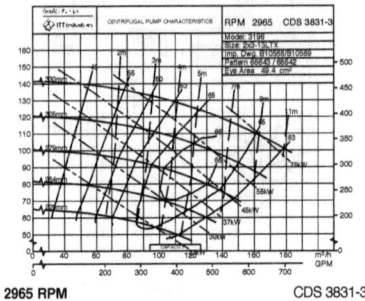

MTX/LTX
2 x 3-13
A30

2965 RPM   CDS 3831-3

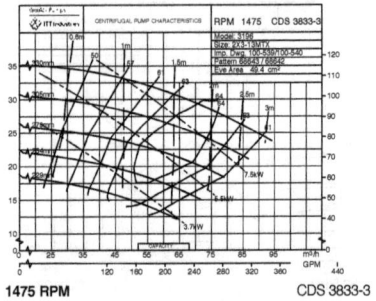

1475 RPM   CDS 3833-3

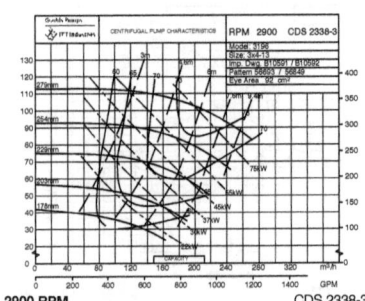

MTX/LTX
3 x 4-13
A40

2900 RPM   CDS 2338-3

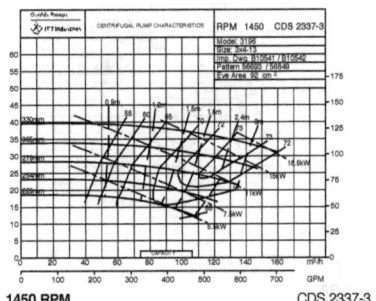

1450 RPM   CDS 2337-3

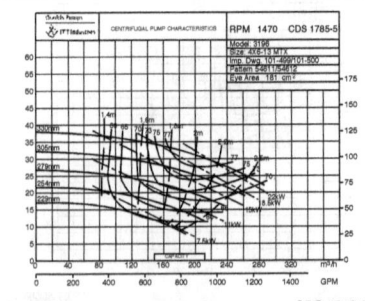

MTX
4 x 6-13
A80

1470 RPM   CDS 1785-5

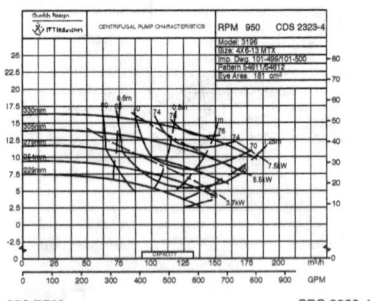

950 RPM   CDS 2323-4

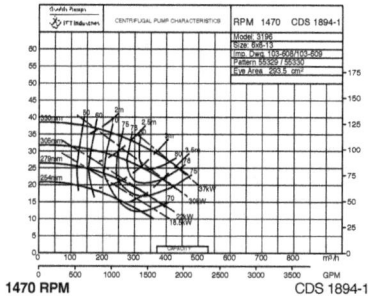

XLT-X
6 x 8-13
A90

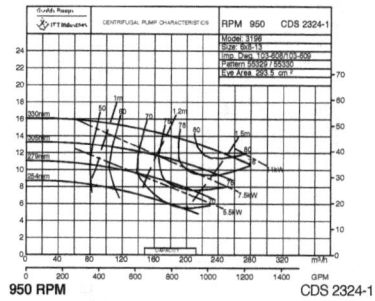

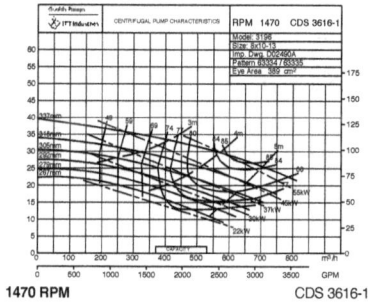

1470 RPM — CDS 1894-1

950 RPM — CDS 2324-1

XLT-X
8 x 10-13
A100

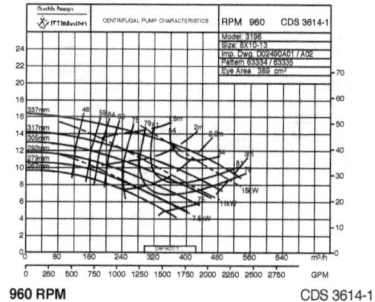

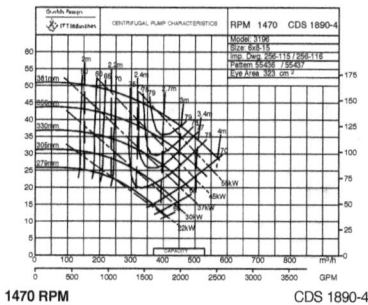

1470 RPM — CDS 3616-1

960 RPM — CDS 3614-1

XLT-X
6 x 8-15
A110

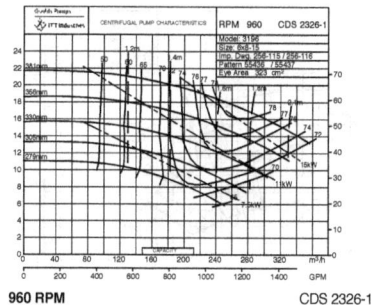

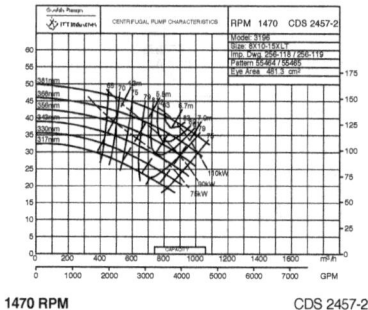

1470 RPM — CDS 1890-4

960 RPM — CDS 2326-1

XLT-X
8 x 10-15
A120

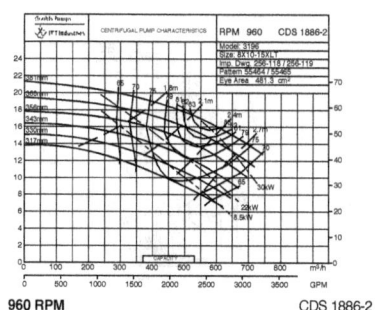

1470 RPM — CDS 2457-2

960 RPM — CDS 1886-2

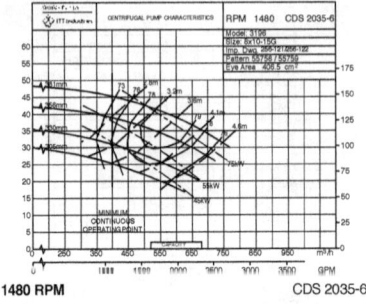

XLT-X
8 x 10-15G
A120

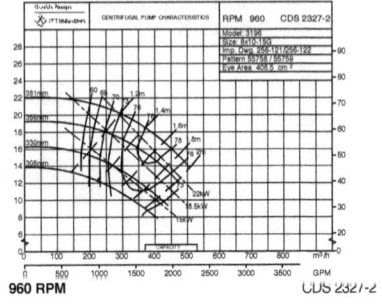

1480 RPM   CDS 2035-6   **960 RPM**   CDS 2327-2

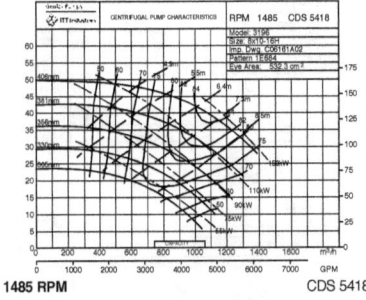

X-17
8 x 10-16H
A120

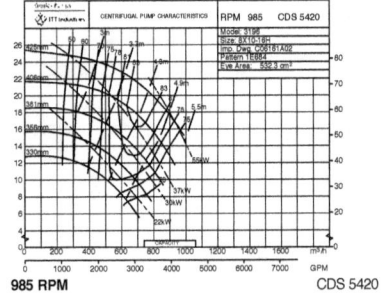

1485 RPM   CDS 5418   **985 RPM**   CDS 5420

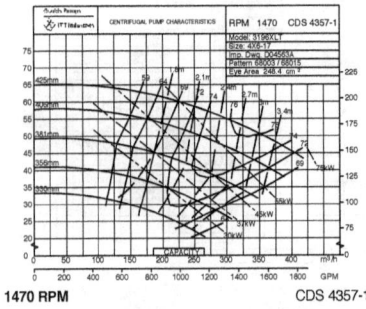

X-17
4 x 6-17
A120

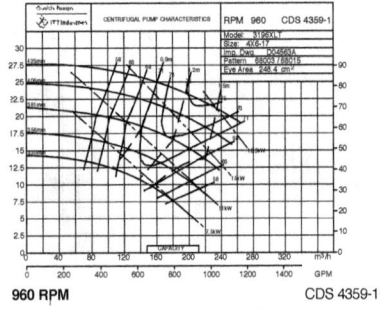

1470 RPM   CDS 4357-1   **960 RPM**   CDS 4359-1

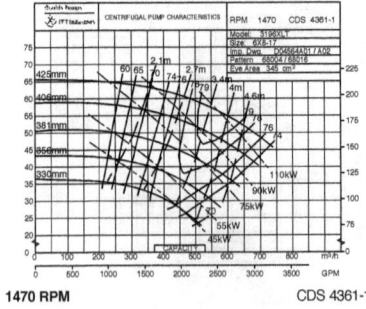

X-17
6 x 8-17

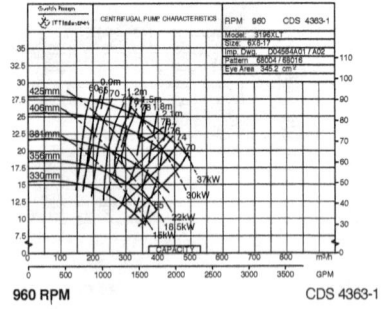

1470 RPM   CDS 4361-1   **960 RPM**   CDS 4363-1

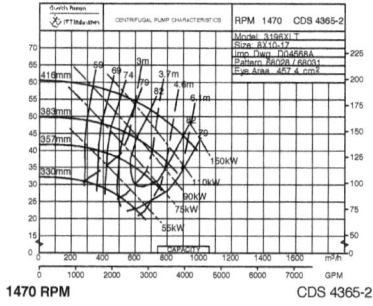

X-17
8 x 10-17

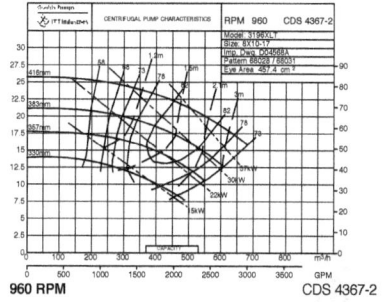

1470 RPM   CDS 4365-2

960 RPM   CDS 4367-2

F Minor Loss Coefficient

The table on the minor loss coefficient presented here has been adapted from the following reference:

Ronald Darby, Ron Darby, and Ravi P. Chhabra. *Chemical Engineering Fluid Mechanics*. CRC Press, 2017.

DOI: 10.1201/9781003049272-F

Table F.1
Minor Loss Coefficient

Item	K, Typical Value	Typical Range
Pipe inlets		
Inward projecting pipe	0.78	0.5–0.9
Sharp cornerflush	0.50	—
Slightly rounded	0.20	0.04–0.5
Bell mouth	0.04	0.03–0.1
Expansions[a]	$(1 - A_1/A_2)^2$(based on V_1)	
Contractions[b]	$(1/C_c - 1)^2$ (based on V_2)	

A_2/A_1	0.1	0.2	0.3	0.4	0.5	0.6	0.7	0.8	0.9
C_c	0.624	0.632	0.643	0.659	0.681	0.712	0.755	0.813	0.892

Item	K, Typical Value	Typical Range
Bends[c]		
Short radius, $r/d = 1$		
90	—	0.24
45	—	0.1
30	—	0.06
Long radius, $r/d = 1.5$		
90	—	0.19
45	—	0.09
30	—	0.06
Mitered (1 m)		
90	1.1	—
60	0.50	0.40–0.59
45	0.3	0.35–0.44
30	0.15	0.11–0.19
Tees	c	—
Diffusers	c	—
Valves		
Check valves		
Swing check[d]	1.0	0.29–2.2
Tilt disk[d]	1.2	0.27–2.62
Lift check[d]	4.6	0.85–9.1
Double door[d]	1.32	1.0–1.8
Full-open gate valve	0.15	0.1–0.3
Full-open butterfly valve	0.2	0.2–0.6
Full-open globe valve	4.0	3–10

G Sample Project Topics

Examples for possible thermal system design project topics are presented here. For more details, readers can refer to the Further Reading provided at the end. The sample project topics can also be expanded for different applications.

1. Cooling of battery banks of an electric car using combined radiator and evaporator system
2. Cooling of battery banks of an electric car using phase change material
3. A micro gas turbine power plant for campus use
4. Centralized air conditioning plant for student activity center
5. Supercomputer/Data center cooling system
6. Compact heat exchanger for electronic cooling
7. High power laser cooling system
8. Piping network for residential units
9. Cold storage system design
10. Heat exchanger system for waster heat recovery
11. Solar energy based heating system
12. Solar power generation system using low boiling point fluid
13. Design of sports car structures
14. Design of Stirling refrigerator for space shuttle
15. Solar thermoelectric generator
16. Thermoelectric cooling for mobile devices/laser
17. Thermoelectric cooling helmet
18. Helmet cooling using phase change materials
19. Thermoelectric power generation from exhaust
20. Automotive thermoelectric air conditioner from exhaust
21. Radioisotope thermoelectric generator
22. Car seat climate control
23. Thermoelectric air heating/cooling system
24. Wearable thermoelectric cooler
25. Thermoelectric power generation from body heat

FURTHER READING

L. Aichmayer, J. Spelling, B. Laumert, and T. Fransson. Micro gas-turbine design for small-scale hybrid solar power plants. *Journal of Engineering for Gas Turbines and Power*, 135(11), 2013. doi: 10.1115/1.4025077.

L. Cao, J. Han, L. Duan, and C. Huo. Design and experiment study of a new thermoelectric cooling helmet. *Procedia Engineering*, 205:1426–1432, 2017. doi: 10.1016/j.proeng.2017.10.339.

A. Capozzoli and G. Primiceri. Cooling systems in data centers: State of art and emerging technologies. *Energy Procedia*, 83:484–493, 2015. doi: 10.1016/j. egypro.2015.12.168.

M. Cosnier, G. Fraisse, and L. Luo. An experimental and numerical study of a thermoelectric air-cooling and air-heating system. *International Journal of Refrigeration*, 31(6):1051–1062, 2008. doi: 10.1016/j.ijrefrig.2007.12.009.

Y. Du, K. Cai, S. Chen, H. Wang, S. Z. Shen, R. Donelson, and T. Lin. Thermoelectric fabrics: Toward power generating clothing. *Scientific Reports*, 5(1), 2015. doi: 10.1038/srep06411.

A. Elarusi, A. Attar, and H. Lee. Optimal design of a thermoelectric cooling/heating system for car seat climate control (CSCC). *Journal of Electronic Materials*, 46(4):1984–1995, 2016. doi: 10.1007/s11664-016-5043-y.

M. N. H. M. Hilmin, M. F. Remeli, B. Singh, and N. D. N. Affandi. Thermoelectric power generations from vehicle exhaust gas with TiO2 nanofluid cooling. *Thermal Science and Engineering Progress*, 18:100558, 2020. doi: 10.1016/ j.tsep.2020.100558.

T. C. Holgate, R. Bennett, T. Hammel, T. Caillat, S. Keyser, and B. Sievers. Increasing the efficiency of the multi-mission radioisotope thermoelectric generator. *Journal of Electronic Materials*, 44(6):1814–1821, 2014. doi: 10.1007/s11664-014-3564-9.

Z. Z. Jiling Li. Battery thermal management systems of electric vehicles. Master's thesis, Chalmers University of Technology, 2014.

R. Jilte and R. Kumar. Numerical investigation on cooling performance of Li-ion battery thermal management system at high galvanostatic discharge. *Engineering Science and Technology, an International Journal*, 21(5):957–969, 2018. doi: 10.1016/j.jestch.2018.07.015.

H. Jouhara, N. Khordehgah, S. Almahmoud, B. Delpech, A. Chauhan, and S. A. Tassou. Waste heat recovery technologies and applications. *Thermal Science and Engineering Progress*, 6:268–289, 2018. doi: 10.1016/j.tsep.2018.04. 017.

S. S. Katoch and M. Eswaramoorthy. A detailed review on electric vehicles battery thermal management system. *IOP Conference Series: Materials Science and Engineering*, 912:042005, 2020. doi: 10.1088/1757-899x/912/4/042005.

J. Kim, J. Oh, and H. Lee. Review on battery thermal management system for electric vehicles. *Applied Thermal Engineering*, 149:192–212, 2019. doi: 10.1016/j.applthermaleng.2018.12.020.

R. A. Kishore, A. Nozariasbmarz, B. Poudel, M. Sanghadasa, and S. Priya. Ultra-high performance wearable thermoelectric coolers with less materials. *Nature Communications*, 10(1), 2019. doi: 10.1038/s41467-019-09707-8.

D. Laing, C. Bahl, T. Bauer, D. Lehmann, and W.-D. Steinmann. Thermal energy storage for direct steam generation. *Solar Energy*, 85(4):627–633, 2011. doi: 10.1016/j.solener.2010.08.015.

Y. Lee, E. Kim, and K. G. Shin. Efficient thermoelectric cooling for mobile devices. In *2017 IEEE/ACM International Symposium on Low Power Electronics and Design (ISLPED)*. IEEE, 2017. doi: 10.1109/islped.2017.8009199.

H. Manchanda and M. Kumar. Study of water desalination techniques and a review on active solar distillation methods. *Environmental Progress & Sustainable Energy*, 37(1):444–464, 2017. doi: 10.1002/ep.12657.

J. B. Marcinichen, J. A. Olivier, and J. R. Thome. On-chip two-phase cooling of datacenters: Cooling system and energy recovery evaluation. *Applied Thermal Engineering*, 41:36–51, 2012. doi: 10.1016/j.applthermaleng.2011.12.008.

G. J. Marshall, C. P. Mahony, M. J. Rhodes, S. R. Daniewicz, N. Tsolas, and S. M. Thompson. Thermal management of vehicle cabins, external surfaces, and onboard electronics: An overview. *Engineering*, 5(5):954–969, 2019. doi: 10.1016/j.eng.2019.02.009.

A. R. Mukaffi, R. S. Arief, W. Hendradjit, and R. Romadhon. Optimization of cooling system for data center case study: PAU ITB data center. *Procedia Engineering*, 170:552–557, 2017. doi: 10.1016/j.proeng.2017.03.088.

M. Olsen, E. Warren, P. Parilla, E. Toberer, C. Kennedy, G. Snyder, S. Firdosy, B. Nesmith, A. Zakutayev, A. Goodrich, C. Turchi, J. Netter, M. Gray, P. Ndione, R. Tirawat, L. Baranowski, A. Gray, and D. Ginley. A high-temperature, high-efficiency solar thermoelectric generator prototype. *Energy Procedia*, 49:1460–1469, 2014. doi: 10.1016/j.egypro.2014.03.155.

F. Tan and S. Fok. Cooling of helmet with phase change material. *Applied Thermal Engineering*, 26(17–18):2067–2072, 2006. doi: 10.1016/j.applthermaleng.2006.04.022.

X. Wang, W. Dai, J. Zhu, S. Chen, H. Li, and E. Luo. Design of a two-stage high-capacity stirling cryocooler operating below 30 K. *Physics Procedia*, 67:518–523, 2015. doi: 10.1016/j.phpro.2015.06.069.

Index

For Product Safety Concerns and Information please contact our EU
representative GPSR@taylorandfrancis.com
Taylor & Francis Verlag GmbH, Kaufingerstraße 24, 80331 München, Germany